Lecture Notes in Physics

For information about Vols. 1–110, please contact your bookseller or Springer-Verlag.

Lecture Notes in Physics

Edited by H. Araki, Kyoto, J. Ehlers, München, K. Hepp, Zürich
R. Kippenhahn, München, H. A. Weidenmüller, Heidelberg
and J. Zittartz, Köln

193

Cool Stars, Stellar Systems, and the Sun

Proceedings of the Third Cambridge Workshop
on Cool Stars, Stellar Systems, and the Sun
Held in Cambridge, Massachusetts
October 5–7, 1983

Edited by Sallie L. Baliunas and Lee Hartmann

Springer-Verlag
Berlin Heidelberg GmbH 1984

Editors

Sallie L. Baliunas
Lee Hartmann
Center for Astrophysics, Harvard College Observatory
Smithsonian Astrophysical Observatory
60 Garden Street, Cambridge, MA 02138, USA

ISBN 978-3-540-12907-3   ISBN 978-3-540-38785-5 (eBook)
DOI 10.1007/978-3-540-38785-5

2153/3140-543210

TABLE OF CONTENTS

3. Results from the Infrared Astronomical Satellite

4. Active, Evolved Stars

Preface

This meeting marks the Third Cambridge Workshop devoted to aspects of Cool Stars. The Organizing Committee structured the meeting with the theme of stellar evolution - beginning with pre-main sequence stars, discussing main sequence evolution and then addressing the post-main sequence red giant phase. This continuity was enriched with the participation of scientists from many subdisciplines of astronomy and astrophysics - all bearing new results, analysis, and conjectures!

October of 1983 was a good time for a meeting because of the influx and analysis of new results from space experiments - the IUE and HEAO-2 ("Einstein") began the deluge of exciting new measurements five years ago. Ground-based techniques have made similar substantial strides with photon counting detectors and dedicated observing programs. All of these measurements allow a true confrontation of observations with theory and strong interaction between solar and stellar physics. Highlights of the meeting included a summary of measurements of stars with IRAS as well as an intensive discussion focused on a cool star prototype - Alpha Orionis (Betelgeuse) - a star that is accessible to observation in many frequency domains and with varied techniques.

It was appropriate too that the annual James Arthur Lecture of the Smithsonian Astrophysical Observatory was given by Jack Eddy during this meeting. His elegant lecture, "The Ancient Sun", gave a very real perspective on new solar and stellar observations of luminosity variations.

During the meetings, science and discussions were intensive, but an historical highlight also accompanied these events. After the banquet, we were entertained by the Harvard singing group "Modified Rapture" who performed selections from the Harvard Observatory Pinafore. This parody of the Gilbert and Sullivan version was written by Winslow Upton in 1879 and contained both astronomers of the Harvard College Observatory and schemes and lore of the day. We were reminded that astronomer's work may have differed from ours - being concerned solely with "photometree" - but the problems of equipment failure, long hours, and low pay for the love of the science ring true now, as they did over a century ago!

Our meeting benefited greatly from the generous support of the Smithsonian Institution and its James Arthur Fund and Langley-Abbot program, the Center for Astrophysics, and the National Aeronautics and Space Administration. We are also grateful to the unsung backstage heroines of this meeting Sara Yorke and Stephanie Deeley, to the "volunteers" confronting the practical problems: N. Burnham, R. Hewett and W. Toth, and to the anonymous, hard-working referees of the manuscripts of the Workshop.

Andrea Dupree
for the Organizing Committee:
S. Baliunas, A. Dupree, L. Hartmann, R. Noyes,
R. Rosner, D. Soderblom, G. Withbroe

Recent Observations of T Tauri Stars

Suzan Edwards and Stephen E. Strom

The Five College Astronomy Department;

Smith College and the University of Massachusetts

Abstract

The nature of the peculiar emission properties of the T Tauri stars is examined in the light of contributions from regions designated as photospheres, chromosphere coronae, and extended envelopes which are often characterized by mass loss. Although some manifestations of T Tauri activity can be linked to vigorous solar-stellar phenomena characteristic of cool stars of low mass, other characteristics of T Tauri behavior are exhibited by a wide range of young stellar objects.

Introduction

The T Tauri stars are generally regarded as low mass pre-main sequence objects on the basis of (1) their location in the vicinity of molecular cloud complexes and optical nebulosity, (2) their location on quasi-hydrostatic equilibrium tracks in the HR diagram, and (3) the strength of their Li Iλ6707 absorption, which is as yet undiminished by convective depletion. A distinctive set of optical spectral characteristics distinguishes T Tauri stars from other irregularly variable bright-line stars in nebulous regions such as dMe or flare stars (Herbig, 1962). Late-type photospheric spectra range from G to M and can occasionally be totally obscured by overlying continuum emission. Characteristic emission lines include strong, broad Balmer lines of hydrogen and Ca II H and K, and fluorescent lines of Fe Iλ4063 and λ4132. These fluorescent lines have been found only in T Tauri spectra and in solar flares (Cowley and Marlborough, 1969). Additional low excitation metallic lines, He I, He II emission, and forbidden lines of O I and S II are often present.

A representative HR diagram for the T Tauri stars in the Taurus-Auriga complex (Cohen and Kuhi, 1979) is shown in Figure 1. The positioning of T Tauri stars in this diagram is somewhat unreliable due to contamination of the photosphere by overlying emission and uncertain extinction corrections. Typical maximum luminosities are about 50 $L_\odot$ and maximum effective temperatures are about 6000 K. Recent observations of faint members of NGC 2264 (Adams, Strom and Strom, 1983)

show probable members of the T Tauri class with luminosities as low as 0.01 L_o
and effective temperatures of 2500 K. A theoretical birthline for low mass stars,
marking the optical emergence of a star as it enters the phase of quasi-hydrostatic
contraction along a conventional Hayashi track, has been estimated by protostellar
collapse calculations of Stahler, Shu and Taam (1980, 81). The birthline is found
to agree well with the observed locus of points of the T Tauri stars in HR diagrams
such as that in Figure 1 (Stahler, 1983; see also this volume). This allows a
rough determination of ages, masses, and radii for the T Tauri stars, which are
typically 10^5 to 10^6 yr, 0.1 to 2.0 M_o, with radii 2.5 to 5 times their
eventual main sequence values.

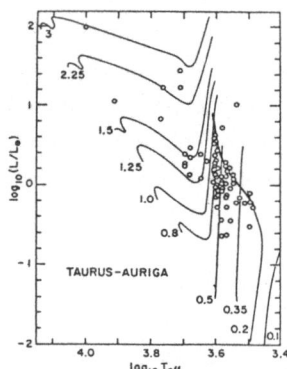

Figure 1 - The locus of T Tauri stars from Tau-Aur in the HR Diagram, with the
Stahler (1983) birthline for stars of $\leq 1M_\odot$. (Reprinted from the Astrophysical
Journal, Cohen and Kuhi, (1979).

Indirect evidence that emission activity in the T Tauri stars, whatever its
source, declines with age comes from correlations of He I, Fe II, and [O I] line
flux with L_{bol} (Cohen and Kuhi, 1979). If the stars are in fact evolving down
nearly vertical convective Hayashi tracks, then this is readily interpreted as a
decline in emission activity with age. The large scatter seen in Hα flux with
L_{bol} implies that multiple regions probably contribute to the Balmer line emission
(Cohen and Kuhi, 1979). The T Tauri stars do not, however, obey the Skumanich
(1972) relation of Ca II K line flux declining with (age)-1/2 exhibited by older
stars with ages from 4×10^7 to 1×10^{10} yr (Kuhi, 1981; Duncan, 1981).

An investigation of rotational velocities among T Tauri stars (Vogel and Kuhi,
1981) demonstrated that these stars are not rapid rotators and that the bulk of the
original angular momentum present at the onset of protostellar collapse is dissipated by the time low mass stars become optically visible. Insufficient data are
available to determine whether the T Tauri stars obey the Skumanich rotational

velocity relation adapted to declining angular momentum with (age)-1/2, since the
majority of vsini determinations by Vogel and Kuhi (1981) were upper limits with
rotational velocities < 25 km s^{-1}. It is important to realize that these conclusions are based on a relatively small sample of T Tauri stars, none of which
exhibit extreme emission characteristics.

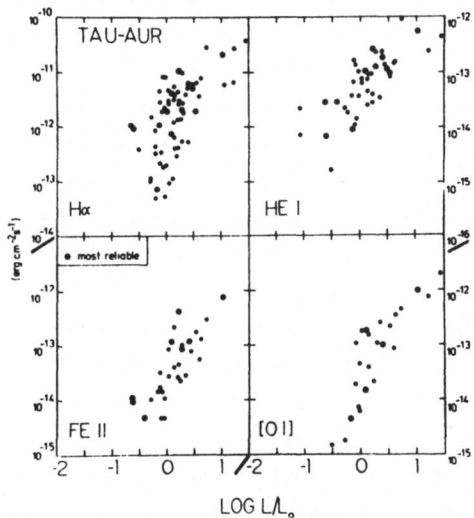

Figure 2 - The relation between bolometric luminosity and the emission flux of Hα ,
He I, Fe II, and [O I] to T Tauri stars in Tau-Aur. (Reprinted from the Astrophysical Journal, Cohen and Kuhi, 1979).

In this review we will describe some recent observational advances at optical,
infrared and radio wavelengths that have begun to resolve long standing issues in
the interpretation of the T Tauri phenomenon. These include the behavior and origin
of the irregular light variations, the extent of the chromospheric contribution to
the emission properties and the nature of the mass loss, in particular its
previously unsuspected highly collimated characteristics.

Photometric Variations

Irregular brightness variations over timescales ranging from minutes to decades
characterize nearly all members of the T Tauri class. It is almost certain that
several mechanisms contribute to the observed variation within the class and even
for a given T Tauri star. Irregular short time scale variations on the order of
minutes with low amplitudes (<0.1 magnitude at visible wavelengths) may be related
to flaring phenomena analogous to solar events. Semi-regular, moderate amplitude

(0.1 $\leq \Delta V \leq$ 3.0 mag) variations could arise from the rotationally modulated appearance and disappearance of spots and active regions. Irregular, large amplitude (>3.0 mag) variations may arise from a number of sources. These include 1) thermal instabilities resulting from rapid rotation (Larson, 1980) 2) irregular infall of envelope material resulting in changes in the relative contribution to the total system light of either the inner regions of an accretion disk (Lynden-Bell and Pringle, 1974) or the shock heated boundary between an accreting envelope and the stellar photosphere (Bertout, 1977), 3) variable obscuration by circumstellar dust or 4) variations in the photospheric effective temperature.

Insight into the intrinsic nature of moderate amplitude photometric variations is provided by Herbst, Holtzman and Klasky (1983) who interpret their observations as due to the growth and decay of bright or dark spots on the stellar surface. They find that while most T Tauri stars become redder as they fade in brightness, the weak emission and the strong emission stars exhibit different characteristics in their Hα emission variations. In the weak emission stars, such as CO Ori, the H α flux is observed to remain constant as the star becomes fainter and the variations are interpreted as arising from the variable presence of cool photospheric spots. In the strong emission stars, such as RW Aur, the Hα emission flux correlates with the continuum brightness and this is attributed to variable surface coverage by bright plages and flare-like events. A recent theoretical model examining the effects of brightness variations in T Tauri stars induced by surface magnetic fields indicates that a decrease in V magnitude of up to 3.5 mag can result from the emergence of cool photospheric spots covering 50% of the stellar surface (Appenzeller and Dearborn, 1983) The reduction in brightness results both from the presence of the cooler spotted area and the spot-free surface, where the convective energy flow is inhibited through-out the star by the presence of surface magnetic fields. Confirmation of these spot theories as the source of T Tauri variability awaits a detailed spectroscopic investigation.

To date, most attempts to search for periodic variations in T Tauri stars have proved unsuccessful. One remarkable exception has been demonstrated by Schaefer (1983). In an exhaustive study of the Harvard plate collection covering 1901 to 1982, Schaefer found evidence for a 6.129 day periodicity in the brightness variations in SY Cha from 1970-72, which he interpreted as rotational modulation from a long lasting bright spot on the stellar surface.

Periodic brightness variations in seven low mass weak emission pre-main sequence stars that occupy the same portion of the HR diagram as the T Tauri stars have been interpreted as rotational modulation by cool photospheric spots (Rydgren and Vrba, 1983a, Rydgren, private communication). These stars have rotational velocities comparable to or greater than the T Tauri stars, which implies that the T Tauri emission properties cannot be attributed solely to rotation.

4

Photospheres

Aside from the recognition of their late spectral types, the underlying photospheric absorption spectra of T Tauri stars have not been subjected to the close observational scrutiny that has been directed toward the emission features. These regions may be from normal, however, as evidenced by the variation in spectral type with wavelength noted in BM And (Herbig, 1977; Walker, 1981), YY Ori (Walker 1978) and the Ae star HK Ori (Strom and Strom 1983). A question frequently posed but largely unresolved is whether photospheric variations accompany the dramatic and irregular light variations common to T Tauri stars. An example of a T Tauri star that has enhanced photospheric absorption at times of brightness minima has been reported by Schwartz and Heuermann (1981), but this phenomenon is not well documented as representative T Tauri behavior.

Chromospheres

The similarity of the low excitation metallic lines in the optical spectra of T Tauri stars to the flash spectrum from the solar chromosphere was first pointed out by Joy (1945). More recently, ultraviolet spectra from the IUE telescope and x-ray data from the Einstein Observatory have strengthened the chromospheric interpretation of the optical spectra by demonstrating that the atmospheres of T Tauri stars include regions with temperatures analogous to the solar transition region and the corona. These results will be discussed elsewhere in this volume.

Chromospheric models have progressed from Herbig's (1970) suggestion that the temperature rise must begin deep in the atmosphere at $\tau(5000) \sim 1$ to fully quantitative models that reproduce many of the observed emission properties of the T Tauri stars. The most successful chromospheric model to date is that of Calvet, Basri and Kuhi (1983). Such chromospheric models account for the total fluxes in the Ca II K and Mg II K lines, the general presence of numerous low excitation lines such as Fe I, Fe II and Na I (Cram 1979), the optical continuum energy distribution, and the ultra-violet lines in T Tauri spectra.

Observational confirmation of differential veiling of the photospheric spectrum by chromospheric emission has been demonstrated in T Tau by Davis (1983). The subtraction of a normal photospheric spectrum of a K0 III star from the observed spectrum of T Tau (K0 IV-V), which shows relatively few of the metallic emission lines common to T Tauri stars, is shown in Figure 3. The resulting spectrum reveals the presence of incipient emission lines, precisely those which are attributed to chromospheric emission in the richest T Tauri emission spectra.

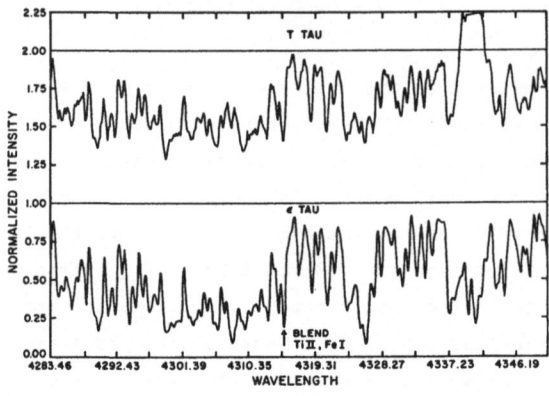

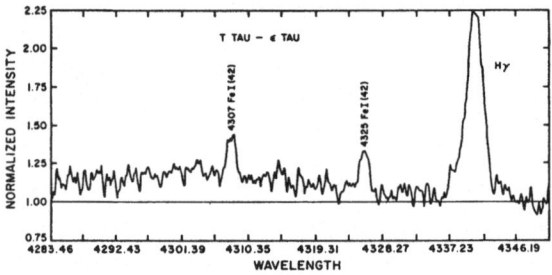

Figure 3 - The upper panel shows 5 A/mm KPNO echelle spectra of ε Tau (KO III) and T Tau (KO IV-V). The spectrum of ε Tau has been rotationally broadened to match the vsini of T Tau. The lower panel shows the difference spectrum with emission features labelled.

Spectroscopic evidence for inhomogeneous surface coverage by deep chromospheric regions has come from near infrared observations of the Ca II triplet by Herbig and Soderblom (1980). The triplet intensity ratio is comparable to the optically thick values found in solar flares and the lines are found to remain saturated even when their emission equivalent width is quite low. This is attributed to variations in the size of the emitting area, which is always optically thick, rather than from changes in the column density of the emitting material.

Extended Envelopes and Mass Loss

The chromospheric models, although very successful in accounting for many T Tauri emission features, cannot explain numerous other emission phenomena that signify the presence of an additional atmospheric component characterized by an extended, and in many cases expanding, circumstellar envelope. These include 1)

6

the intensity and the profile structure of the Hα emission line (Cram 1979;
Heidemann and Thomas 1980; Calvet, Basri and Kuhi 1983; Hartmann Edwards and
Avrett, 1982), 2) the steepness of the Balmer decrement (Cram 1979, Hartmann
Edwards and Avrett, 1982), 3) the optical continum emission in the most extreme
emission stars, 4) the infrared excesses (Calvet Basri and Kuhi 1983; Rydgren and
Vrba 1983), 5) the radio continuum spectra (Cohen, Bieging and Schwartz 1982), 6)
the blueshifted low excitation forbidden lines (Schwartz 1975, Jankovics,
Appenzeller and Krautter 1983), 7) the highly collimated, sometimes bipolar motion
of Herbig-Haro (HH) Objects directed away from some T Tauri stars and low
luminosity embedded sources (Cudworth and Herbig 1979; Herbig and Jones 1981, 1983;
Jones and Herbig 1982; Mundt and Fried 1983; Strom, Strom and Stocke 1983; Graham
and Elias 1983), and 8) the spatially extended high velocity molecular gas
expanding away from some T Tauri stars and low luminosity embedded sources (Snell,
Loren and Plambeck 1980; Edwards and Snell 1982, 1983, 1983a and references
there-in; Kutner et al. 1982). In the limited space available here, we will
examine those observations which provide the most compelling evidence for energetic
winds from T Tauri stars.

The Hα and the Na D line profiles are suggestive of mass loss in many T Tauri
spectra. Representative Hα profiles may be found in Ulrich and Knapp (1979).
Kuhi (1978) has estimated that at least 70% of all T Tauri stars show some sort of
P Cygni structure at Hα. Estimates of mass loss rates from T Tauri stars based on
analysis of Hα lines range from 10^{-8} to 10^{-9} $M_\odot$ yr^{-1} (Kuhi 1964; Hartmann
Edwards and Avrett 1982). Mass loss estimates are about an order of magnitude
lower in the Alfvén wave-driven wind model of Hartmann Edwards and Avrett, which
predicts transverse velocities in the optically thick Hα emitting region that are
of comparable magnitude to the expansion velocities, thereby greatly enhancing line
strengths and widths in comparison to the ad-hoc, optically thin Sobolev-type
approach of Kuhi's earlier model. De Campli (1981) has pointed out the ambiguity
inherent in estimating mass loss rates based on Hα profiles alone, and has shown
that the maximum mass loss rates to be expected for a typical T Tauri star, based
on standard wind models, are $\leq 3 \times 10^{-8}$ $M_\odot$ yr^{-1}.

Spectroscopic evidence for multiple eruptive mass loss events in T Tauri stars
has been provided by Mundt (1981). He finds multiple narrow blueshifted shell
absorption components with velocity shifts of -50 to -150 km s^{-1} superposed on
the broad Na D emission features. These components are found to be constant in
velocity and intensity while Hα intensities and profile structure are seen to
vary, and are similar in appearance to the Na D profiles in FU Ori stars, which are
known to have undergone recent episodes of short term, energetic mass loss.

The detection of radio continuum free-free emission from a small number of T
Tauri stars (Cohen, Beiging and Schwartz, 1982) has been interpreted as the

signature of ionizing stellar winds from these stars. Although the large observed range in spectral indices for these sources precludes a uniform estimate of mass loss rates, it appears likely that those sources detected with the VLA are preferentially those found in the vicinity of HH Objects (Beiging, Cohen and Schwartz, 1983).

Indirect evidence for extremely energetic winds from several T Tauri stars and embedded infrared sources of low bolometric luminosity, which are likely to be young stellar objects in a pre-T Tauri phase, comes from observations of stellar winds interacting with the molecular surroundings of these objects. One optical manifestation of this is the presence of shock excited HH Objects in the vicinity of five visible, rather normal, T Tauri stars (Schwartz, 1983). A classic example of this phenomenon is found around T Tau itself where Schwartz (1975) showed that both Burnham's emission nebula and a small section of Hind's reflection nebula have HH-like spectra with large negative radial velocities, which he interpreted as arising from mass motion away from T Tau at velocities of about 100 km s^{-1}. In some instances (references listed in the opening paragraph of this section), radial velocity and proper motion data indicate that HH Objects are small, high density knots moving away from a T Tauri star at velocities of 200 to 300 km s^{-1} in a highly collimated and sometimes bipolar allignment. One current model suggests that "interstellar bullets" originally accelerated by the wind close to the star are currently impacting with surrounding stationary molecular material (Norman and Silk, 1979). An alternative model envisions the stellar wind as currently impacting an initially stationary clump in the interstellar medium, which is in the process of being accelerated up to wind speed (Schwartz and Dopita, 1980). Mass loss estimates for T Tauri stars providing the momentum for these high velocity HH Objects exceed 1×10^{-5} $M_{\odot}$ yr^{-1} if isotropic winds are assumed (Hartmann and Raymond, 1983).

Additional evidence for highly collimated energetic mass outflows from some T Tauri and pre-T Tauri objects is provided by CCD images of optically emitting jets with HH Object-like spectra. These jets may extend 10^{16} cm from the star and in some instances blueshifted and redshifted radial velocities from jets on opposing sides of the star indicate bipolar ejection (Mundt and Fried, 1983; Mundt, Stocke and Stockman 1983; Strom, Strom and Stocke 1983). An illustration of highly collimated, shock excited gas projecting from the 30 $L_{\odot}$ embedded infrared source L1551 IRS-5 is shown in Figure 4.

Those T Tauri stars showing optical evidence for the interaction of an energetic wind and the surrounding interstellar environment as well as pre-T Tauri objects in the vicinity of optical HH Objects (Edwards and Snell 1982, 1983, 1983a) frequently are found to be surrounded by largescale expanding lobes of molecular gas. Radio spectral line profiles with broad assymetric wings at the ^{12}CO J=1-0

transition signify highly supersonic mass outflow from these young stars. The
spatial extent of this cold (T ~ 20 K) expanding molecular gas may reach as far as
.5 to 1 pc from its source. An excellent example of a highly collimated molecular
outflow is shown in Figure 5 for L 1551 IRS-5 (Snell and Schloerb 1983.) This
outflow is also accompanied by high velocity HH Objects (Cudworth and Herbig 1979),
and optical and radio jets (Cohen Bieging and Schwartz 1982; Mundt and Fried, 1983).

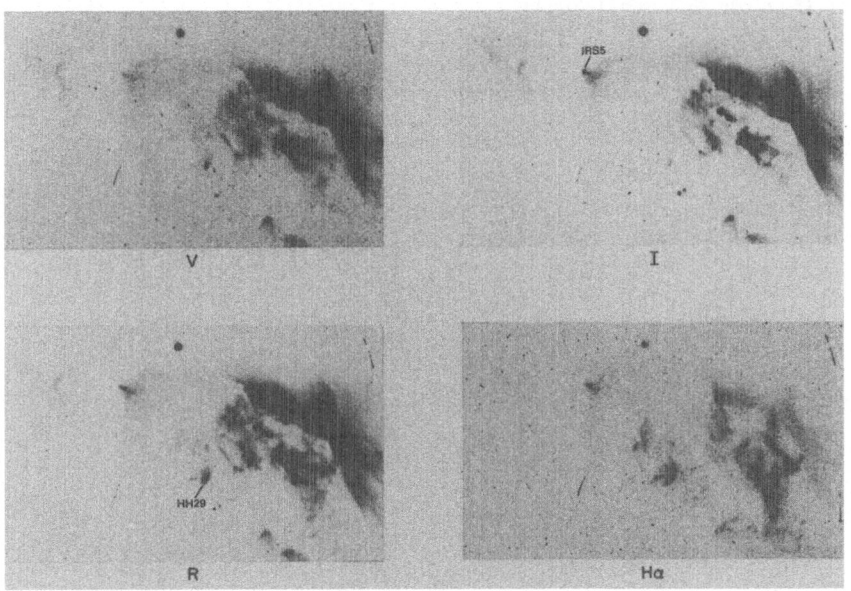

Figure 4 - Optical images of the L1551 IRS-5 region through V, R, I and Hα
(Δλ=38 A) filters using the RCA CCD on the KPNO 0.9m telescope. The V and I
filters probe the scattered light component in an HH complex, while the Hα and R
filters can be combined to define an index of shock excited emission in an HH
object.

The energy and momentum in these molecular outflows must be provided by a source
with mass loss rates ranging from 10^{-5} to 10^{-7} $M_{\odot}$ yr^{-1}, if the observed momentum
has been deposited by a steadily flowing wind over the dynamical timescale of the
outflow (Edwards and Snell, 1983a). These estimates are considerably larger than
those which a low luminosity, slowly rotating, relatively high surface gravity T
Tauri star would appear to be capable of sustaining by a continuously outflowing
stellar wind. Moreover, a mass loss rate of 10^{-5} $M_{\odot}$ yr^{-1} in a star of a few
$M_{\odot}$ must necessarily be a short-lived phenomenon, of considerable shorter duration
than the typical T Tauri stage. Dynamic timescales for the molecular outflows are
estimated to be about 10^4 yr, and thus may be consistent with short bursts of
episodic mass loss rather than a continuous wind. In actuality only a small number
of optically identified stars with standard T Tauri characteristics are believed to

be the source of energetic molecular outflows or of highly collimated optically emitting shock excited gas. The majority of these energetic mass loss sources with luminosity $<100L_\odot$ are embedded infrared objects and might not resemble a typical T Tauri star. The observed frequency of such sources is, however, consistent with a phase of energetic mass loss of $\dot{M} > 10^7$ $M_\odot$ yr^{-1} characterizing at least all stars of 1 $M_\odot$ or greater, if each source is assumed to have one mass loss episode (Edwards and Snell 1983a).

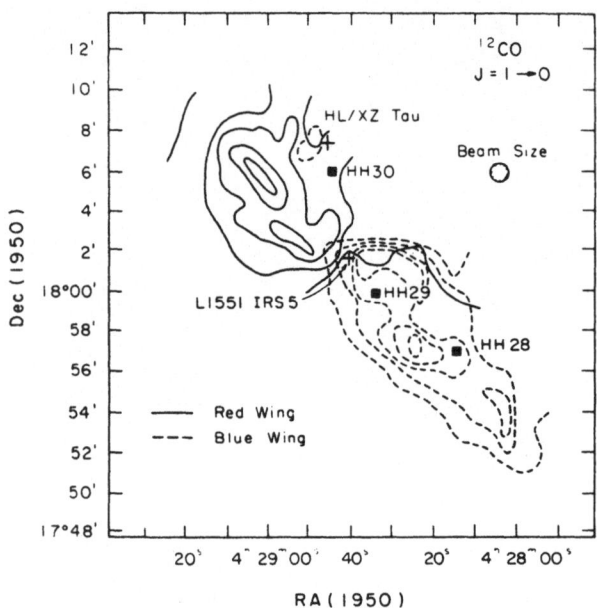

Figure 5 - Contour map of the ^{12}CO integrated intensity of blueshifted and redshifted high velocity molecular gas in the vicinity of L1551 IRS-5. (Snell and Schloerb, 1983).

The recent discovery of the low luminosity infrared companion to T Tau (Dyck, Simon and Zuckerman, 1982; Schwartz et al., this volume,), which is a stronger source of radio continuum emission than T Tau itself, raises the possibility of whether the embedded (evolutionarily younger) object is in fact the source of the molecular mass outflow in this region (Edwards and Snell, 1982). If so, then embedded low luminosity companions may be present in other regions where a typical T Tauri star has been cited as the source of an energetic molecular outflow or of collimated, outflowing shock-excited gas.

The role of circumstellar disks in the collimated ejection of material from T Tauri stars has yet to be determined. Indirect evidence for the presence of an edge-on disk around HL Tau, including a large far infrared flux, a unique near infrared spectrum, and a large, optical linear polarization, has been summarized by Cohen (1983). The presence of a disk or a non isotropic density distribution around a T Tauri star has been suggested as the collimating mechanism for the optical jets and molecular outflows (Snell, Loren, and Plambeck, 1980; Konigl, 1982). Furthermore, mass accretion from a circumstellar disk could provide a convenient source of gravitational energy to power energetic winds as well as provide a source for some of the excess continuum at optical and infrared wavelengths. The extensive work on mass outflow from T Tauri stars has perhaps overwhelmed the observations which suggest that infall of material may play an important role in T Tauri energetics. The presence of infalling material around some T Tauri stars is well documented in high Balmer series members and occasionally at Na D, although it is always accompanied by simultaneous mass outflow (Walker, 1972; Ulrich 1978, Edwards, 1979; Bertout et al. 1982; Hartmann, this volume). An alternate view would attribute the infall to the return of plumes of material ejected from the star that never reach escape velocity.

Conclusions and Questions

   The recent optical, infrared and radio observations of T Tauri stars are best interpreted in terms of a multi-component emission region, which includes a vigorous, deep chromosphere inhomogeneously covering the stellar surface and an extended envelope undergoing mass loss, which in some cases appears to be highly collimated. The chromospheric activity, decaying with stellar age, will eventually assume the character of activity seen in a main sequence, low mass star. The energetic winds, which in the most extreme cases probably represent a phase of short duration, are found to be common to stars of a very young evolutionary age, independent of luminosity or mass, as evidenced by the highly collimated flows around young objects of considerably greater luminosity than any T Tauri star (Bally and Lada, 1983). This phase of highly collimated mass loss is likely associated with the configuration of a young stellar object and its surrounding placental material shortly after the gravitational collapse to a stellar state. The mass loss exhibited by the great majority of T Tauri stars, as evidenced by $H\alpha$ and Na D lines, can be accounted for by standard wind models, notably those powered by magnetic waves.

Acknowledgements

We would like to thank W. Herbst, E. Rydgren, and R. L. Snell for providing us with material in advance of publication.

References

Adams, M.T., Strom, K.M., and Strom, S.E. 1983, Ap.J. Supple. (in press).

Appenzeller, I. and Dearborn, D.S.P. 1983, preprint.

Bally, J. and Lada, C.J. 1983 Ap.J. $\underline{265}$, 824.

Bieging, J.H., Cohen, M., and Schwartz, P. 1983, preprint.

Bertout, C. 1977, Ast. and Astr. $\underline{58}$, 153.

Bertout, C., Carros L., Mundt, R., and Wolf, B. 1982 Ast. and. Astr. Sup. $\underline{47}$, 419.

Calvet, N., Basri, G., and Kuhi, L.V. 1983, preprint.

Cohen, M. 1983, Ap.J. Lett. $\underline{270}$, L69.

Cohen, M., and Kuhi, L.V. 1979, Ap.J. Supple. $\underline{41}$, 743.

Cohen, M., Beiging, J.H., and Schwartz, P. 1982, Ap.J. $\underline{253}$, 707.

Cowley, C. and Marlborough, J.M. 1969, Ap.J. $\underline{158}$, 803.

Cram. L. 1979 Ap.J. $\underline{234}$, 949.

Cudworth, K.M., and Herbig, G.H. 1979, A.J. $\underline{84}$, 548.

Davis, R.E. 1983, Ph.D. thesis, Univ. of Arizona.

De Campli, W. 1981, Ap.J. $\underline{244}$, 124.

Duncan, D.K. 1981, Ap.J. $\underline{248}$, 651.

Dyck, H.M., Simon, T. and Zuckerman, B. Ap.J. $\underline{255}$, L103.

Edwards, S. 1979 P.A.S.P. 91, 329.

Edwards, S. and Snell, R.L. 1982, Ap.J. $\underline{261}$, 151.

Edwards, S. and Snell, R.L. 1983, Ap.J. $\underline{270}$, 605.

Edwards, S. and Snell, R.L. 1983a, Ap.J. submitted

Graham, J.A. and Elias, J.H. 1983, Ap.J. $\underline{272}$, 615.

Hartmann, L., Edwards, S. and Avrett, E. 1982, Ap.J. $\underline{261}$, 279.

Hartmann, L. and Raymond, J. 1983, C.F.A. preprint.

Heidemann, N. and Thomas, R.N. 1980, Ast.and Astr. $\underline{87}$, 36.

Herbig, G.H. 1962, Adv. in Ast. and Astroph. $\underline{1}$, 47.

_____ 1970, Mem. Soc. Roy. Sci. Liege $\underline{19}$, 13.

_____ 1977, Ap.J. $\underline{214}$, 747.

Herbig, G.H. and Jones, B.F. 1981, A.J., $\underline{86}$, 1232.

_____ 1983, A.J., $\underline{88}$, 1040.

Herbig, G.H. and Soderblom, D.R. 1980, $\underline{Ap.J.}$, 242, 628.

Herbst, W., Holtzmann, J.A. and Klasky, R.S. 1983, preprint.

Jankovics, I. Appenzeller, I. and Krauter, J. 1983, preprint.

Jones B.F., and Herbig, G.H. 1982, A.J. $\underline{27}$, 1223.

Joy, A.H. 1945, Ap.J. $\underline{102}$, 168.

Konigl, A. 1982, Ap.J. $\underline{261}$, 115.

Kuhi, L.V. 1964, Ap.J. $\underline{140}$, 1409.

_____ 1978, Protostars and Planets, ed. T. Gehrels, U. of A. Press. p. 708.

_____ 1981, Second Cambrdige Workshop on Cool Stars, Stellar Systems and the Sun,
    Ed. M Giampapa and L. Golub, S.A.O. Special Report #392, p. 141.

Kutner, M.L., Leung, C.M., Machnik, D.E., and Mead, K.N. 1982 Ap.J. $\underline{259}$, L35.

Larson, R. 1980, M.N.R.A.S. $\underline{190}$, 335.

Lynden-Bell, D. and Pringle, J.E. 1974, M.N.R.A.S. $\underline{168}$, 603.

Mundt, R. 1981, Second Cambridge Workshop on Cool Stars, Stellar Systems, and
    the Sun, Ed. M. Giampapa and L. Golub, S.A.O. Special Report #392, p. 181.

Mundt, R. and Fried, J.W. 1983, preprint.

Mundt, R., Stocke, J. and Stockman, H.S. 1983, preprint.

Norman, C. and Silk, J. 1979, Ap.J., $\underline{228}$, 197.

Rydgren A.E. and Vrba, F.J. 1983, A.J. 88, 7.
_____ 1983a, Ap.J. 267, 191.
Schaefer, B.E. 1983, Ap.J. Let., 266, L45.
Schwartz, R.D. 1975, Ap.J., 195, 631.
_____ 1983, Ann. Rev. Astr. Ap., 21, 209.
Schwartz, R.D. and Heuermann, R.W. 1981, A.J. 86, 1526.
Schwartz, R.D. and Dopita, M.A. 1980, Ap.J. 236, 543.
Skumanich, A. 1972, Ap.J. 171, 565.
Stahler, S.W. 1983, preprint.
Stahler, S.W., Shu, F.H. and Taam, R.E. 1980, Ap.J. 241, 637.
_____ 1981, Ap.J. 248, 727.
Snell, R.L., Loren, R.B. and Plambeck, R.L. 1980, Ap.J. Let., 239, L17.
Snell, R.L. and Schloerb, F. P. 1983, in preparation.
Strom, S.E. and Strom, K.M. 1983, in preparation.
Strom, K.M., Strom, S.E. and Stocke, J. Ap.J. Let., 271, L23.
Vogel, S.N. and Kuhi, L.V. 1981, Ap.J., 245, 960.
Ulrich, R.K. 1978 in Protostars and Planets, ed. T. Gehrels, U. of Arizona Press, p. 718.
Ulrich, R.K. and Knapp, G. 1979, Ap.J. Let., 230, L99.
Walker, M.F. 1972, Ap.J. 175, 89.
Walker, M.F. 1979, Ap.J. 224, 546.
_____ 1981, P.A.S.P., 92, 66.

Results From Ultraviolet Observations of T Tauri Stars

Mark S. Giampapa
National Solar Observatory
P. O. Box 26732
Tucson, Arizona 85726-6732/USA

Introduction

The advent of the International Ultraviolet Explorer (IUE) satellite observatory has made new wavelength regions accessible to analysis and, consequently, new thermal regimes in the atmospheres of T Tauri stars can be investigated. In particular, direct diagnostics of plasma at temperatures and heights that are intermediate (in main sequence-like atmospheres) between those of the stellar upper photosphere and the corona become available. The analysis and interpretation of the ultraviolet spectrum of T Tauri stars will be naturally guided by our experiences derived from parallel investigations of the Sun and other late-type stars. The implicit assumption in this approach is that the physical origin of T Tauri atmospheric activity is fundamentally similar to that of main sequence and post-main sequence stars which also exhibit chromospheric and coronal activity, although the relative scales of any such activity (i.e., net radiative cooling, mass loss rates, surface magnetic flux, etc.) can be vastly different. Indeed, I consider it incumbent upon those who would examine the T Tauri stars as an isolated class that exhibit atmospheric characteristics that have no relation to similar manifestations in more evolved stars, to identify the stellar type in the H-R diagram that would represent the transition between "T Tauri atmospheric physics" and "solar-stellar physics." Conversely, it is clearly important to recognize that the T Tauri stars are pre-main sequence objects and thus are not in hydrostatic equilibrium; their line spectra are complex and display indications of the simultaneous occurrence of both mass outflow and mass infall; the stars are often embedded in circumstellar material; and the T Tauri stars rotate more rapidly than the Sun.

In this review I will discuss the principal results that have been deduced from ultraviolet observations of T Tauri stars as obtained with the IUE satellite in the 1150Å-3200Å wavelength range. In particular, I will review the basic aspects of their UV line spectra, including atmospheric properties as inferred from density sensitive line diagnostics and emission measures, observations of UV variability and manifestations of multicomponent atmospheres.

Finally, I will summarize the major results and present questions which should be addressed by future investigations of the T Tauri stars as observed in the ultraviolet.

Far Ultraviolet Line Spectra

I display in Table 1 the far ultraviolet (1150Å–2000Å) line surface fluxes, relative to solar line surface fluxes, for several T Tauri stars as given by Imhoff and Giampapa (1982b). A more extensive compilation with some revised values of these relative fluxes will be presented by Imhoff and Giampapa (1983). Nevertheless, the basic conclusions from Table 1 remain unchanged, namely, that the far UV fluxes in T Tauri stars are distinctly enhanced by typical factors of $\sim 10^{2-3}$ relative to the quiet Sun. The total contribution of the far UV line emission to radiative cooling can be as high as 0.1% of the stellar bolometric luminosity (Imhoff and Giampapa 1982b).

TABLE 1
Ratios of Stellar to Solar Far-Ultraviolet
Line Surface Fluxes for Eight T Tauri Stars

| | | | | | | T Tauri Stars | | | | | | RS CVn Stars | |
	T Tau	DR Tau	RW Aur 1978	RW Aur 1979	GW Ori	CoD -35 10525	RU Lup	AS 205 I	AS 205 II	S CrA	UX Ari	HR 1099
NV 1240	5400:			<420	3300:	5100	350				240	67
OI,SI 1304	2800		<120		980	1400	2500	950			110	63
CII 1335	1700		100	100	610	1400	2800				96	80
SiIv 1400	4800	2500	380	1300	3300	1700	4800	14000	120	2400	226	76
CIV 1550	3100	4000	150	330	2100	2400	1700	4800	50	810	110	91
HeII1640	3100	2700		<160	2800	2200	<850	6500	77	<770	270	192
SiII,SI 1813	620	870	120		150	230	750	3000	30	560	22	19

Note to Table 1: AS 205 Case I: KO, $A_v = 3\overset{m}{.}02$
Case II: MO, $A_v = 1\overset{m}{.}3$

I display in Figures 1 and 2 examples of ultraviolet spectra of a small sample of T Tauri stars. The spectra in Figure 1 are taken from Appenzeller et al. (1980) while the spectrum in Figure 2 is from Imhoff and Giampapa (1980). As in the case of many late-type stars, lines of transition region ions such as C IV, Si IV, N V and He II can appear in the far UV spectra of T Tauri stars. In addition, upper chromospheric lines of SiI, CII and OI are present while FeII features can be identified throughout their spectra. Shell absorption or emission lines occur in the far UV spectra of T Tauri stars with no or little photospheric contribution in this wavelength range (Imhoff and Giampapa 1981; Appenzeller and Wolf 1980; Mundt et al. 1981). Moreoever, far ultraviolet molecular H_2 emission has been identified by Brown et al. (1981) in the direction of the recently discovered infrared companion of the prototype T Tau. Furthermore, the UV

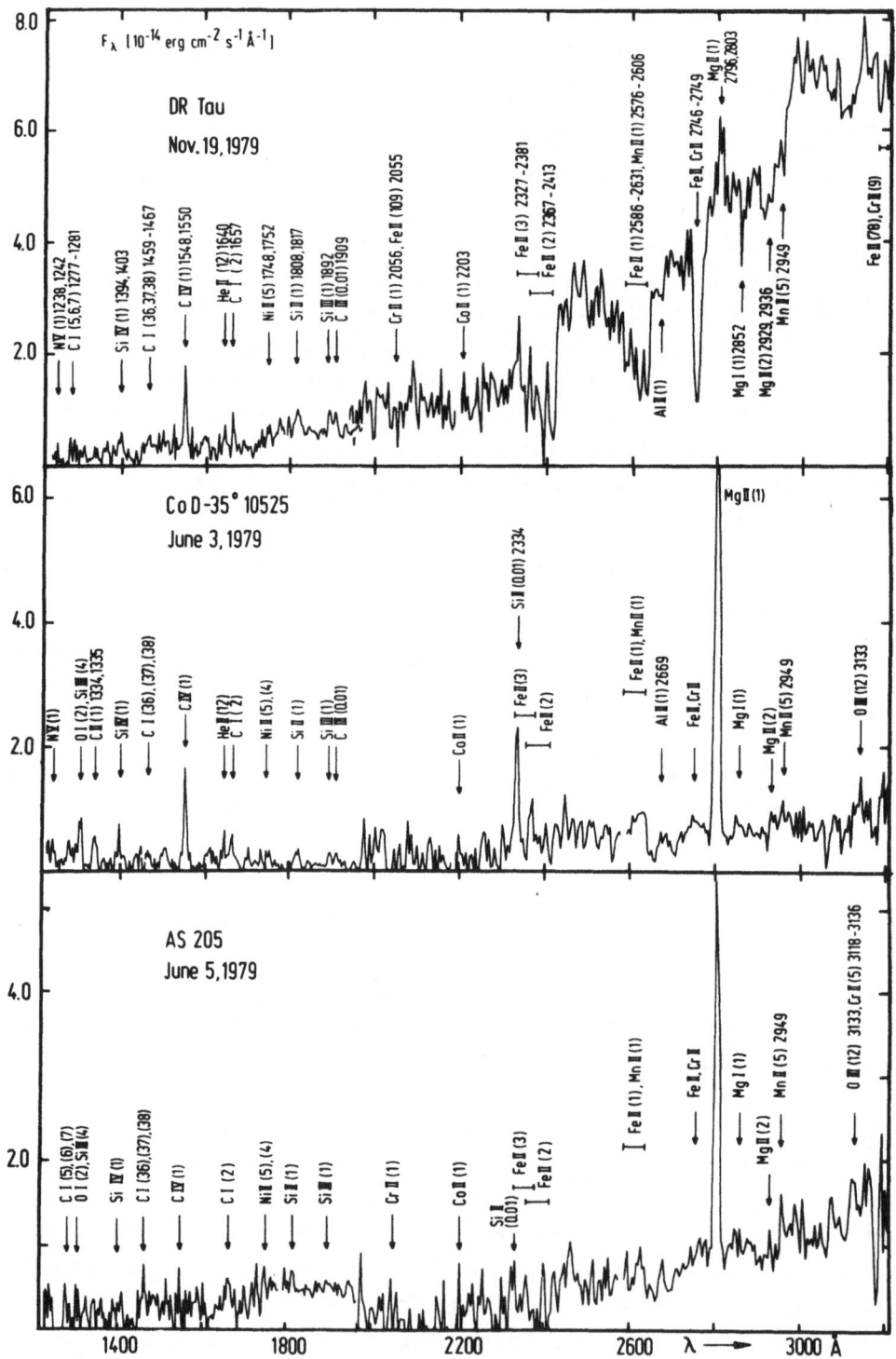

Figure 1 The observed ultraviolet spectra of DR Tauri, CoD -35°10525, and AS 205

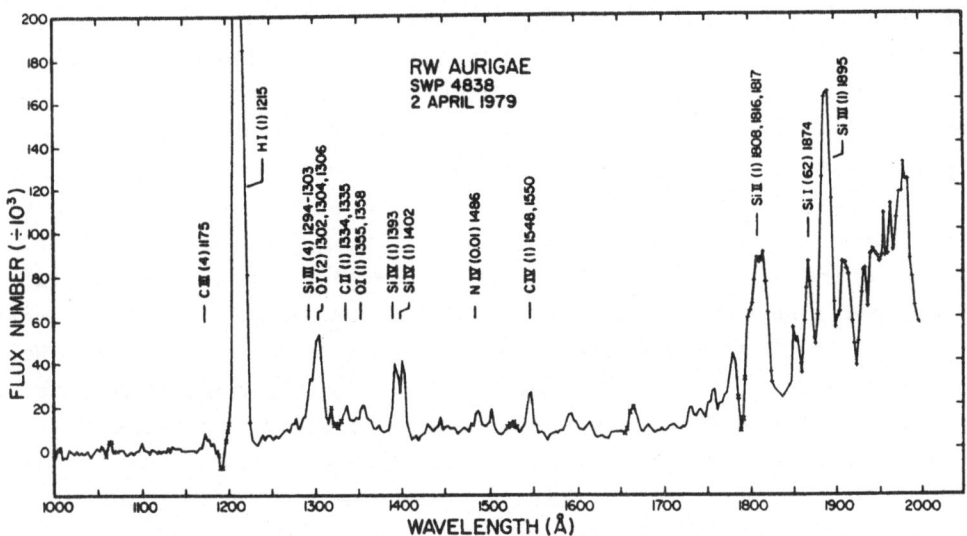

Figure 2.   The far ultraviolet spectrum of RW Aurigae.

continuum appears to be dominated, in many instances, by free-free and
hydrogen recombination radiation emanating from an extended region
that is hotter than the underlying photosphere.   A cool stellar
photospheric energy distribution (corresponding to the known spectral
type of a T Tauri star as inferred from optical and infrared
observations) often does not satisfy the observed ultraviolet spectral
energy distribution (Imhoff and Giampapa 1980; Appenzeller and Wolf
1980).   Clearly, the photospheric properties of the T Tauri stars
cannot be productively studied in the ultraviolet.   Finally,
Appenzeller et al. (1980) claim that there is a correlation between
strong UV continuum emission and optical "blue veiling" indicating
that the so-called veiling can be the result of envelope emission seen
against a cool photospheric background, as opposed to deep
photospheric heating at the onset of a steep chromospheric temperature
gradient.   However, exceptions are noted, such as RU Lupi, which are
"strongly veiled" in the optical but shows a relatively weak
ultraviolet continuum in near simultaneously acquired optical and UV
spectra (Appenzeller et al. 1980).   The blue veiling is likely the
result of both an extended envelope contribution and deep photospheric
heating.   The relative importance of these processes in the veiling
phenomenon is yet to be determined.

    The far ultraviolet spectra of T Tauri stars invites comparisons
with those of active chromosphere, main sequence stars such as the M

dwarf stars. Representative far UV spectra of dMe and dM stars are given by Linsky et al. (1982; their Figure 1). Inspection of these spectra reveals that the C IV $\lambda 1550$ resonance doublet is the most prominent emission line feature in the far UV spectra of dMe stars. Moreover, lines of high temperature species, such as N V, Si IV and He II $\lambda 1640$ (which may result from recombination following X-ray photo-ionization), are clearly evident. By contrast, the far UV spectra of non-dMe stars are dominated by lines arising from low excitation neutral metal species. The C IV $\lambda 1550$ feature is only weakly present, if at all, in dM stellar spectra while lines of N V, Si IV, and He II are not detected. Inspection of Table 1 and Figures 1 and 2 herein reveals that the T Tauri stars can exhibit a similar pattern in their UV line spectrum. In particular, Imhoff and Giampapa (1980, 1982b, 1983) originally noted the relative weakening of far UV lines of ions such as C IV, Si IV, He II and the absence of N V $\lambda 1240$. These investigators interpreted this observation as evidence that the maximum temperature, T_{max}, of plasma in these particular T Tauri stars is less than coronal temperatures ($T_{cor} \sim 10^{6-7}$K).

In the case of the M dwarf stars, the difference between the dMe and dM stars can be attributed to the relative degrees of nonradiative heating present in their atmospheres which is, in turn, likely related to the difference in the amount of magnetic flux on the stellar surface. Thus chromospheric, transition region and coronal emissions are each systematically reduced in the dM stars compared to the dMe stars (Giampapa et al. 1981, 1982; Linsky et al. 1982; Vaiana et al. 1981). One may be tempted to offer the same explanation for those T Tauri stars that exhibit relatively weakened transition region line emission. However, this explanation is not applicable to the T Tauri stars. In particular, the T Tauri stars are characterized by strongly enhanced emission line fluxes that are indicative of the highest degree of nonradiative heating present in the atmospheres of these stars as compared to any other class of single, late-type stars (Giampapa et al. 1981b). Hence the relative weakening, or even absence, of high temperature transition region line emission in some T Tauri stars must be due to a source of enhanced nonradiative cooling. More specifically, mass loss becomes the dominant component in the energy balance of the outer atmospheres of some T Tauri stars as opposed to radiative cooling via coronal X-ray emission. While the degree of mass loss is uncertain, estimates are generally in the range of 10^{4-7} times greater than the solar mass loss rate.

Imhoff and Giampapa (1980, 1982a,b, 1983) originally proposed this explanation utilizing ultraviolet and optical observations of T Tauri stars. In particular, these investigators noted that the optically "weak emission" T Tauri stars (that is, those stars with relatively less strong Balmer line and Ca II H and K line emission in the optical) show C IV, N V, Si IV and He II emission lines in normal relative strengths and were detected in the X-ray by Einstein (HEAO-B). The optically "strong emission" T Tauri stars, which are presumably characterized by higher mass loss rates, displayed both a relative weakening of high temperature lines in the UV and were not detected in the X-ray by Einstein. Theoretical corroboration of this explanation has recently been advanced by Hartmann et al. (1982) in their Alfvén wave driven mass loss models for T Tauri stars. More specifically, this model predicts that the local Alfvén wave heating rate increases more slowly with wind density than does the radiative cooling rate. Hence there is a decline in the wind temperature, T_{max} with increasing mass loss rate. In other words, those regions of highest Alfvén wave flux (and therefore highest heating rates) will also have the largest mass flux rates and will generally be cooler. This favors low-temperature emission over high-temperature emission. Thus coronal temperatures are never attained in those T Tauri stars characterized by a sufficiently high Alfvén wave flux and correspondingly high mass loss rate. Thus the far ultraviolet can actually be utilized to ascertain the T_{max} of plasma in the outer atmospheres of X-ray quiet T Tauri stars if the far UV emission generally arises from wind regions.

Interestingly, Gahm (1980) and Walter and Kuhi (1981) proposed that all T Tauri stars possess coronae and the soft X-ray emission from those T Tauri stars that were not detected by Einstein was simply extinguished by overlying gas and dust. However, the aforementioned arguments demonstrate that the most natural explanation that is consistent with the ultraviolet (and optical) data is that coronal temperatures are never attained in X-ray quiet T Tauri stars and the "smothered coronae" hypothesis (Walter and Kuhi 1981) is not correct (additional arguments that illustrate the inconsistencies of this hypothesis have already been delineated by Imhoff and Giampapa [1982b]). However, I note that in those T Tauri stars that were detected in the X-ray, the level of X-ray emission is not enhanced to the degree that would be expected from an extrapolation of the far UV fluxes observed in these stars combined with the UV and X-ray scaling relations given by Ayres et al. (1981).

In summary, the ultraviolet spectra of T Tauri stars are characterized by intense line and continuous emission indicative of a high degree of nonradiative heating occurring in the atmospheres of these stars. There exists T Tauri stars that do not possess coronae even though strong chromospheric line emission is present. In particular, the relative weakening or absence of high temperature UV transition region lines reveals that the maximum plasma temperature attained in these stars is only $T_{max} \sim 10^5$ K. Mass loss appears to become the dominant atmospheric cooling mechanism and this, in turn, constitutes corraborative evidence for the applicability of Alfvén wave-driven mass loss models for T Tauri stars. <u>This represents a fundamental conclusion that could only have been deduced from far ultraviolet observations.</u>

<u>Emission Measures and Densities</u>

The distribution of emission measures with temperature can provide insight on the operative heating mechanisms in the atmospheres of T Tauri stars. Moreover, emission measures combined with estimates of electron density can yield information on the depth (or geometric extent) of a line source region and the degree to which the stellar atmosphere can be regarded as "plane-parallel and homogeneous". Cram <u>et al</u>. (1980) find that the temperature distribution of emission measures for RU Lup and two observations of RW Aur is similar to the solar case but at a considerably enhanced level (i.e., solar x 5 x 10^4). Cram <u>et al</u>. (1980) utilized the density sensitive ratios of Si III $\lambda 1892$/C III] $\lambda 1908$ and Si IV $\lambda 1403$/C III] $\lambda 1908$ to obtain a crude estimate of the electron density in the source region of RU Lup. These investigators find a mean value of log N_e = 10.2±0.6 but emphasize that the line ratios are highly sensitive to uncertain atomic parameters and atmospheric models. The propagation of uncertainties in estimates of emission measure and N_e lead to a range of 0.08 R_* - $4R_*$ for the source depth of SiIV emission in RU Lup (Cram <u>et al</u>. 1980). This range is compatible with both a relatively compact emission region and a hot, extended region.

Hartmann <u>et al</u>. (1982) computed the flux ratio for Si IV/C III] within the context of an Alfvén wave model for mass loss in T Tauri stars. These investigators find Si IV/C III] = 0.94 which is to be compared to the observed mean value <Si IV/C III]> = 2.5±1.5 and the range Si IV/C III] = 0.4-4.9 for seven T Tauri stars (Imhoff and Giampapa 1983). By contrast, the static, deep chromospheric, nonspecific model of T Tauri stars constructed by Cram (1979) yields

Si IV/C III] = 2.61. While the observed mean value for this line flux ratio appears more consistent with the static, deep chromosphere model, the range of observed values is consistent with both models. I regard this as evidence for the presence of both compact regions and extended regions in the atmospheres of T Tauri stars. Interestingly, Brown et al., (1983) find, in their detailed examination of the spectrum of T Tau, discrepant values of the electron density as inferred from different line diagnostics formed in similar thermal regimes. These investigators claim that this apparent discrepancy can only be reconciled by postulating a two-component atmosphere for T Tau (see also Giampapa et al. 1981). In particular, Brown et al. (1983) claim that the observed flux ratios and inferred electron densities can be understood if the permitted lines and X-ray emission are predominantly formed in a high pressure, hydrostatic region while the semiforbidden lines (C II], C III], Si II], etc.) are formed in an extended, low density component of the atmosphere. Moreover, the emission measure of the low density region must be comparable, but not greater than, the emission measure of the high density region (Brown et al. 1983; see also Bertout 1983 for a single-component parameterization of T Tauri atmospheres).

In summary, the aforementioned investigations of T Tauri stars combine to demonstrate that a realistic model of their atmospheres requires two components; a high density, compact region analogous to closed magnetic field structures observed on the Sun, and open regions, analogous to solar coronal holes, that are characterized by a relatively strong mass loss rate and a T_{max} ~10^5 K. Hence the observed transition region line emission is composed of a wind, or "extended region" contribution, and a compact region contribution. The X-ray emission can only arise from the compact component since $T_{max} < T_{cor}$ in the wind. I suggest that this may explain why X-ray emitting T Tauri stars appear somewhat underluminous in X-rays than would be expected on the basis of the observed strong enhancement of their far UV emission. The relative proportion of each contribution likely depends upon the extent and kind of magnetic field configurations that exist on the stellar surface at the time of observation.

Ultraviolet Variability

The T Tauri stars are characterized by both rapid and long-term variability at practically all wavelengths. The UV variability of RW Aur has been described in detail by Imhoff and Giampapa (1981) and I update those results herein. In particular, the visual brightness of RW Aur increased by a factor of 2.5 between July 1978 and April 1979. During this time, the UV continuum at 2700Å±100Å and 2900±100Å increased by a factor of 5. The resonance lines of C IV and Si IV each increased by 4-5 times while the lower excitation lines of C II and Mg II (h and k) increased by factors of 2.5 and 1.5, respectively. In addition, the UV shell spectrum changed from emission to absorption. This latter observation implies that the enhanced UV continuum emission was interior to the shell.

The origin of the variability is unknown. However, the degree and pattern of the UV continuous and line emission enhancement is reminiscent of violent flare activity. Violent outbursts and flare-like activity has been observed in the optical (Worden et al. 1981; see also Mundt and Giampapa 1982 and Mundt 1983) and the X-ray (Feigelson and De Campli 1981). Of course, temporal and spatial variability in the wind structure can affect the far UV high temperature emission (e.g. see Mundt and Giampapa 1982). The ultraviolet variability of T Tauri stars clearly merits further investigations with IUE and Space Telescope using improved temporal resolutions.

The Mg II h and k Resonance Lines

The spectra of T Tauri stars in the 2000Å-3200Å range is dominated by the resonance lines of Mg II. The results of an extensive, low resolution study of the Mg II lines in a sample of T Tauri stars has been presented by Giampapa et al. (1981b). According to these investigators, the Mg II (h and k) line fluxes indicate that the atmospheres of T Tauri stars are characterized by the most extreme degree of nonradiative heating among the class of single late-type stars (Giampapa et al. 1981b, their Figure 3). In fact, the Mg II lines, the Ca II lines, the far UV emission combined with any X-ray emission can comprise 1% or more of the stellar bolometric luminosity. The total outer atmospheric radiation losses (including Balmer and Lyman line emission) in the specific case of T Tau have been estimated to be 8% of its bolometric luminosity (Brown et al. 1983).

High resolution observations of both the Mg II h and k lines and the Ca II K line reveal P Cygni type profiles indicative of the presence of an expanding chromospheric region (Giampapa et al. 1983). Moreover, the Ca II K line profile and Mg II k line profile have similar shapes. Hence these lines arise from similar regions in

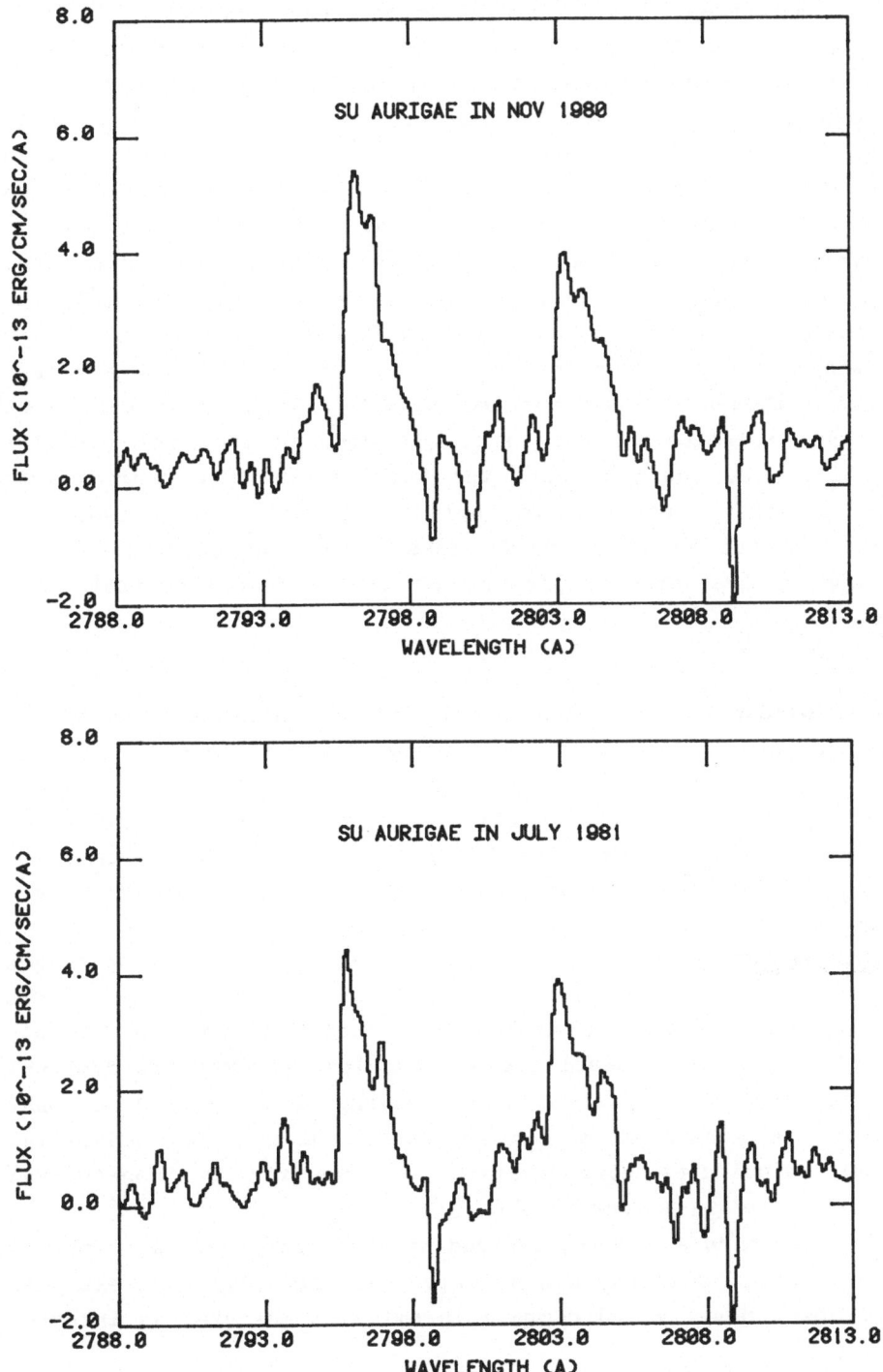

Figure. 3 The Mg II h and k line profiles
in the spectrum of SU Aurigae.

a T Tauri atmosphere. This observation is consistent with the fact
that the Mg II and Ca II resonance lines share similar formation
characteristics. Furthermore, the velocity displacement of the P
Cygni absorption from line center reveals that the Mg II and Ca II
resonance line profiles can be utilized as diagnostics of the wind
acceleration region (Giampapa et al. 1983). Mundt (1983) infers a
similar result for the Na I D lines. The Mg II h and k lines exhibit
profile variability as well as variability in their total flux (e.g.
see Brown et al. 1983). I display in Figure 3 two high resolution
observations of the Mg II h and k lines in SU Aur. The variability in
the blue wings of these profiles is evident. Thus the mass loss rate
is variable although a quantitative estimate of both the mass loss
rate and the variation in this rate, as deduced from the profiles
given in Figure 3, would require the application of a hydrodynamic
radiative transfer code. In brief summary, high resolution
observations of the Mg II h and k lines reveal the existence of an
accelerating chromospheric component characterized by a variable mass
loss rate.

Summary

   In the following I summarize the principal results that have thus
far emerged from ultraviolet observations of T Tauri stars:
1. The ultraviolet chromospheric and transition region emission lines
are strongly enhanced ($\sim 10^{2-3}$) relative to the quiet Sun and other
late-type stars. The outer atmospheric line and continuous emission
luminosity can, in total, be several percent of the stellar bolometric
luminosity.
2. The relative weakening or absence of lines of high temperature
species, such as N V and C IV, in X-ray quiet T Tauri stars, reveals
that the maximum temperature attained in their atmospheres is only
$T_{max} \sim 10^5$ K. In these T Tauri stars, mass loss becomes the dominant
component in the outer atmospheric energy balance and coronal
temperatures are consequently not attained. This is a fundamental
result that constitutes corroborative evidence for the non-thermal
origin of mass loss in T Tauri stars.
3. The T Tauri stars are characterized by multi-component atmospheres
composed of static, compact, dense regions and geometrically extended
regions of lower density and comparable emission measure, as inferred
from the far UV.
4. High resolution observations of the Mg II (and Ca II) resonance
lines reveal the presence of a chromospheric component that is
expanding. Moreover, these features can be utilized as probes of the

wind acceleration region in T Tauri stars. The line profiles are variable, particularly in the blue wings, and indicative of a variable mass loss rate.

5. The UV variability of T Tauri stars is correlated with their visual brightenings and "blue veiling". The pattern of UV line and continuous emission features is reminiscent of violent flare out-bursts. Shell emission and absorption lines are present in the spectrum but no photospheric features are seen. While the UV can be utilized as a probe of the extended regions, it is not a useful wavelength region for the investigation of the photospheric properties of T Tauri stars.

In the following I pose some of the many questions that future investigations of T Tauri stars should address. In particular:

1. What are the initial conditions that lead to a mass flux and associated cooling rate in some T Tauri stars such that $T_{max} < 10^6$ K? Are we witnessing the evolution of stellar coronae?

2. Are there evolutionary trends (in outer atmospheric properties) within the class of pre-main sequence late-type stars? What is the evolutionary relationship of the T Tauri stars to the so-called "post-T Tauri" stars?

3. What is the range of variability in the mass loss rate as inferred from ultraviolet data?

4. Is there evidence in the UV for infalling as well as outflowing material?

5. Is the circumstellar material still primordial or the result of mass loss and episodic outbursts (see Mundt 1983)?

6. Is there a rotation-activity relation for the T Tauri stars?

Finally, a recent investigation by Canuto et al. (1983) utilized ultraviolet observations of T Tauri stars to formulate a model of the UV irradiance from the young Sun. These investigators then studied the effect of this UV radiation in the disassociation of oxygen from outgassed CO_2 to produce free oxygen and an ozone layer in a model of the prebiological paleoatmosphere of the Earth. Thus ultraviolet observations of the T Tauri stars may eventually assist in explaining how we can be here to contemplate the evolution of the stars.

References

Appenzeller, I. et al. 1980, Astron. Astrophys., 90, 184.

Appenzeller, I. and Wolf, B. 1979, Astron. Astrophys., 75, 164.

Ayres, T. R., Marstad, N. C. and Linsky, J. L., Ap. J., 247, 545.

Bertout, C. 1983, "T Tauri Stars: Not One But Two Transition Regions", in <u>Proceedings of the Third European IUE Conference</u>, in press.

Brown, A., Ferraz, M. C. de M. and Jordan, C. 1983, <u>M.N.R.A.S.</u>, in press.

Brown, A. <u>et al.</u> 1981, <u>Nature</u>, <u>290</u>, 34.

Canuto, V. M. <u>et al.</u>, <u>Nature</u>, <u>305</u>, 281.

Cram, L. E. 1979, <u>Ap. J.</u>, <u>234</u>, 949.

Cram, L. E., Giampapa, M. S. and Imhoff, C. L. 1980, <u>Ap. J.</u>, <u>239</u>, L115.

Feigelson, E. D. and De Campli, W. M. 1981, <u>Ap. J. (Letters)</u>, <u>243</u>, L89.

Gahm, G. 1980, <u>Ap. J. (Letters)</u>, <u>242</u>, L163.

Giampapa, M. S. <u>et al.</u> 1981a, <u>Ap. J.</u>, <u>246</u>, 502.

Giampapa, M. S. <u>et al.</u> 1981b, <u>Ap. J.</u>, <u>251</u>, 113.

Giampapa, M. S., Worden, S. P. and Linsky, J. L. 1982, <u>Ap. J.</u>, <u>258</u>, 740.

Giampapa, M. S. <u>et al.</u> 1983, in preparation.

Hartmann, L., Edwards, S. and Avrett, E. H. 1982, <u>Ap. J.</u>, <u>261</u>, 279.

Imhoff, C. L. and Giampapa, M. S. 1980, <u>Ap. J. (Letters)</u>, <u>239</u>, L115.

_____. 1981, "The Ultraviolet Variability of the T Tauri Star RW Aurigae", in <u>the Universe at Ultraviolet Wavelengths: The First Two Years of IUE</u>, ed. R. D. Chapman, (NASA Conference Publication 1981), p. 185.

_____. 1982a, "Far Ultraviolet and X-ray Evidence Concerning the Chromospheres and Coronae of the T Tauri Stars", in <u>The Second Cambridge Workshop on Cool Stars, Stellar Systems, and the Sun</u>, eds. M. S. Giampapa and L. Bolub, SAO Special Report <u>392</u>, p.175.

_____. 1982b, "Chromospheres and Coronae in the T Tauri Stars", in <u>Advances in Ultraviolet Astronomy: Four Years of IUE Research</u>, eds. Y. Kondo, J. M. Mead and R . D. Chapman (NASA Conference Publication 2238, 1982), p.456.

Imhoff, C. L. and Giampapa, M. S. 1983, <u>Ap. J.</u>, to be submitted.

Linsky, J. L. <u>et al.</u> 1982, <u>Ap. J.</u>, <u>260</u>, 670.

Mundt, R. 1983, <u>Ap. J.</u>, in press.

Mundt, R. 1981, <u>Astron. Astrophys.</u>, <u>93</u>, 412.

Mundt, R. and Giampapa, M. S. 1982, <u>Ap. J.</u>, <u>256</u>, 156.

Vaiana, G. S. <u>et al.</u> 1981, <u>Ap. J.</u>, <u>245</u>, 163.

Walter, F. M. and Kuhi, L. V. 1981, <u>Ap. J.</u>, <u>250</u>, 254.

Worden, S. P. <u>et al.</u> 1981, <u>Ap. J.</u>, <u>244</u>, 520.

X-ray Emission from Pre-Main Sequence Stars

Eric D. Feigelson
Department of Astronomy
The Pennsylvania State University
University Park, PA 16802

ABSTRACT

The status and implications of recent X-ray studies of pre-main sequence (PMS) stars are reviewed in detail. Images of nearby star formation regions including the Orion, Taurus-Auriga, ρ Ophiuchi and Chamaeleon clouds obtained with the Einstein X-ray Observatory have revealed more than 150 X-ray luminous PMS stars to date. Some are classical strong emission line T Tauri stars while others are late-type PMS stars with weak or absent lines. The soft X-ray luminosities range from $\lesssim 10^{29}$ erg/s to several 10^{31} erg/s, or $\sim 10^3$ times that of local main sequence stars. The X-ray emission is highly variable; most show large amplitude variations on timescales of days, and a few exhibit flares on timescales of 10^2-10^4 seconds. The rapid variations, hard spectra, and lack of correlation with wind-produced optical emission lines all point to a flare origin of the enhanced X-ray emission. Solar-type stars thus exhibit their most violent surface activity during their PMS stage of evolution. Implications for the conditions in PMS stars, stellar activity-rotation age relation, for the chemistry and evolution of molecular clouds, and for various chemical and isotopic anomalies in the ancient solar system are discussed.

It has been known for over a decade that stellar activity is inversely correlated with stellar age (Skumanich 1972). However, it has proved difficult to extend the traditional activity indicators, such as Ca II optical and ultraviolet emission lines, to the youngest pre-main sequence (PMS) stars. Spectroscopic evidence for chromospheric activity in classical T Tauri stars is confused by the presence of dense envelopes or winds, and there has been no reliable procedure for finding PMS stars in which surface activity could be directly studied independent of wind properties. A new tool has emerged to address this problem. X-ray images of nearby star forming clouds have been obtained with the Imaging Proportional Counter (IPC) and High Resolution Imager (HRI) on board the Einstein X-ray Observatory (HEAO-B satellite) between 1979 and 1981. They show dozens of X-ray luminous PMS stars, many of them without T Tauri emission line spectra. Their X-ray emission can be confidently attributed to flare-like events, firmly establishing that 10^6 yr old stars are characterized by very strong surface activity. I review here the observational results on PMS X-ray emission and discuss the implications for stellar physics and related topics. This field has been more briefly reviewed by Vaiana (1981) and Stern (1983) in the context of main sequence X-ray emission, by Kuhi (1983) in light of other properties of T Tauri stars, and by Zahnle and Walker (1982) in terms of solar system evolution.

I.   Pre-Einstein History

Prior to the advent of imaging X-ray telescopes, there were several theoretical speculations and one detection of X-ray emission from low-mass PMS stars. Bisnovatyi-Kogan and Lamzin (1977) considered the possibility that T Tauri winds are thermally driven, requiring a dense corona at $\sim 10^6$ °K at the base of the wind. The resulting X-ray luminosities of $\gtrsim 10^{34}$ erg/s are orders of magnitude greater than those observed, and were already precluded by early X-ray sky surveys. Impact of an outflowing wind on the surrounding interstellar cloud   (Schwartz, 1978) or of infalling material on the surface of YY Orionis-type PMS stars (Ulrich 1976; Mundt 1981; Vaiana 1981) could also produce X-ray emission. The difficulty here is that T Tauri winds have velocities $\lesssim 300$ km/s which, upon thermalization, produce X-rays around 0.1 keV rather than the observed ~ 1 keV. Relating to stellar flares, Gurzadyan (1973) calculated that $\sim 10^{31}$ erg/s of X-rays will be produced when mono-energetic MeV electrons penetrate dust grains and produce transition radiation. This prediction proved to be quite accurate, despite the lack of wide acceptance of Gurzadyan's 'fast electron' theory.

The sole detection of X-rays from PMS stars by non-imaging instruments was the report from ANS and SAS-3 satellite observations that most of the Orion nebula X-ray source 4U0531-05, with L_x = 4 x 10^{33} erg/s (1-8 keV), is about half a degree in extent (den Boggende et al. 1978; Bradt and Kelley 1979). A compact component is produced by the Trapezium OB stars, but the extended component was tentatively attributed to luminous coronae around a few T Tauri stars (den Boggende et al.) or to a diffuse hot plasma (Bradt and Kelley).

II.   Einstein X-ray Observatory Results

a)   Orion Nebula. The Orion nebula was the first star formation cloud observed with the Einstein Observatory. An exposure with the low resolution IPC, shown in Figure 1, reveals more than 20 discrete sources within $\sim 30'$ of the Trapezium (Ku and Chanan 1979). The X-ray sources are spatially associated with 'nebular variable' stars and have 0.5-4.5 keV luminosities ranging from 1 x 10^{31} erg/s to 5 x 10^{31} erg/s. A follow-up observation with the HRI shows 58 statistically significant sources (Ku et al. 1982). Almost all of these stars were numbered by

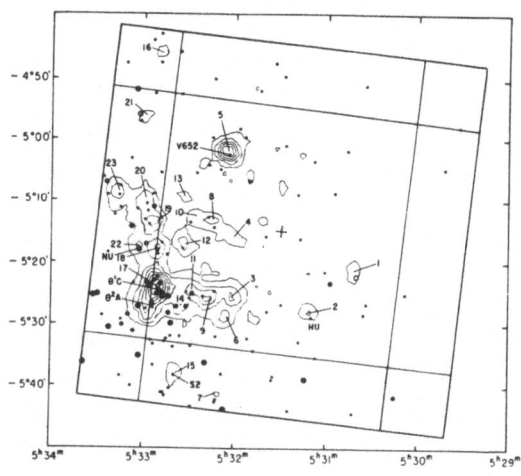

Figure 1. Contour map of the Einstein IPC X-ray image of the central region of the Orion nebula (Ku and Chanan 1979). The bright X-ray source in the southwest corner is due to the Trapezium OB stars. The remaining ∿20 sources are low mass PMS stars.

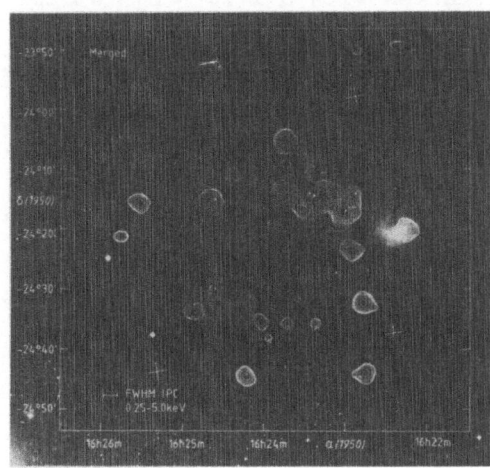

Figure 2. IPC contour map of the central region of the ρ Ophiuchi cloud superposed on the red Palomar Sky Survey print (Montmerle et al. 1983). This map is the sum of exposures taken on six separate days (see Figure 4c).

Figure 3. A mosaic of three overlapping IPC images of the Chamaeleon I star formation cloud (Feigelson and Kriss 1983). Of the 23 labeled locations, 11 are associated with previously known Hα-emission stars, 8 with X-ray identified PMS stars, and 2 are statistical fluctuations.

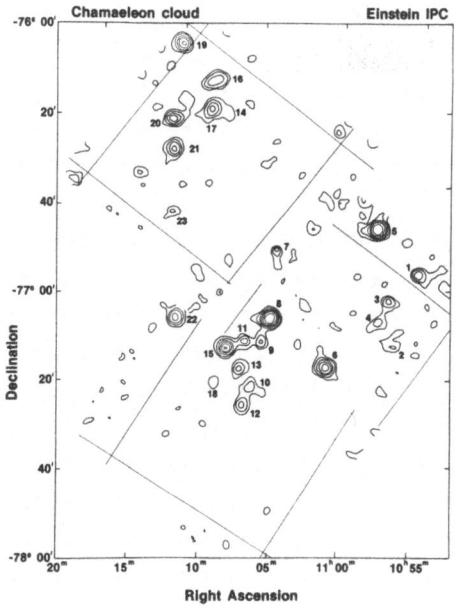

Parenago, half are variables in Kukarkin's catalog, but only five appear in Herbig and Rao's (1972) catalog of PMS stars. Spectral types range from K to O. The rich population of X-ray luminous PMS stars is not limited to the core of the Orion nebula: an IPC image pointed 1.5° south of the Trapezium showed another $\sim$20 stellar X-ray sources (Pravdo and Marshall 1981). It is reasonable to believe that a sufficiently complete and sensitive survey of the Orion complex would reveal some hundreds of stars with $L_x \gtrsim 10^{30}$ erg/s.

b) Taurus-Auriga Cloud Complex. Since this region is more than 20° in extent only portions were examined with Einstein. Thirty PMS stars from the catalog of Herbig and Rao (1972) lie in 18 IPC fields reported by Gahm (1980), Feigelson and DeCampli (1981), and Walter and Kuhi (1981 and 1983). Of these 30 sources, 12 to 14 (the uncertainty is due to two unresolved double T Tauri stars) are detected with $5 \times 10^{29} \leq L_x \leq 8 \times 10^{30}$ erg/s. Five stars that were not found in optical Hα surveys are also detected. The latter group has been intensively studied: all have K7/M0 spectral types, lie $1-2^m$ above the ZAMS, have no significant infrared excess, exhibit weak and variable Hα emission (1-5 Å equivalent width), strong Li 6707 Å absorption (0.7 Å equivalent width), low level quasi-sinusoidal photometric variability (<0.1 to 0.5m), and moderately rapid surface rotation (<10 to 30 km/s) (Rydgren, Schmelz, and Vrba 1982; Feigelson and Kriss 1981 and 1983a; Mundt et al. 1983, Rydgren and Vrba 1983; Walter, this volume). These X-ray PMS stars, which are spectro-scopically similar to the least active optically identified PMS stars (Cohen and Kuhi 1979), may be some of the mysteriously rare 'post-T Tauri' stars or simply T Tauri stars that currently have no winds. One of these stars, 1E0429+1755, is the first known PMS double-lined spectroscopic binary (Mundt et al. 1983). It is comprised of nearly identical K7/M0Ve stars with orbital period 3.9 d, zero eccentricity, and estimated separation of 5 stellar radii.

Several further characteristics of the stars detected in the Tau-Aur clouds were noted (Gahm 1980; Feigelson and DeCampli 1981; Walter and Kuhi 1981 and 1983). First, only T Tauri stars brighter than $\sim 13^m$ ($M_v \lesssim 7$) are detected in short IPC exposures, suggesting a correlation between L_{bol} and L_x. Second, considering the various predictions that T Tauri winds would produce X-rays (section I above), it was surprising that many of the T Tauri stars with the strongest optical emission lines (e.g. RW Aur, UZ Tau, XZ Tau) were not detected while many of those which are X-ray luminous had very weak lines. This could imply either that the heaviest T Tauri winds absorb X-rays produced near the surface, or that wind and X-ray production are intrinsically anticorrelated. Third, the

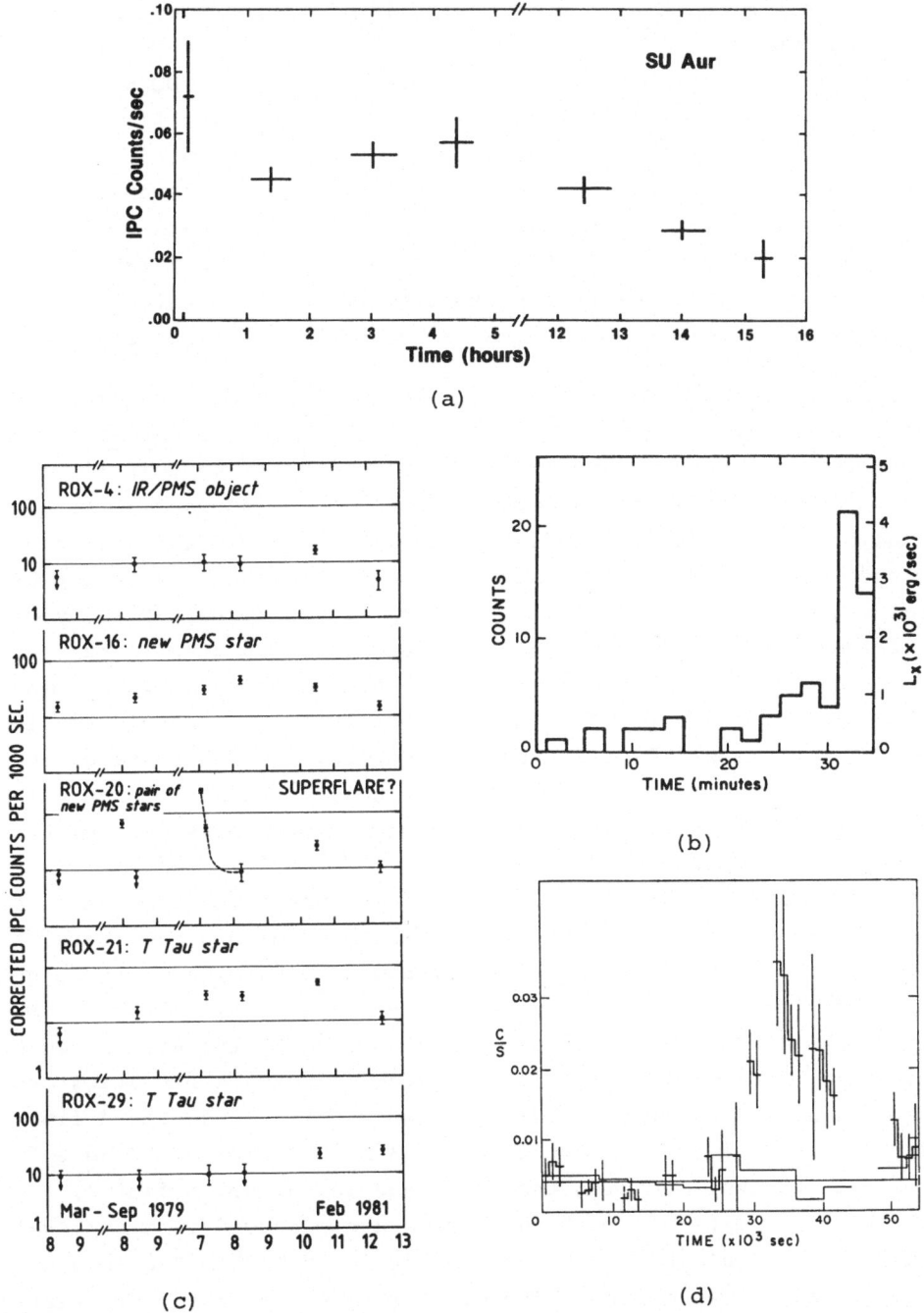

(a)

(b)

(c)

(d)

Figure 4. X-ray light curves of rapid variations seen in PMS stars:
(a) SU Aur shows a mild decline in 16 hours (Sanders and Hartmann, in
preparation);   (b) DG Tau exhibits the most rapid flare seen, rising
a factor >6 in ∿ 4 minutes (Feigelson and DeCampli 1981);   (c) several
typical and one extraordinary (ROX20) variations from the ρ Ophiuchi
PMS stars are shown (Montmerle 1983b);   and (d) a complete X-ray event
is seen in AS205 within 4 hours (Walter and Kuhi 1983).

majority of stars which were observed during two epochs exhibit variability and two stars changed brightness within a single observation. The X-rays from SU Aur decreased by 50% in 16 hours and DG Tau showed a spectacular increase from $<6 \times 10^{30}$ erg/s (3σ upper limit) to 4×10^{31} erg/s in only 10^2 seconds. These light curves are shown in Figure 4a and 4b. Fourth, spectral fits to the IPC pulse height channel distributions indicate that the emission process is either nonthermal or thermal from plasma with $T \gtrsim 1 \times 10^7$ K. In either case, the X-rays are absorbed below ∼1 keV to the degree expected from the observed visual absorption assuming cosmic abundances in the obscuring material.

c) ρ Ophiuchi Cloud. Motivated by the coincidence of the γ-ray source 2CG353+16 with the ρ Oph star formation cloud, Montmerle et al. (1983a) obtained two IPC exposures separated by 6 months of the central region in 1979. Finding more than a dozen variable X-ray sources, several additional exposures spaced 1-2 days apart were obtained, along with a mosaic of the entire $2° \times 2°$ cloud at two epochs. A merged X-ray map of the central region is shown in Figure 2. Forty-seven certain X-ray sources with luminosities ranging from 5×10^{29} erg/s to 2.5×10^{31} erg/s are found. All but 6 have optical stellar counterparts with visual magnitudes ranging from 4.9 to 18.5. These include 9 of 11 previously identified T Tauri stars, 15 of 19 near-IR sources, several B stars, and at least 4 new weak-Hα PMS stars similar to those found in the Tau-Aur clouds. Up to 20 additional X-ray sources of low statistical significance may also be present.

The extensive temporal coverage of the ρ Oph cloud demonstrates that short timescale variability is not restricted to a few stars (Montmerle et al. 1983a and b). Twenty-two of 31 sources with adequate data are seen to vary within 5 days, and virtually all 47 sources vary within 2 years. One X-ray source, designated ROX20 and associated with a previously unnoticed visual binary system of weak-Hα PMS stars, exhibited the most extreme variability (Figure 4c). It declined from 2×10^{31} erg/s to a 'quiescent' level of 8×10^{29} erg/s with an e-folding time of 105 minutes. Scaling from the behavior of solar flares, Montmerle et al. estimate that the X-rays are produced in magnetically confined arches $\gtrsim 10^{11}$ cm in size containing $3-6 \times 10^{10}$ cm^{-3} plasma at 10^7 K undergoing radiative cooling. The similarity to solar and dMe flares is also suggested by a power law distribution of L_x(max)/L_x(min) values for the ρ Oph X-ray stars.

Walter and Kuhi (1983) have examined a region containing two T Tauri stars several degrees from the ρ Oph cloud during two epochs. One of them, AS205, exhibited a flare lasting 4 hours with a peak luminosity of $\sim 1.5 \times 10^{31}$ erg/s (after a factor of 5 correction for absorbing material) and quiescent level >10 times fainter (Figure 4d). This is the only X-ray variation for which both the rise and fall is observed. Other PMS sources in the vicinity are also found.

d) <u>Chamaeleon Cloud</u>. Shortly before the Einstein Observatory ceased functioning in 1981, IPC observations of the Chamaeleon cloud were obtained (Feigelson and Kriss 1983b). This cloud, 1° x 2° in size, is less well known than other star formation regions because of its extreme southern declination (-77°). Thirteen of the 45 optically selected Hα emission line stars are detected, along with 8 additional stars with weak or absent emission lines (Figure 3). As in the Tau-Aur complex, X-ray emission is correlated with optical magnitude. Statistical associations between X-ray and emission line or UBVRI colors are not apparent.

IV. DISCUSSION

a) <u>Origin of the X-ray Emission</u>. As mentioned in section I, there are several possible locations where X-rays could be generated: at the surface of the star, in a hot fraction of the stellar wind, or at the interface of the wind and its environment. Three facts argue persuasively that winds are not involved in X-ray production: (i) many of the X-ray luminous stars have weak or absent optical emission lines, which are essential indicators of dense winds or envelopes; (ii) the rapid variability seen in ROX20, AS205, SU Aur and particularly DG Tau requires densities exceeding 10^{10} cm^{-3} (if the emission mechanism is thermal), which could only be found close to the stellar surface; and (iii) the available X-ray spectra (with one apparent exception, 1E0500+2518; Feigelson and Kriss 1981) point to emission temperatures substantially higher than the thermalization temperature of PMS winds.

Once the proximity to the stellar surface is established, the more subtle distinction between a 'coronal' and 'flare' origin can be sought. The spectral hardness of PMS stars clearly differentiates them from less luminous ($L_x = 10^{27}$-10^{29} erg/s) main sequence coronae. PMS temperatures can not be as low as seen in main sequence F to M stars (kT = 0.1-0.5 keV; Vaiana 1981, Figure 15) even when foreground absorption is taken into account. However, temperatures observed in more active evolved stars, such as RS CVn systems (Swank <u>et al</u>. 1981) and dMe stars (Haisch <u>et al</u>. 1980), are similar to those in PMS stars. The high X-ray temperature of PMS may be a manifestation of the L_x-T correlation predicted in theoretical models of magnetically-confined coronal plasma (Rosner,

Tucker and Vaiana 1978). The extremely rapid (10^2-10^4 sec) and luminous (peak $L_x \sim 10^{31}$ erg/s, total $E_x \sim 10^{34}$-10^{36} erg) variations seen in several PMS stars suggest that explosive flaring rather than quasi-static coronae are responsible for much of the PMS X-ray emission. These events are too rapid to be produced, for instance, by the rotational emergence or disappearance of an active region behind the stellar limb. However, the distinction between an enhanced "coronal" model (e.g. Walter and Kuhi 1981) and a "superflare" model (e.g. Montmerle et al. 1983a) is probably not very great. PMS X-ray emission, like RS CVn and dMe star emission, is probably produced in large loops of magnetically-confined plasma which can evolve on both short (10^2-10^4 sec) and long ($\geq 10^5$ sec) timescales.

It should be mentioned here that there is a growing body of infra-red, optical and ultraviolet evidence for extreme surface activity in classical T Tauri stars. Faint rapid flickering in the U band (Worden et al. 1981), long-term UBVRI variation (Herbst et al. 1982, Schmelz 1983), saturation of the CaII infrared triplet (Herbig and Soderblom 1980) and ultraviolet emission lines seen with IUE (Gaimpapa et al. 1981) have all been interpreted as manifestations of an active and inhomogeneous photosphere-chromosphere-transition region combined with frequent flares. Similarities between T Tauri and flare stars have been noted for many years (see Haro 1968, Gurzadyan 1980 and references therein). The 'chromospheric' model of T Tauri stars has recently been placed on a more quantitative footing with detailed model atmospheres that account for many of their spectroscopic and photometric properties (Calvet, Basri and Kuhi 1983; Hartmann, this volume). The principal open question in linking the X-ray properties with this interpretation is the possible anti-correlation between L_x and T Tauri stars with the strongest chromo-sphere and transition region lines. This could be due to absorption or 'smothering' of the X-rays by the dense wind/envelope in these stars (Gahm 1980; Walter and Kuhi 1981), a failure for these stars to develop an X-ray emitting corona (Imhoff and Giampapa 1981), or the presence of a second transition region between the wind and circumstellar gas (Bertout 1982). Relevant observations include the observed rise in Hα emission accompanying a drop in X-ray emission in AA Tau (Walter and Kuhi 1983) and a statistical anti-correlation of L_x and Hα equivalent width in Tau-Aur and ρ Oph PMS stars. This L_x-Hα anti-correlation is expected in the wind absorption model, but is also probably consistent with the alternative suggestions.

b) <u>The Stellar Activity-Age-Rotation Relation</u>. Given the preceeding discussion, there is no reason why X-ray emission from PMS stars can not be considered in the context of the well-established correlations

between stellar activity, youth, and rotation. Generally speaking, PMS
X-ray luminosities follow the expected trend: local low mass disk stars
have typical $L_x \sim 10^{27}$-10^{28} erg/s, Hyades stars are $\sim$30 times stronger
(Stern et al. 1981), and PMS stars are $\sim$30 times stronger again (Vaiana
1981, figure 18). A more quantitative estimate can be made by combining
the empirical deceleration law, v sin i = 8 $(t/10^9 yr)^{-1/2}$ km/s, with the
two suggested X-ray/rotation relations, $L_x \sim 1 \times 10^{27}$ (v sin i)^2 erg/s
and $L_x \sim 3 \times 10^{28}$v erg/s, giving $L_x \sim 6 \times 10^{31}$ $(t/10^6 yr)^{-1}$ erg/s and
$L_x \sim 6 \times 10^{30}$ $(t/10^6 yr)^{1/2}$ erg/s respectively (Feigelson 1982 and references therein). Most of the PMS stars examined with Einstein are
estimated to be about 2-3 x 10^6 yrs or younger in age (Cohen and Kuhi
1979). The approximate median X-ray luminosity of the detected PMS
stars is $\sim$8 x 10^{30} erg/s in Orion, $\sim$2 x 10^{30} erg/s in Taurus-Auriga,
$\sim$5 x 10^{30} erg/s in ρ Ophiuchi, and $\sim$0.5 - 2 x 10^{30} erg/s in Chamaeleon
(the range here is due to uncertainty in its distance). The true median
is undoubtedly lower by at least a factor of 2 because an uncertain
fraction of the PMS population is undetected and because the time
averaged luminosity of a given star may be substantially lower than the
detected level if flares have a short duty cycle. The observed average
L_x levels are thus several fold lower than the predicted values, suggesting that between stellar ages of 10^6 and 10^8 years either the ro-
tation/age of L_x/rotation correlations is weaker than inferred from
older stars. It seems likely that the v $\propto$ t$^{-1/2}$ relation fails for PMS
stars: optical rotation surveys of very young stars indicate that the
majority have v sin i $\lesssim$ 20 km/s (Vogel and Kuhi 1981; Smith, Beckers
and Barden 1983). The X-ray rotation relation is directly investigated
by Smith, Pravdo and Ku (1983) for 22 young Orion G stars with both
rotational velocities and X-ray luminosities or upper limits. Their
data indicate that a correlation between L_x/L_{bol} and v sin i is present
(the statistical probability is 95% based on a 2 x 2 contingency table
analysis of their Figure 1), but the dependence of L_x on rotation appears
to be less steep than the (v sin i)^2 or v^1 relation found in more evolved
stars. In summary, the available evidence suggests that PMS are more
active than open cluster and local disk stars but the age rotation
relation, and perhaps also the X-ray rotation relation, is weaker than
predicted by extrapolation from older stars. However, the difficulties
created by the uncertain PMS population and source variability preclude
certain conclusions.

c) <u>Flares on PMS Stars</u>. The rapid and large amplitude X-ray variations
seen in DG Tau, ROX20, AS205 and SU Aur argue strongly for the existence
of 'superflares' on PMS stars. X-ray variability is perhaps the only

TABLE 1

PMS and Comparison X-ray Flare Parameters

Star	Rise time	Fall time	Quiescent L_x (erg/s)	Peak L_x (erg/s)	Total L_x (ergs)	Ref.
DG Tau	2×10^2s	–	$<6\times10^{30}$	4×10^{31}	$\geq 1\times10^{34}$	1
AS 205	$<5\times10^3$	1×10^4	$<1\times10^{30}$	1.5×10^{31}	1.5×10^{35}	2
ROX20	–	6×10^3	8×10^{29}	$\geq 2\times10^{31}$	$\geq 8\times10^{35}$	3
Sun-subflare	10^2	1×10^3	10^{26}-10^{27}	1×10^{25}	10^{29}	4,5
Sun-Class 3	10^3	1×10^4	10^{26}-10^{27}	1×10^{27}	10^{31}-10^{32}	4,5
Prox Cen	3×10^2	1×10^3	1.5×10^{27}	7×10^{27}	2×10^{31}	6
YZ CMi	1×10^2	5×10^2	2×10^{28}	8×10^{28}	4×10^{31}	7
HD27130	$<2\times10^3$	2×10^3	1×10^{29}	$> 1\times10^{31}$	$> 3\times10^{34}$	8

References: 1=Feigelson and DeCampli (1981), 2=Walter and Kuhi (1983), 3=Montmerle et al. (1983a), 4=Moore et al. (1980), 5=Dere et al. (1973), 6=Haisch et al. (1980), 7=Kahler et al. (1982), 8=Stern et al. (1983).

---

way individual flares, rather than the collective effects of many flares, can be studied. Table 1 summarizes the observed parameters of the PMS flares along with parameters for small and large solar flares, events seen in the dMe flare stars Prox Cen and YZ CMi, and an X-ray flare seen in the Hyades binary system HD27130. Similar tables are given by Montmerle et al. (1983a) and Stern et al. (1983). The PMS flares are basically similar to solar and dMe flares except for their much greater strength. The total X-ray energies, which (at least in solar flares) constitute a substantial fraction of the total flare energy, are $\sim 10^3$ times those seen in large solar or dMe flares. The X-ray rise and fall times, as well as the inferred temperature of the emitting plasma ($1-3 \times 10^7$K), are all consistent with those seen in other stars. Calculations of the physical parameters of the PMS flares indicate that densities are several x 10^{10}cm^{-3} in regions several x 10^{10}cm in size. Radiative cooling, rather than conduction, is the principal cooling mechanism. The enormous energies of PMS flares can be attributed to the large volume of the emitting region rather than higher densities, temperatures or longer duration compared to solar flares. The emitting loops may be $\sim\frac{1}{2}R_*$ in size. It should be noted that if other properties of solar flares scale upward with X-ray emission in PMS flares, 10^{35} erg of optical and ultraviolet emission, 10^{36} erg in shock waves, and 10^{20}g of ejected material should be present (Feigelson 1982). Extrapolation from solar flares specifically predict 10^{33} ergs increase in 'white light' and 10^{33} ergs increase in Hα emission on timescales of

4-20 minutes in PMS flares, which are consistent with U band and Hα variations seen in some T Tauri stars.

d) Effects on the Molecular Cloud. Theorists have recently calculated that the X-rays emitted by PMS stars embedded in dense cores of star forming clouds could have a number of important effects on the chemistry and dynamics of the surrounding gas. A steady X-ray source with $T \sim 10^7$K and $L_x \sim 10^{30}$ erg/s will create a series of nested 'Stromgren' spheres: a coronal region with radius $\sim 10^{13}$cm at X-ray temperatures, an HII region at 10^4°K, a hot HI region, and finally a cool H_2-dominated region starting $\sim 10^{14}$cm from the star (Lepp and McCray 1983). The photoabsorbed X-rays will be reradiated in the far infrared by dust grains and in the near infrared by H_2 vibrational and rotational transitions. Krolik and Kallman (1983) find that the integrated X-ray flux of the Orion PMS stars may be responsible for the ionization of molecules in the core. The traditional attribution to unobservable low energy cosmic rays may not be necessary. X-ray ionization may also account for the unexplained high abundance of C, CN, and reaction products of $C_2H_2^+$ seen in molecular clouds. Silk and Norman (1983) further point out that X-ray ionization in the vicinity of PMS stars will couple the local magnetic field to the gas, thereby inhibiting collapse leading to further star formation. They derive a slowly increasing star formation rate of $10^{-7}pc^{-3}yr^{-1}$, which is consistent with the X-ray determined PMS population in the ρ Ophuichi cloud. PMS X-ray emission may thus provide a crucial negative feedback mechanism that regulates star formation rates in the galaxy.

e) Efficiency of Star Formation. In addition to the physical effects on the cloud collapse described by Silk and Norman, the X-ray results may affect empirical estimates of star formation efficiency. This efficiency, estimated to be several percent, is usually determined by comparing the mass of the molecular gas with the population of PMS stars formed within it (e.g. Myers 1982). We can now add the X-ray luminous PMS stars to the numbers of spectroscopically identified stars, infrared sources and stars with proper motions coincident with the cloud. In three well-studied Taurus clouds, the total PMS population is about 2 times the T Tauri population in these clouds (Feigelson and Kriss 1983a). In the ρ Oph and Orion clouds, the ratio of X-ray PMS to optical T Tauri populations is considerably higher. The X-ray survey of ρ Oph revealed 40-60 PMS stars, where 11 Tauri stars and 19 embedded infrared sources were known. It appears that X-ray observations reveal a PMS population that is sometimes (e.g. Taurus and Chamaeleon clouds) comparable to and sometimes (e.g. the Orion and Ophiuchi clouds) considerably more numerous than spectroscopically identified PMS stars. It is interesting to note that there is little overlap between the ~80 X-ray detected Orion stars and 176 Orion stars known to undergo rapid optical flaring (Haro 1968). The total population of flaring PMS stars is thus probably larger than either X-ray or optical techniques reveal alone.

f) <u>Effects on the Early Solar System</u>. As discussed in sections III b and c above, the X-ray activity of solar mass stars decreases with stellar age approximately as $t^{-1/2}$ between $10^6 \leq t \leq 10^9$ yr. If our Sun exhibited this trend, and had flares as powerful as those seen in DG Tau or ROX20 during its PMS phase, then nearby planets or nebular material would have been exposed to very high fluxes of energetic particles and radiation. This is important because it may elucidate a number of long-standing controversies in lunar and meteoritic studies (Feigelson 1982). High VVH flare particle track densities in lunar highland rock and certain meteorites, high abundances of Xe and other solar wind implanted elements in the lunar regolith, and most importantly, the ^{26}Al and other isotopic anomalies in carbonaceous chondritic meteorites have all suggested - but not conclusively demonstrated - that solar activity was higher in the past than seen today. These findings could all be consequences of the $t^{-1/2}$ trend and enhanced flaring during the PMS phase. The 'local proton irradiation' model of the ^{26}Al anomaly is now more viable than previously considered (e.g. Clayton et al. 1977), due partly to the discovery of PMS X-ray flares and partly to the recent revision of the ^{25}Mg (p,γ) ^{26}Al cross-section at low proton energies (Champagne et al. 1983). The ^{26}Al and other anomalies are usually attributed to nearby supernova explosion rather than solar activity. Flare protons may also be responsible for the destruction of ice grains in the vicinity of PMS stars (Strazzulla, Pirronello and Foti 1983). The higher ultraviolet solar flux that would have accompanied enhanced solar activity in the past would also have important effects on the photochemistry of primordial planetary atmospheres (Sekiya et al. 1981; Zahnle and Walker 1982; Canuto et al. 1982) and possibly may have influenced the origins of life on Earth (Gaustad and Vogel 1982; Whitehouse 1983).

IV. PROSPECTS

A considerable number of Einstein observations of PMS objects have yet to be reported. These include a time series of the central Orion region, images of dark clouds (the NGC2264/Cone Nebula region, B68, B72, B92, and Lynds dark nebulae 134, 409, 434, 669, 1517, 1529, 1590 and 1605), Herbig-Haro objects (H-H 26, 28-30, 34 and 43), and a variety of regions of intermediate- and high-mass star formation (OB and R associations, HII regions). These images are sure to contain dozens of additional X-ray luminous PMS stars.

Further optical, infrared, and radio research on the published X-ray stars is also needed. Many of the Orion PMS stars, for instance, lack modern optical spectra. Surface rotation velocities and lithium abundances should be sought for more stars to elucidate L_x-v_{rot}-t relationships. The relation between X-ray flaring and optically flaring PMS stars is not at all clear at present, and the presence of several X-ray luminous non-emission line B8 stars in dark clouds has yet to be explained. Comparison of X-ray and far-infrared PMS populations (Beichman, this volume) might tell whether protostars younger than 10^6 yr can have magnetically active surfaces. A deep radio continuum survey of the Ophiuchi cloud performed recently with the VLA has revealed variable radio emission from a bright X-ray PMS star possessing no optical emission lines (Montmerle and Feigelson, in preparation). This raises the possibility that non-thermal radio emission, as well as thermal radio radiation from the cool outer winds of classical T Tauri stars, may be present in PMS stars.

X-ray astronomy can make unique contributions to the study of PMS stars during the coming decade. The one currently active imaging X-ray satellite, the European Space Agency's EXOSAT launched in January 1983, has already obtained a deep exposure of the ρ Ophiuchi cloud simultaneously with IRAS pointed observations (Montmerle, private communication). A high throughput X-ray instrument with modest angular resolution, such as the proposed LAMAR, could survey dozens of molecular and dust clouds in a few x 10^5 seconds. The all sky survey to be made with Germany's ROSAT, scheduled for launch in 1986-87, could (for instance) produce an X-ray luminosity function for the entire Herbig and Rao (1972) catalog of PMS stars. But perhaps most important for furthering our understanding of the physics of stellar activity in young stars will be detailed variability and spectral studies of the most active stars detected with Einstein. Any of the above-mentioned satellites could obtain extensive time series of the populous Orion, ρ Ophiuchi or Chamaeleon clouds to study the frequency, duration, and amplitude distribution of PMS X-ray flares. The Advanced X-ray Astrophysics Facility (AXAF) planned for a 1991-92 launch can make enormous advances. It will be able to detect 10^{30} erg/s PMS stars up to 1-3 kpc away, and embedded stars with $A_v \lesssim 20$ due to high energy response of its telescope. If it witnessed a flare like that seen in DG Tau, AXAF could measure the temperature evolution, and even the elemental composition, on timescales of minutes. X-ray astronomy can thus make unique contributions to the study of young stars during the coming decade.

V. SUMMARY

(1) Large numbers of low mass pre-main sequence (PMS) stars have been detected in all nearby (d $\leq$ 500 pc) star formation regions observed with the <u>Einstein</u> X-ray Observatory. Studies of the Orion, Taurus-Auriga, ρ Ophiuchi, and Chamaeleon cloud complexes have revealed about 150 X-ray luminous PMS stars (Figures 1-3). X-ray data yet to be reported will contain dozens more PMS stars.

(2) The level of PMS X-ray emission is 10^3 times higher than in typical main-sequence stars, with soft X-ray luminosities ranging from $L_x < 3 \times 10^{29}$ to 4×10^{31} erg/s. Their X-ray spectra are significantly harder than seen in main sequence stars (T $\geq 1 \times 10^7 °$K if thermal bremsstrahlung is the emitting mechanism), and absorption of soft X-rays is often seen at the level expected from measured visual extinctions.

(3) Rapid (10^2-10^5 second timescales) and strong (factors of 1.5 to 30) variability is characteristic of PMS X-ray emission. Individual events on timescales of hours have been seen in SU Aur, ROX20 and AS205, and an extremely rapid rise on a timescale of minutes was detected in DG Tau (Figure 4a-d). These events are interpreted as scaled up solar-type flares (Table 1). PMS flare energies are 10^3 times those of the largest solar flares, 10^2 times those of dMe flare stars, and comparable to the largest events seen in RS CVn binary systems.

(4) PMS X-ray emission is correlated with optical magnitude but not with optical spectroscopic properties. High X-ray emission does not depend on the presence of a dense circumstellar envelope or wind. L_x can be high in stars with weak or absent Hα, and appears to be anomalously low in T Tauri stars possessing the strongest chromosphere/transition region emission lines. The latter finding could be due to absorption of X-rays by the densest winds, or to the failure of closed coronal loops to develop in these stars.

(5) The above four points lead to conclusion PMS X-ray emission is produced by coronal- and flare-like surface activity, and is not related

to the envelopes or winds in classical T Tauri stars. The large population of X-ray luminous stars, sometimes several times greater than the number of classical T Tauri stars, indicates that virtually all low mass stars exhibit their highest level of surface activity during their PMS phase. The level of X-ray emission is higher in PMS stars than Hyades or Pleiades stars, though not quite at the level expected from naive extrapolation of the activity-age-rotation relations.

(6) This high level of activity will have important effects on the environments of young stars, due both to the X-rays themselves and to the high-energy radiation and particle fluxes expected to accompany the X-ray flares. X-rays from PMS stars deeply embedded in molecular clouds will photoionize the surrounding material, producing ionized molecular species and freezing the magnetic field to the gas. If the early Sun exhibited flares like those seen in PMS stars, the enhanced proton flux and wind would produce elemental and isotopic anomalies like those seen in lunar and meteoritic material. The ^{26}Al anomaly in particular could be due to 'local' solar flare protons rather than a nearby supernova.

Acknowledgments. I would like to thank Drs. W.-H. Ku, T. Montmerle, W. Sanders, and F. Walter for permission to reproduce published and unpublished figures.

REFERENCES

Bertout, C. (1982). Proc. 3rd Europ. IUE Conf., Madrid, p. 89.
Bisnovatyi-Kogan, G.S. and Lamzin, S.A. (1977). Sov. Astr. 21, 720.
Bradt, H.V. and Kelley, R.L. (1979). Astrophys. J. 228, L33.
Calvet, N., Basri, G.S. and Kuhi, L.V. (1983). Submitted to Ap.J.
Canuto, V.M., Levine, J.S., Augustsson, T.R., and Imhoff, C.L. (1982). Nature 296, 816.
Champagne, A.E., Howard, A.J. and Parker, P.D. (1983). Ap.J. 269, 686.
Clayton, D.D., Dwek, E. and Woosley, S.E. (1977). Ap.J. 214, 300.
Cohen, M. and Kuhi, L.V. (1979). Ap.J. Suppl. 41, 743.
den Boggende, A.J., Mewe, R., Gronenschild, E.H., and Grindlay, J.E. (1978). Astron. Astrophys. 62, 1.
Dere, K.P., Horan, D.M. and Kreplin, R.W. (1973). In World Data Center A, UAG-28, Part II, NOAA, p. 298.
Feigelson, E.D. (1982). Icarus, 51, 155.
Feigelson, E.D. and DeCampli, W.M. (1981). Ap.J. 243, L89.
Feigelson, E.D. and Kriss, G.A. (1981). Ap.J. 248, L35.
Feigelson, E.D. and Kriss, G.A. (1983a). Astron. J. 88, 431.
Feigelson, E.D. and Kriss, G.A. (1983b). In preparation.
Gahm, G.F. (1980). Ap.J 242, L163.
Gaustad, J.E. and Vogel, S.N. (1982). Origins of Life 12, 3.
Giampapa, M.S., Calvet, N., Imhoff, C.L. and Kuhi, L.V. (1981). Ap.J. 251, 113.
Gurzadyan, G.A. (1973). Astron. Astrophys. 28, 147.
Gurzadyan, G.A. (1980). Flare Stars. NY: Pergamon.

Haisch, B.M., Linsky, J.L., Harnden, F.R., Rosner, R., Seward, F.D., and Vaiana, G.S. (1980). Ap.J. 242, L99.

Haro, G. (1968). In Nebulae and Interstellar Matter (ed. B.M. Middlehurst and L.H. Aller), Stars and Stellar Systems Vol. VIII, p. 141.

Herbig, G.H. and Rao, N.K. (1972). Ap.J. 174, 401.

Herbig, G.H. and Soderblom, D.R. (1980). Ap.J. 242, 628.

Herbst, W., Holtzman, J.A. and Phelps, B.E. (1982). A.J. 87, 1710.

Imhoff, C.L. and Giampapa, M.S. (1981). Proc. 2nd Cool Star Workshop

Kahler, S., et al. (1982). Ap.J. 252, 239.

Krolik, J.H. and Kallman, T.R. (1983). Ap.J. 267, 610.

Ku, W. H.-M. and Chanan, G.A. (1979). Ap.J. 234, L59.

Ku, W. H.-M., Righini-Cohen, G. and Simon, M. (1982). Science 215, 61.

Kuhi, L.V. (1983). Rev. Mexicana Astron. Astrof., in press.

Lepp, S. and McCray, R. (1983). Ap.J. 269, 560.

Montmerle, T., Koch-Miramond, L., Falgarone, E., and Grindlay, J. (1983a). Ap.J. 269, 182.

Montmerle, T., Koch-Miramond, L., Falgarone, E., and Grindlay, J. (1983b). Physica Scripta in press.

Moore, R.L. et al. (1980). In Solar Flares (ed. P. Sturrock), Boulder: Col. U. Press, p. 341.

Mundt, R. (1981). Astron. Astrophys. 95, 234.

Mundt, R., Walter, F.M., Feigelson, E.D., Finkenzeller, U., Herbig, G.H. and Odell, A.P. (1983). Ap.J. 269, 229.

Myers, P.C. (1982). Ap.J. 257, 620.

Pravdo, S.H. and Marshall, F.E. (1981). Ap.J. 248, 591.

Rosner, R., Tucker, W. and Vaiana, G.S. (1978). Ap.J. 220, 643.

Rydgren, A.E., Schmelz, J.T. and Vrba, F.J. (1982). Ap.J. 256, 168.

Rydgren, A.E. and Vrba, F.J. (1983). Ap.J. 267, 291.

Schmelz, J. (1983). A.J. in press.

Schwartz, R.D. (1978). Ap.J. 223, 884.

Sekiya, M., Hayashi, C. and Nakazawa, K. (1981). Prog. Theor. Phys. 66, 1301.

Silk, J. and Norman, C. (1983). Ap.J. 272, L49.

Skumanich, A. (1972). Ap.J. 171, 565.

Smith, M.A., Beckers, J.M. and Barden, S.C. (1983a). Ap.J. 271, 237.

Smith, M.A., Pravdo, S.H. and Ku, W. H.-M. (1983b). Ap.J. 272, 163.

Stern, R.A. (1983). Adv. Space Res., in press.

Stern, R.A., Underwood, J.H. and Antiochos, S.K. (1983). Ap.J. in press.

Stern, R.A., Zolcinski, M.C., Antiochos, S.K. and Underwood, J.H. (1981). Ap.J. 249, 647.

Strazzalla, G., Pironello, V., and Foti, G. (1983). Ap.J. 271, 255.

Swank, J.H., White, N.E., Holt, S.S., and Becker, R.H. (1981). Ap.J. 246, 208.

Ulrich, R.K. (1976). Ap.J. 210, 377.

Vaiana, G.S. (1981). Space Sci. Rev. 30, 151.

Vogel, S.N. and Kuhi, L.V. (1981). Ap.J. 245, 960.

Walter, F.M. and Kuhi, L.V. (1981). Ap.J. 250, 254.

Walter, F.M. and Kuhi, L.V. (1983). Submitted to Ap.J.

Whitehouse, D.R. (1983). Observatory 103, 160.

Worden, S.P., Schneeberger, T.J., Kuhn, J.R. and Africano, J.L. (1981). Ap.J. 244, 620.

Zahnle, K.J. and Walker, J.C.G. (1982). Rev. Geophys. Space Phys. 20, 280.

INFRARED AND MOLECULAR LINE OBSERVATIONS OF YOUNG STARS AND ASSOCIATED DENSE GAS

P. C. Myers[1], P. J. Benson[2], and E. L. Wright[3]

[1]Center for Astrophysics, Cambridge, MA; [2]Wellesley College, Wellesley, MA; [3]University of California, Los Angeles, CA

Recent investigations of dark clouds within a few hundred pc of the Sun indicate the presence of "dense cores," condensations which are small ($\sim$0.1 pc), cold ($\sim$10 K), and dense ($\sim$3x10^4 cm^{-3}), with mass of order 1 $M_\odot$ (e.g. Churchwell, Winnewisser, and Walmsley 1978). Such studies suggest that some dense cores have formed, are forming, or will soon form low-mass stars. To test this idea further, we searched the visually opaque region around each of 25 dense cores reported by Myers and Benson (1983) and by Benson (1983) for emission at 2.2 μm. We found five stars which appear associated with their clouds.

The infrared observations were made with the 1.3 m and 2.1 m telescopes of Kitt Peak National Observatory[a] over the past three years; the detections reported here were made in 1983 January. The observing procedure consisted of (a) scanning the 1.3 m telescope, equipped with the K (2.2 μm) filter, across a $\sim$5' diameter box, centered on the $\sim$2' diameter dense core; (b) returning to any position with intensity $\geq$5σ for a peak-up; if the peak-up gave repeatable results, then (c) peaking up with the 2.1 m telescope; (d) making photometric measurements at the J,H,K,L, and M bands; and (e) taking continuously variable filter (CVF) spectra at 21 equally spaced points between 2.0 and 2.5 μm, to seek the presence or absence of the CO V=2$\rightarrow$1 vibrational band head in absorption. The typical 5σ sensitivity at K was 10.5 mag; the rms position uncertainty was $\leq$3".

We examined the CVF spectra for each detected star to determine whether it is associated with the target cloud. According to Elias (1978a,b,c), $\sim$90% of the field stars at 2 μm are G, K, or M giants, whose continuum should be absorbed by cool CO molecules in their atmospheres at wavelengths longward of 2.3 μm. We used the criterion that a featureless spectrum with flux increasing with wavelength from 2.0 to 2.5 μm is most probably due to a star associated with the cloud. A star having a flat spectrum is probably associated. We also relied on the result of Elias (1978b) that a star with K-L $\geq$1.5 mag is probably associated.

[a]Kitt Peak National Observatory is operated by the Association of Universities for Research in Astronomy, under contract with the National Science Foundation.

We found that stars toward B5, L1489, L1536, and L1582 are associated, and the star toward L1262 is probably associated. Recently these detections were confirmed by detections in all four bands of the Infrared Astronomical Satellite (IRAS) survey photometer for B5 (IRAS Circular No. 1; Beichman et al. 1983), L1489 (IRAS Circular No. 1), and L1536 and L1582 (Beichman 1983). The IRAS detections strengthen our claim of association.

We present in Figure 1 maps of NH_3 $(J,K) = (1,1)$ emission from five dense cores near the associated stars reported here; and also from the dense core near the far infrared source in B335 (Keene et al. 1983). We include this object, although we did not detect it as 2 μm, because it is somewhat similar to the stars reported here, and because its spatial relation to a NH_3 map has not yet been shown. The position of each star is indicated by an asterisk. The maps were made at Haystack Observatory[a] and are described in more detail by Benson (1983). They show that the projected position of the star is approximately at the peak of the NH_3 emission, probably the densest portion of the cloud, for B5, L1582, and B335. For L1489 and L1262 the star is near the outer NH_3 emission contour, and for L1536 the star lies significantly beyond the outer contour. We found a faint, uncatalogued star on the red Palomar print coincident with each of the associated infrared stars in L1536 and L1262. The remaining four stars have no visible counterparts. Therefore each of these four stars is probably embedded in its associated core.

The evolutionary state of the newly detected objects is uncertain, but those two (in B5 and L1489) which have spectral measurements available for 2-100 μm have spectral slope and luminosity ($\sim$3 $L_\odot$) nearly identical to those of HL Tau, which Cohen (1983) estimates to be $\sim$1x10^5 yr old from its position on the HR diagram. Since they are optically invisible, these new sources may be younger than $\sim$10^5 yr. If so, the similarity between their age and the free-fall times of their associated cores ($\sim$2x10^5 yr) suggests that they may have formed in the cores where they are now seen.

References

Beichman, C. 1983, personal communications to P.C.M.
Beichman, C. et al. 1983, in preparation.
Benson, P.J. 1983, unpublished Ph.D. Thesis, M.I.T. Department of Physics.

[a]Radio Astronomy at the Haystack Observatory of the Northeast Radio Observatory Corporation is supported by the National Science Foundation.

Figure 1 - Positions of newly discovered associated stars and of the star in B335 (Keene et al. 1983) (asterisks), superposed on maps of NH_3 (J,K) = (1,1) line emission (Benson 1983). Filled circles indicate NH_3 mapping positions which gave no detection; crosses indicate positions with line intensity $\geq 3\sigma$. The contours of line antenna temperature T_A, in K, are corrected for atmospheric extinction. Dashed contours are extrapolated. The horizontal line below each map indicates 0.1 pc, assuming cloud distance 160 pc for B5; 460 pc for L1582; and other distances as in Myers and Benson (1983). The (0,0) coordinates, epoch 1950, for each map are B5, α = $03^h44^m28\overset{s}{.}7$, δ = $32°44'30"$; L1489, α = $04^h01^m45\overset{s}{.}0$, δ = $26°10'33"$; L1536, α = $04^h30^m26\overset{s}{.}0$, δ = $22°37'10"$; L1582, α = $05^h29^m16\overset{s}{.}0$, δ = $12°29'20"$; B335, α = $19^h34^m33\overset{s}{.}3$, δ = $07°27'00"$; L1262, α = $23^h23^m32\overset{s}{.}2$, δ = $74°01'45"$.

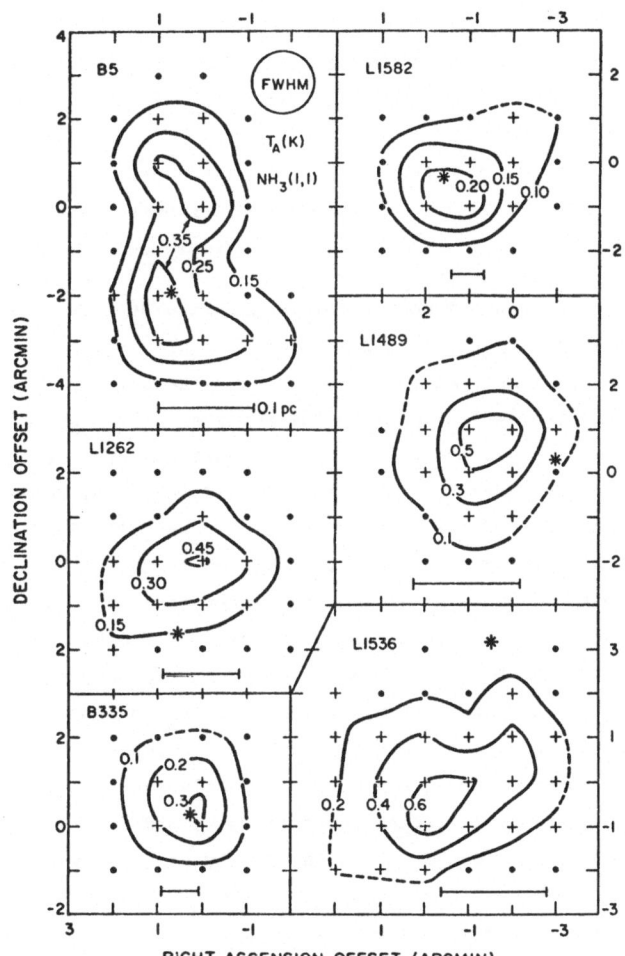

Churchwell, E., Winnesser, G., and Walmsley, C.M. 1978, Astr. Ap., 67, 139.

Cohen, M.L. 1983, Ap. J. (Letters), 270, L69.

Elias, J.H. 1978a, Ap. J., 223, 859.

Elias, J.H. 1978b, Ap. J., 224, 453.

Elias, J.H. 1978c, Ap. J., 224, 857.

Keene, J., Davidson, J., Harper, D., Hildebrand, R., Jaffe, D., Lowenstein, R., Low, F., and Pernic, R. 1983, in preparation.

Myers, P.C., and Benson, P.J. 1983, Ap. J., 266, 309.

PHOTOMETRIC MONITORING OF T TAURI STARS,
RS CVn STARS AND RELATED OBJECTS

William Herbst, Jon A. Holtzman, and Ronald S. Klasky

Van Vleck Observatory, Wesleyan University

Middletown, CT 06457

A program of photometric monitoring of T Tauri stars, Ae-irregular variables, and RS CVn stars is in its third year of operation at Van Vleck Observatory. Results from the first year were published by Herbst, et al. (1982; Paper I), and from the second year will appear in the November 1983 issue of AJ, (Paper II). Stars observed so far include the following: T Tauri (RW Aur, SU Aur, RY Tau, T Tau, CO Ori); Herbig Ae/Be (BD+40° 4124, BD+41° 3731, BD+67° 1283, BD+46° 3471, BD+61° 154, AB Aur, Z CMa); Ae-Irregular (BF Ori, UX Ori, WW Vul); RS Cvn and related (FK Com, DM UMa). On the program this year, in addition to several of the stars listed above, are T Ori, GW Ori and HH Aur. We are very interested in coordinating observations of these or other similar stars with the efforts of spectroscopists, IUE observers, or infrared observers. Anyone wishing to propose a collaborative effort should contact W. Herbst.

Observations are made with a single-channel photometer attached to the 60 cm Boller and Chivens reflector (the Perkin telescope) of the Van Vleck Observatory. A GaAs tube is used permitting UBVRI (Cousin's system) data to be obtained. Narrow and intermediate band filters centered on Hα and the K-line of CaII are also used. Our limiting magnitude, set by the extremely bright sky at our on-campus location, is V ~13. We prefer to work considerably brighter than that. All observations are made differentially. There have been roughly 40 useful nights per 6-month observing season during the past few years. Data handling and reduction are highly automated (through an LSI 11/02 microprocessor linked to Wesleyan's mainframe computer), permitting quicklooks, and an easy, efficient observing environment. An award from the Dudley Observatory to W. Herbst and H.L. Nations has made it possible for us to begin construction of a fully automated photometric system, capable of being operated remotely. Hopefully the new system will be in place by next Fall.

As a single example of results obtained so far from our program, we discuss in this paper, data obtained on five T Ori stars and its interpretation. A T Ori star is one which exhibits large, aperiodic drops in brightness by ~0.5 to 3 mag on timescales of ~1 week, followed by an equally rapid recovery. Between minima, most stars remain close to maximum light with only small amplitude, irregular fluctuations. There are five T Ori stars on our program, the three Ae-irregular variables plus the two G-type T Tauri stars, CO Ori and SU Aur. Last year's light curve for UX Ori is shown in Paper II and is an example of typical T Ori-type behaviour.

Near maximum light, the spectra of all five program stars are predominantly photospheric in nature, with relatively weak Hα emission. To substantially reduce the brightness of the stars, therefore, one must substantially diminish the light reaching us from their photospheres. This may be done by interposing an occulting body between us and the star, or by disrupting the photosphere itself in a manner perhaps analagous to the disruption caused by large spots on RS CVn stars.

Our data on these stars lead us to favor a "spot" model, over the competing "dust obscuration" models for the following reasons:

1. UX Ori and CO Ori are observed to become redder when dimmer in all colors up to a certain point. When very faint, however, the color curves "turn around" (most dramatically at short wavelengths) and the stars become bluer as they fade further. Reddening as a star fades has been used to support the dust obscuration model for the variability of these stars; however, it can just as well be explained in a spot model as we discuss below and as is observed in RS CVn stars. The color turnarounds near minimum are completely unexpected in dust obscuration models but have a natural explanation in spot models- namely, as the photosphere fades beyond a certain point, light from the bluer chromosphere begins to dominate.

2. We also observe a tendency for the Hα equivalent width to increase as these stars fade. Again, rather special geometries must be invoked in dust obscuration models to explain a phenomenon which would naturally be expected in the spot model. As in point (1), the increasing dominance of the chromospheric light naturally accounts for the observations in a spot model.

3. Spot models can reproduce the color variations quite well with reasonable values for spot temperatures (see Paper II). Of course, one requires enormous spot areas - up to 90% of the star - but some RS CVn stars are known to be within factors of 2 or 3 of these values. (See End Note)

4. The timescale for the large brightness drops and subsequent recoveries are roughly consistent with expected convective timescales in these stars, supporting the spot model.

5. Spot models allow one to tie together in a natural way the T Ori-type variation of photosphere-dominated Orion population stars and the more erratic behaviour of chromosphere-dominated stars such as RW Aur, RY Tau, and T Tau. In each case, the fundamental cause of the variability is the same - surface activity. However, the light curve behaviour is different because the photospheres respond differently than the chromospheres to the activity. This speculation suggests that a correlation ought to exist between light curve class and emission class, and it does, as Weaver and Frank (1980) have shown. Related correlations between spectral type and light curve class and emission class and spectral type are demonstrated in Paper II.

6. Quasi-periodic variations are sometimes seen in T Ori and other T Tau objects. These are naturally explained in the spot model as rotational modulation of surface inhomogeneities.

The light curves of T Ori stars, and perhaps all light curve class I T Tauri stars, can be accounted for by a "spot" model, in which the rapid brightness drops are caused by the growth of a "spot" over most of the star's surface. Actually, the term "spot" is probably misleading since essentially the entire star is affected. What we are proposing is that the photospheres of these stars cool during minimum light, presumably on account of the disruption of convection near their surfaces. Spectra of T Ori stars near minimum should reveal, particularly in the red, evidence of this cooling, if our hypothesis is correct. Another consequence of the hypothesis, if correct, is that pre-main-sequence A stars such as UX Ori and BF Ori must transport energy primarily by convection near their surfaces.

References

Herbst, W., Holtzman, J.A. and Phelps, B.E. (1982), Astron. J. 87, 1710. (Paper I).

Herbst, W., Holtzman, J.A., and Klasky, R.S. (1983) Astron. J. 88, (November issue), in press. (Paper II).

Weaver, W.B. and Frank, J.L. (1980) Mon. Not. R. Astron. Soc. 191, 321.

End Note: A preprint by Appenzeller and Dearborn arrived just after completion of this paper. They show that spot areas as small as 50% of a star's surface can cause brightness drops by as much as 3.5 magnitudes in V.

ANGULAR MOMENTUM LOSS DURING PRE-MAIN SEQUENCE CONTRACTION

L. Mestel
Astronomy Centre
University of Sussex
Falmer, Brighton BN1 9QH, England

ABSTRACT

Current ideas on the early stages of star formation suggest strongly that proto-stars enter the pre-main sequence phase rotating rapidly. The subsequent rotational evolution of both single and binary stars is largely determined by angular momentum transport by a magnetically-controlled stellar wind. A massive star with a cool corona but with a modest relic magnetic field may be prevented from fragmenting into a binary system through the braking action of a centrifugally-driven wind. A late-type pre-main sequence star will generate by dynamo action a magnetic field of surface strength B_0 that depends on the angular velocity Ω. The consequent braking can be powerful enough to cause spin-down in spite of the contraction. The predicted $\Omega(t)$ law depends on the $B_0(\Omega)$ relation, on whether the wind is thermal, thermo-centrifugal, or magneto-centrifugal, and on the fraction of the magnetosphere with open field-line. Angular momentum loss from a tidally locked binary pair can lead to the formation of late-type contact binaries with rapidly rotating members, and ultimately to their coalescence into one rapid rotator.

1.   OBSERVATIONAL BACKGROUND

Among the more recent observational reports in the literature, the following are of particular interest for theorists interested in angular momentum loss.

(A)   Vogel and Kuhi (1981) have examined the rotational velocities of pre-main sequence (PMS) stars. The previously reported large rotations of solar-type PMS stars are withdrawn. Other PMS stars have rotational velocities up to 200 km/sec, corresponding to a surface ratio of centrifugal acceleration to gravitation $V^2R/GM \simeq .2\bar{R}/\bar{M}$ (barred quantities in solar units, V in km/sec.), but three-quarters have values of V below the observational limit of 25-35. The evolutionary track of an $\bar{M}=1.5$ star is a convenient divider: when $\bar{M} < 1.5$, Vsini < 25, and when $\bar{M} > 1.5$, $25 < $ Vsini < 200. The interpretation is that the low-mass stars have spent a longer time in regions of the H-R diagram where a strong braking torque is exerted. On radiative tracks V increases with M. On convective tracks most stars have V < 35. No T Tauri star has V > 76 (corresponding to $V^2R/GM \simeq 3 \times 10^{-2}\bar{R}/\bar{M}$), and most are slow rotators (Vsini < 35). This suggests that a significant angular momentum loss has already occurred during the epoch of hydrostatic contraction, otherwise one would see rapid rotators on the convective tracks.

(B)   Gray (1981) had earlier found evidence for a sharp quasi-discontinuous drop in the rotations of giant stars at type G5III. He suggested that this is due to the sudden turn-on of a strong dynamo mode as convection develops, which ceases when V ~ 5. He subsequently produced evidence for a qualitatively similar effect for lower main sequence stars (1982). In a diagram with V plotted against spectral type, he finds a hyperbola-type curve for the upper bound to V. He interprets this curve as indicating the point where the strong braking process ceases as a function of spectral type: subsequently, the stars suffer the weaker "classical" slow braking by the magnetically controlled stellar wind, e.g. as described by the (1972) Skumanich $t^{-1/2}$ law (cf. Section 5 below). However, one must note that there are some reservations among other observers.

(C)   Soderblom et al (1983) and Stauffer et al (1984) have reported rapid rotations among Pleiades dwarf stars. The latter study found eighteen such stars with 25 < Vsini < 140, most of them being confined to a narrow range of colour, corresponding to spectral types K2-K6. Photometric observation leads to their identification as heavily spotted, rapidly rotating MS stars. Skumanich and Young reported at the meeting the discovery of a rapidly rotating field M-dwarf.

These observations are just a part of the rapidly expanding field aptly called the "solar-stellar connection." They have served to revive interest in the study of magnetic braking both during the early and late phases of star formation, and in the PMS and MS stages. I shall not give the definitive solution to any problem, but will rather survey the present state of magnetic braking theory, indicating especially where it is lagging behind what is needed for a convincing interpretation of observation.

2.   ANGULAR MOMENTUM LOSS DURING STAR FORMATION

A fundamental question to be faced by any theoretical picture of star formation is the familiar "angular momentum problem." An interstellar cloud forming from gas partaking in the rotation about the galactic centre would normally have such a high angular momentum about its mass-centre that subsequent contraction with angular momentum conservation would be halted at densities far below either PMS or MS values. The difficulties are still greater if a realistic estimate is made for the angular momentum associated with the interstellar turbulence. One requires some efficient means of removing angular momentum from a contracting cloud to the background medium, and likewise from a condensing fragment to the rest of the cloud. But equally one must not prove too much: as emphasized by Bodenheimer (1977), Mouschovias (1977) and others, fragments must retain sufficient angular momentum to account for the existence of binary stellar systems and solar systems, as well as for the individual PMS stars observed to be rapidly rotating.

We summarise here the conclusions of recent studies on the early stages of star formation in the presence of the local quasi-uniform galactic magnetic field B_o (Mestel and Paris 1979, 1983; Mouschovias 1983). A simple picture of cloud formation has cooled gas accumulating by flow down the galactic field-lines. As the mass approaches the <u>critical mass</u> $M_c \simeq F/3G^{1/2}$ defined in terms of its magnetic flux F, self-gravitation causes contraction in all three dimensions, and the frozen-in magnetic field is locally distorted. As long as $M < M_c$, the forces exerted by the distorted poloidal field B_p can balance gravity, and the cloud evolves through a sequence of states in near magneto-thermo-gravitational equilibrium. Further, the tendency of the cloud to spin up as it contracts is off-set by efficient angular momentum transport by the magnetic field. The rotational shear between cloud and surroundings generates a toroidal field component B_t, maintained by a poloidal current density $j_p = c\nabla \times B_t/4\pi$ which in turn yields a force $j_p \times B_p/c$ that brakes the rotation of the cloud. When M is less than but close to M_c, the timescale of braking by these torsional Alfvén waves is comparable with but longer than the free-fall time; hence if the mass M and the flux F are changing moderately slowly, the magnetic torques keep the cloud in approximate corotation with the background angular velocity Ω_o.

The physics changes radically if and when the ratio of charged to neutral particles becomes so low that the bulk of the magnetic flux leaks out of the cloud (Mestel and Spitzer 1956). Both the magnetic force opposing gravity and the braking torque become negligibly small, and the cloud can now collapse gravitationally, conserving the angular momentum it had at the start of this collapse phase. If, as is likely, rapid flux loss occurs at the molecular cloud phase (Mouschovias 1977), then the ratio $\Omega_o^2 R^3/GM$ of centrifugal force to gravity at the beginning of collapse will be a small number, and the spin-up during the collapse will not yield centrifugal forces comparable with gravity until densities are reached well into the opaque, PMS phases. Thus in this picture the bulk of the angular momentum problem is resolved through the magnetic stresses being able to maintain corotation with the background up to moderately high, molecular cloud densities, but enough angular momentum is left to yield rapidly rotating PMS stars.

The picture as outlined is idealized at a number of points. For example, if strict flux-freezing holds during cloud formation, then the accumulation length of gas down the galactic field is uncomfortably long. A more plausible model may involve violent cloud formation with an associated turbulent diffusion across the field; however, this does not affect the analysis of cloud behaviour in terms of the critical mass $M_c(F)$. More sophisticated treatments distinguish between the central cores of clouds and their outer regions (Mouschovias 1983; Shu, this volume), and focus attention on possible changes in field topology following field-line reconnection (Mestel and Ray, in preparation). But there is no reason to question the general conclusion that initially slightly sub-critical clouds can be the loci for the condensation of low angular momentum masses that nevertheless are inevitably rapid rotators by the time they reach the PMS stage. No clear prediction of a mass spectrum seems yet to have emerged, and we shall merely assume that PMS stars of different masses are able to form, all normally with rapid rotations.

A different picture results for clouds formed with super-critical masses $M > M_c(F)$. Magnetic braking again removes angular momentum, but now the magnetic stresses alone are unable to prevent contraction, and the appropriate models normally have centrifugal forces comparable with gravity. Rapid flux-loss at the molecular cloud phase now leaves weakly magnetic, rapidly rotating clouds of moderate density. The dynamical evolution of such clouds will depend on a gas-dynamical process for angular momentum redistribution. How the end products will be related to observations of PMS stars is as yet unclear.

3.   CENTRIFUGAL WIND FROM A PMS STAR WITH A RELIC MAGNETIC FIELD

It is plausible that a PMS star will have retained some of the flux of the galactic magnetic field that threaded the region from which it was born. Let us consider now a moderately massive, rapidly rotating PMS star with a relic dipolar field of fixed total flux, but without the hot corona that would drive a Parker-type thermal wind. The magnetic field will tend to maintain isorotation along individual field-lines, so that the ratio of centrifugal acceleration to gravity will normally increase outwards. Along field-lines that leave the star at high magnetic latitudes, one therefore expects the magnetically maintained strong centrifugal acceleration to drive a <u>centrifugal wind</u>, with the energy coming essentially from the kinetic energy of the stellar rotation. If the surface ratio $\Omega^2 R^3/GM$ is well below unity, then because of the small thermal scale-height the density at the point of centrifugal balance will have exponentiated to a low value, and the centrifugal wind will be weak, with only a small fraction of the star's magnetosphere taking part in the wind, and angular momentum loss will be correspondingly small. Equally, as the star contracts and spins up the centrifugal acceleration will exceed gravity at photospheric densities, and the consequent angular momentum loss may be sufficient to halt the increase in $\Omega^2 R^3/GM$. One does not expect such a wind to convert the star into a slow rotator, but it may keep the contracting star always rotating near the centrifugal limit.

The process may be of cosmogonical importance. There is as yet no definitive answer as to what happens to a contracting star which <u>conserves</u> its angular momentum. It is perhaps doubtful whether a compressible, centrally-condensed body will undergo the essentially divergence-free "fission" process as discussed e.g. by Jeans (1929) for incompressible bodies (even allowing for the inevitable dissipative processes which will make **irrelevant** counter-arguments based on the reversibility of strictly conservative systems (Lyttleton 1953)). However, if angular momentum is conserved, the proto-star would – other things being equal – try to approach a disk-like isothermal structure, with centrifugal balance holding in two dimensions and an isothermal pressure gradient balancing gravity in the third. Such a system would inevitably be unstable to the appropriate generalization of Jeans's gravitational instability (Ledoux 1951; Ebert 1964), and one can plausibly argue that a massive proto-star which conserves its angular momentum has no option but to develop into a binary or multiple system, with the bulk of the angular momentum now appearing in the orbital motion of the components, which being of lower mass may each evolve rather differently, especially if they generate dynamo-built fields (cf. below). But equally, it may be that a massive proto-star which has retained a significantly large primeval flux will lose enough angular momentum via a centrifugal wind to prevent such fragmentation. It is therefore of interest to try and estimate how much flux is needed for the time-scale of braking by a centrifugal wind to be shorter than the contraction time.

The problem is a convenient means of introducing the basic ideas of the rotating, magnetically-controlled stellar wind (Mestel 1966, 1967, 1968; Weber and Davis 1967). For simplicity we assume the field symmetric about the rotation axis. Bernoulli's equation for the outstreaming gas becomes

$$\tfrac{1}{2}v^2 + \tfrac{1}{2}\Omega^2\tilde{\omega}^2 - \Omega^{*}\Omega\tilde{\omega}^2 + \int dp/\rho - GM/r = \text{constant on field-streamlines}, \quad (1)$$

where $\tilde{\omega}$ is distance from the axis, Ω is the local angular velocity, Ω^{*} that of the star, and v the wind velocity (parallel to the poloidal field). Thus the wind kinetic energy per gram includes the rotatory part $\tfrac{1}{2}\Omega^2\tilde{\omega}^2$, but the same magnetic torque that is giving angular momentum to the outstreaming gas is also doing work on the gas: the term $-\Omega^{*}\Omega\tilde{\omega}^2$ comes from this term $(j \times B/c).v$. The steady state torque and hydromagnetic integrals enable us to introduce the Alfvénic surface S_A, defined by

$$v^2 = B_p^{\ 2}/4\pi\rho \tag{2}$$

where B_p is the poloidal field. Well within this surface the gas is kept in approximate corotation with the star, so $\Omega \simeq \Omega^{*}$. Well beyond the sonic point but still within S_A, Bernoulli's equation (1) yields $v \simeq \Omega^{*}\tilde{\omega}$; the work done by the magnetic torque is twice that required to supply the energy of corotation, leaving the balance available to drive the centrifugal wind. By the time the gas reaches S_A its angular velocity has dropped well below Ω^{*}, but the angular momentum carried to infinity jointly by the wind and the Maxwell stresses is exactly equivalent to that found by assuming corotation maintained out to S_A.

The remaining problem is to determine the structure of the magnetic field in the star's surroundings, in particular to estimate the relative proportions of the magnetosphere forming the "wind zone", with field-lines extending to S_A and beyond, and the "dead zone", with loops that close within S_A (cf. Section 4). For illustration, we adopt the extreme case with radial external field-lines $(B_p \propto 1/r^2)$, so maximalising the braking efficiency. From continuity of flow along B_p with the centrifugal wind speed written typically as $v = \Omega r$ (the asterisk now being dropped), we have

$$\rho v/B_p = \rho\Omega r/B_p = \rho^* a/B^* \tag{3}$$

where a is the sound speed (taken for simplicity as uniform), ρ^* is the density at the sonic point, assumed to be near the photosphere, and B^* is a typical field strength at the star. At an Alfvénic radius r_A

$$\rho v^2 \simeq \rho\Omega^2 r^2 = B^2/4\pi = B^{*2}(R/r_A)^4/4\pi \tag{4}$$

whence

$$(r_A/R)^3 = F^2/\pi^2 R^{7/2}(4\pi\rho^* a)(GM)^{1/2} \tag{5}$$

with $F = \pi B^* R^2$ is the relic flux, supposed conserved during the contraction, and the star is given the maximum angular velocity $\Omega = (GM/R^3)^{1/2}$. The equation to the braking of a star of radius of gyration kR

$$- d(k^2 R^2 M\Omega)/dt \simeq (4\pi\rho_A v_A r_A{}^2)\Omega r_A{}^2 \simeq (4\pi\rho^* a R^2)\Omega r_A{}^2 \tag{6}$$

yields a characteristic braking time

$$\tau_b = \pi k^2 M^{4/3} G^{1/3} R^{1/3}/(4\rho^* a)^{1/3} F^{4/3} \tag{7}$$

$$\simeq 10^7 \overline{R}^{1/3} \ \overline{M}^{4/3}/(T/10^4)^{1/6} \ (\rho^*/10^{-7})^{1/3} \ (B^*\overline{R}^2/10^2)^{4/3} \text{ years.}$$

Note the weak dependence on the temperature T and on the density ρ^* at the sonic point. With ρ^* taken as a photospheric density $\simeq 10^{-7}$, then τ_b will be $\simeq 10^7$ years if the relic flux corresponds to $B^* = 10^2$ at a solar radius, a modest requirement. Although we emphasize that the assumption of a radial external field is probably too optimistic, the deduction of a braking time of 10^7 years does suggest that the postulated relic flux – though far less than that originally threading the primeval stellar material – may be cosmogonically important in the way outlined during the comparatively relaxed PMS time-scales.

4.   THE EFFECT OF MAGNETIC FIELD STRUCTURE ON THE ANGULAR MOMENTUM LOSS

    The problem of greatest interest is the rotational evolution of late-type PMS stars with strong outer convection zones and with consequent dynamo-maintained magnetic fields. Prima facie, one expects a more rapid rotation to yield a stronger surface field; we first ask how the angular momentum loss depends on the surface field and hence on the angular velocity. We assume a steady wind, postponing the question of its energy source, interacting with the field emanating from the star as outlined in the last Section. There is a hidden paradox in the theory of magnetic braking by a wind: we want the magnetic stresses to control the rotation of the outflowing gas, but simultaneously we demand that the field does not prevent the outflow. This manifests itself in the structure of the magnetosphere, which should contain both wind zones with open field-lines as well as zones of closed field-lines, where the field energy density is strong enough to prevent outflow. Within the wind zone, continuity of flow along the meridional field $\underset{\sim}{B}_p$ yields

$$\rho v/B_p = \rho_A v_A/B_{pA} \tag{8}$$

the suffix A referring to the point where the field-streamline considered cuts S_A. Then

$$4\pi\rho v^2/B_p^2 = (4\pi/\rho)(\rho v/B_p)^2 = (4\pi/\rho)(\rho_A v_A/B_{pA})^2 = \rho_A/\rho \qquad (9)$$

by definition of S_A; and as ρ decreases outwards in any reasonable model, we expect $B_p^2 \gg 4\pi\rho v^2$ near the star. The structure of $B_{\sim p}$ is fixed by the component of the equation of motion perpendicular to B_p which balances $(\nabla \times B_{\sim p}) \times B_{\sim p}/4\pi$ against the density of thermal, gravitational and inertial forces, and the force density $(\nabla \times B_{\sim t}) \times B_{\sim t}/4\pi$ exerted by the toroidal field B_t associated with the non-uniform rotation generated by the wind. Well within S_A it is known that $B_p \gg B_t$, and $B_p^2/8\pi \gg \rho v^2$, ρa^2 and $\rho GM/r$, whence we can argue that $\nabla \times B_{\sim p} \simeq 0$ near the star; a marked deviation from the curl-free structure would imply a magnetic force density much larger than the non-magnetic force densities. This approximation becomes progressively less good the nearer to S_A, and well beyond S_A where $4\pi\rho v^2 > B_p^2$, we may expect the dominant wind energy to pull out the field-lines to be nearly radial. The simplest approximations take the dipolar field to be curl-free all the way between the star and S_A, and to be radial beyond (Mestel 1968). This indeed predicts the division of the magnetosphere into a wind zone and a dead zone. The subsequent solar X-ray observations showing the wind to be associated with the cooler "coronal hole" regions support this general picture.

It is important to have a reasonably accurate field structure in order to see how the rate of braking varies with the strength of the field at the stellar surface, which will itself be given by the appropriate dynamo model as a function of Ω. A stronger field implies that S_A is further from the star, so that gas flowing across S_A carries more angular momentum per gram, but at the same time the amount of flux defining the wind zone will decrease, so reducing the mass flux. In the first computations (Mestel 1968) these two effects were found to be almost compensating, so that the rate of angular momentum loss increased only weakly with the surface polar field strength B_o. These models were rather crude, with a discontinuity in the field at the transition on S_A from a curl-free to a radial field; but as long as S_A is nearly spherical, these models with surface currents at S_A are tolerable as preliminary simulations of the real structure. As the gas approaches S_A, the inertial forces acting on flow along the curving field-lines are no longer negligible, and require a magnetic body force $j_t \times B_{\sim p}/c$ to balance them; the surface currents present in these models may thus be thought of as a local concentration of the actual volume currents $j_{\sim t}$.

More recently Rowse and Roxburgh (1981, 1983) have modified the model by assuming S_A to be spherical, and constructing the curl-free dipolar field between the star and S_A that is continuous in both components with the radial dipolar field beyond:

$$(B_r, B_\theta) = (B_o R^3/r^3)\left[(1 + r^3/2r_A^3)\cos\theta, \ (1 - r^3/r_A^3)\sin\theta/2\right] . \qquad (10)$$

If the field strength B_p is approximated by $B_o R^3/r^3$ then (8) and the Alfvénic condition yield

$$v_A(r_A/R)^3 = B_o^2/4\pi\rho_o v_o \qquad (11)$$

as the relation between quantities on S_A and at a convenient point near the star's surface where $B \simeq B_o$. The star's rate of loss of angular momentum H - as already noted, given accurately by assuming that corotation holds out to S_A - is then approximately

$$-dH/dt \simeq (8\pi/3)\rho_o v_o \Omega R^4 (r_A/R). \qquad (12)$$

This has a stronger dependence on (r_A/R) than that found in Mestel (1968), but is much weaker than the result

$$- dH/dt \simeq (8\pi/3) \rho_0 v_0 \Omega R^4 (r_A/R)^2 \qquad (13)$$

given by the Weber-Davis radial field model (1967).

I conjecture that – at least for stars with moderately rapid rotations – the actual dependence will be $\propto (r_A/R)^m$ with $0 < m < 1$. The argument leading to (10) takes the curl-free approximation somewhat too seriously, applying it strictly all the way to S_A, whereas in fact $\nabla \times B$ becomes steadily more significant as S_A is approached; and intuitively the consequent changes in the field structure would appear to reduce the amount of flux that crosses S_A. We shall in fact use (12) (with the associated relation (11)) as a provisional approximation when we illustrate how the different possible wind models and variations in the parametrization of the $B_0(\Omega)$ relation affect the predicted braking laws. For comparison, we shall also include the consequences of assuming the radial field result (13), together with its analogue of (11)

$$v_A (r_A/R)^2 = B_0^2/4\pi\rho_0 v_0. \qquad (14)$$

5.   APPLICATIONS TO SINGLE LATE-TYPE STARS

Suppose first that the star is such a slow rotator that neither of the terms in Ω in Bernoulli's equation (1) is important, but that the corona is hot enough for the sonic point to be close to the star, with the associated density consequently high enough to yield a powerful thermal wind (in contrast to the case studied in Section 3). It is known from the Parker thermal wind theory that once the gas has passed through the sonic point ($v = a$) it subsequently accelerates slowly, so that typically $v_A \simeq 2a$. Thus (11) and (12) yield for the curl-free field model

$$r_A/R \propto B_0^{2/3}, \quad -dH/dt \propto \Omega B_0^{2/3} R^4, \qquad (15)$$

whereas (13) and (14) yield for the radial field model the stronger B_0-dependence

$$r_A/R \propto B_0, \quad -dH/dt \propto \Omega B_0^2 R^4. \qquad (16)$$

A rapidly rotating star with a hot corona will emit a thermo-centrifugal wind: the rotatory terms in (1) have only a small effect on the position on each field-streamline of the sonic point and so on the mass-loss rate, but at S_A the wind speed is again better approximated by $v_A = \Omega r_A$ (cf. Section 2). Now (11) and (12) predict

$$r_A/R \propto B_0^{1/2}/\Omega^{1/4} R^{1/4}, \quad -dH/dt \propto \Omega^{3/4} B_0^{1/2} R^{15/4}, \qquad (17)$$

while (13) and (14) predict

$$r_A/R \propto B_0^{2/3}/\Omega^{1/3} R^{1/3}, \quad -dH/dt \propto \Omega^{1/3} B_0^{4/3} R^{10/3} \qquad (18)$$

The centrifugal acceleration can yield much higher asymptotic speeds, but by making the gas reach the Alfvénic speed earlier it reduces somewhat the angular momentum loss.

We now make a link-up with phenomenological dynamo models which relate B_0 and Ω by

$$B_o \propto \Omega^p. \tag{19}$$

Any such relation represents the saturation of kinematic dynamo action by a dynamical back-reaction of the magnetic stresses on either or both of the turbulence and the differential rotation, both essential ingredients of the "$\alpha\omega$" dynamos supposed active in the convective envelopes of solar-type stars. Arguments can be advanced for the choices $p = 1$ or $p = 2$; we shall consider both cases. Thus for each model - dipolar or radial field structure, thermal or thermo-centrifugal wind - we substitute for B_o from (19), ending finally with an equation analogous to (6) for the braking of a single star.

As we are interested in cases where braking rather than contraction is primarily responsible for the evolution of Ω, as indeed seems to be implied by observations of late-type PMS stars, we integrate the equations keeping R constant. The results are summarised as follows:

(1) Radial field in whole magnetosphere (mass-loss rate independent of B_o)

(a) thermal wind

$p = 1 \qquad \Omega \propto t^{-1/2} \qquad$ (Skumanich law (1972))

$p = 2 \qquad \Omega \propto t^{-1/4}$

(b) thermo-centrifugal wind

$p = 1 \qquad \Omega \propto t^{-3/2}$

$p = 2 \qquad \Omega \propto t^{-1/2}$

(2) Curl-free field within S_A (mass loss rate declines with increasing B_o)

(a) thermal wind

$p = 1 \qquad \Omega \propto t^{-3/2}$

$p = 2 \qquad \Omega \propto t^{-3/4}$

(b) thermo-centrifugal wind

$p = 1 \qquad \Omega \propto t^{-4}$

$p = 2 \qquad \Omega \propto t^{-4/3}$

The most striking conclusion is the wide variety of laws that can be derived. Certainly the $t^{-1/2}$ law is in no sense a general deduction, and one is interested to know just how compelling is the observational support for the index - 1/2. On the theoretical side, we are faced with the incompleteness of dynamo theory, which does not yet tell us how B_o depends on Ω, and the uncertainties in the structure of the external field already discussed. For the thermal wind case we have argued for the curl-free rather than the radial approximation well within S_A; however, one wonders whether in the more relevant thermo-centrifugal case, the growth towards the equator of the centrifugal pull may not act so as to increase markedly the fraction of the flux crossing the Alfvénic surface, making the field structure closer to the radial field model. The field within S_A would adjust to being curl-free except near the equator, where the magnetic pressure would need to be balanced by an increased thermal pressure. It may be significant that a $t^{-1/2}$ law results also for a thermo-centrifugal wind with a radial field, provided $p = 2$.

A further uncertainty comes from the possibility of a non-dipolar structure for the dynamo-generated field. For example, Roxburgh (1983) has shown that the curl-free quadrupolar field analogous to (10) has a larger fraction of the magnetosphere "dead" - i.e. consisting of closed field-lines - than the dipolar field, with an angular momentum loss that is independent of r_A and so of B_o. He has suggested that the Vaughan-Preston gap could be associated with a sudden increase in angular momentum loss consequent on a sudden change from a quadrupolar to a dipolar dynamo mode. However, the evidence cited by Noyes et al (1983) casts doubt on this type of explanation. (Also, one cannot help wondering whether the Vaughan-Preston gap will itself have survived the observational surveys under way between now and the next Workshop).

The different $\Omega(t)$ relations all depend on the braking time's being short compared with the PMS contraction time. Also, Gray's studies (cf. Section 1) on giants and on dwarf MS stars require a very efficient braking process at high Ω. A fairly strong $B_o(\Omega)$ dependence and a possible preference at high Ω for a more nearly radial field structure both work in this direction. As an example, assume a thermo-centrifugal wind with a radial field structure. The estimated braking time is $\simeq 10^7 \tau_m^{1/3}/\overline{R}^{5/3}(B_o/1)^{4/3}$ years, where τ_m is the mass-loss time measured in years. With reasonable mass-loss rates assumed, this suggests that B_o needs to be $\simeq 10^2$ gauss in order to make this time short compared with giant or PMS evolution times.

The discussion has been in terms of steadily flowing winds in which the magnetic field plays an important or even dominant role in the wind dynamics, but indirectly, through its control of the rotation field. We have seen in Section 3 that if the star's corona is cool (i.e. the coronal scale-height is small compared with the radius), then a strong centrifugal wind will flow only in a star rotating near the centrifugal limit. However, the weak thermal pressure may be replaced by the effective pressure $\varepsilon = <\delta B^2>/8\pi$ of small wavelength magnetic field fluctuations generated in the envelope through interactions with the turbulence (Belcher 1971; Hollweg 1973; Lago 1979; Hartmann et al 1982). If there were no damping of the waves, then WKBJ theory predicts that ε varies with height according to

$$\varepsilon M_A (1 + M_A)^2 = \text{constant} \tag{20}$$

where $M_A = v/(B_p/(4\pi\rho)^{1/2})$ is the local Alfvénic Mach number. For damping to be treated as a small correction, it is necessary that $\varepsilon \ll 1$; and in fact if ε were $\simeq 1$ near the star, the acceleration of the gas would be so violent that the speed would become Alfvénic too quickly and the associated angular momentum loss would be small. Relation (20) predicts that even without dissipation ε decreases outwards with the steady increase in M_A. Thus the most realistic description of a T Tauri wind may be "magneto-centrifugal", i.e. with the mass efflux fixed near the star by magnetic fluctuations acting as a pressure, but with the asymptotic speed determined by the magnetically controlled centrifugal acceleration that persists out to the Alfvénic surface.

6.   BINARY STARS

We have been concerned with angular momentum loss from a single PMS star, efficient enough to cause spin-down in spite of the contraction. If the star is a member of a binary system, the braking would then make the spin Ω_s lag behind the orbital angular velocity Ω_o. However, if the stars are sufficiently close, synchronization of Ω_s and Ω_o will in fact be maintained (e.g. by the classical process of tidal friction): orbital angular momentum is fed into the spin, so that a powerful braking torque on the individual stars leads to their approaching each other, with an accompanying increase in their spins. The presence of extensive outer convective zones assists synchronization, by supplying a strong turbulent drag on the laminar tidal motions present in unsynchronized systems. Magnetic field-lines linking the two stars may also assist in dissipating the tidal energy.

The picture was originally introduced (Mestel 1966, 1968) to account for the existence of the rapidly rotating late-type W.Ursae Majoris contact binaries. Rather than having to find a reason why these stars have not suffered the same braking as single stars of the same type, it is simpler to argue that magnetic braking has indeed occurred, but their rapid rotations are just an inevitable consequence of the process by which they have remained components of a "close" binary system (with their mutual separation $d \simeq 2R$) in spite of their contraction to the main sequence. It is encouraging to see that Baliunas and Guinan (this Workshop) report the presence of four contact binaries in the old open cluster NGC 188, arguing that they have probably likewise evolved from detached or semi-detached binary systems by angular momentum loss; otherwise it is difficult to account for the apparent over-representation of the comparatively short-lived W UMa systems in an old cluster. And one is led to conjecture that some recently discovered rapidly rotating dwarf stars in the Pleiades and in the field (Section 1) are the logical result of the continuous braking that must lead to the mutual merging of the components of a close binary to form a rapidly rotating single late-type star, which will in time evolve through further angular momentum loss into a normal slow rotator.

Note that the angular momentum to be removed by the wind is much more than that in two non-coupled rapid rotators. Consider for simplicity two equal stars of mass M moving in a circular orbit of diameter d. To a high approximation $\Omega_0^2 d^3 = 2GM$, and the total orbital angular momentum is $2M\Omega_0(d/2)^2 = (GM^3 d/2)^{1/2}$. The contribution of the spin angular momentum is $2k^2 MR^2 \Omega_s$, and with $\Omega_s \simeq \Omega_0$ this is much smaller even for a contact binary $(d/R = 2)$ since in a centrally condensed PMS star $k^2 \simeq 1/8$ or less. Thus if the braking process were independent of the rotation, it might be difficult to produce a model which could simultaneously yield acceptable results for both single and binary systems. However, with B_0 given by (19) and with $p \gtrsim 1$, then there is the possibility that the continual spin-up of the individual stars as their mutual distance decreases will yield an angular momentum loss rate that is an order of magnitude higher than for individual stars, especially if the magnetic field in a thermo-centrifugal wind has a structure closer to the radial rather than to the curl-free model, so that formula (18) rather than (17) applies. Again, the discussion highlights the need for a better understanding of stellar winds from rotating magnetic stars.

My final remark is on the implicit assumption of Sections 3 – 6 that the star as a whole feels instantaneously the effect of the surface magnetic stresses. One expects the convective envelope to adjust rapidly to the steady-state variation of Ω with latitude and depth, given by one or the other theory of the interaction of rotation and turbulence (e.g. the balance between a macroscopic anisotropic viscous force and advection of angular momentum by laminar circulation). There are a number of estimates of the expected lag between the central radiative regions and the base of the convective envelope; for example the steadily developing shear is pictured as being limited by a weak turbulence that develops as soon as the rotation law has become marginally unstable. These studies have been given an impetus by tentative evidence of non-uniform rotation in the sun from rotational splitting of solar oscillation frequencies, and oblateness measurements. I would just ask that the assumption of negligible magnetic coupling between core and envelope be made explicit, and be quantified. One requires that the time of travel of a torsional Alfvén wave across the core be long compared with a characteristic braking time. A large-scale relic field of strength as low as 10^{-4} gauss is not negligible in this context. Until one has incontrovertible evidence of rotational decoupling between solar core and envelope, it is reasonable to assume uniform rotation in the present studies, especially in the light of the gross uncertainties in the theory already stressed.

REFERENCES

Belcher, J.W.: 1971, Astrophys. J. 168, p. 509.
Bodenheimer, P.: 1977, in "Star Formation", eds. T. de Jong and A. Maeder, IAU
    Symposium 75.
Ebert, R.: 1964, Habilitationschrift, Un. Frankfurt-am-Main.
Gray, D.F.: 1981, Astrophys. J. 251, p. 155.
Gray, D.F.: 1982, Astrophys. J. 261, p. 259.
Hartmann, L.W., Edwards, S., and Avrett, E.H.: 1982, Astrophys. J. 261, p. 279.
Hollweg, J.V.: 1973, Astrophys. J. 181, p. 547.
Jeans, J.H.: 1929, "Astronomy and Cosmogony" (CUP).
Lago, T.: 1979, D. Phil dissertation, Un. of Sussex.
Ledoux, P.: 1951, Ann. d'Astrophys. 14, 438.
Lyttleton, R.A.: 1953, "The Stability of Rotating Liquid Masses" (CUP).
Mestel, L.: 1966, Liège Symposium.
Mestel, L.: 1967, Stellar Magnetism, in "Plasma Astrophysics", ed. P.A. Sturrock
    (Academic Press).
Mestel, L.: 1968, Mon. Not. R. astr. Soc. 138, p. 359.
Mestel, L.: 1969, Liège Symposium.
Mestel, L., and Paris, R.B.: 1979, Mon. Not. R. astr. Soc. 187, p.337.
Mestel, L., and Paris, R.B.: 1983, Astron. Astrophys., submitted.
Mestel, L., and Ray, T.P.: 1984, in preparation.
Mestel, L., and Spitzer Jr., L.: 1956, Mon. Not. R. astr. Soc. 116, p.503.
Mouschovias, T.C.: 1977, Astrophys. J. 211, p. 147.
Mouschovias, T.C.: 1983, in "Solar and Stellar Magnetic Fields", ed. J.O. Stenflo,
    IAU Symposium 102.
Noyes, R.W., Hartmann, L.W., Baliunas, S.L., Duncan, D.K., and Vaughan, A.H.:
    1983, preprint.
Rowse, D.P.,and Roxburgh, I.W.: 1981, Solar Phys. 74, p. 165.
Roxburgh, I.W.: 1983, in "Solar and Stellar Magnetic Fields", ed. J.O. Stenflo,
    IAU Symposium 102.
Skumanich, A.: 1972, Astrophys. J. 171, p. 565.
Soderblom, D.R., Jones, B.F., and Walker, M.F.: 1983, Astrophys. J. 273.
Stauffer, J.R., Hartmann, L., Soderblom, D.R., and Burnham, N.:
    1984, Astrophys. J. 278.
Vogel, S.N., and Kuhi, L.V.: 1981, Astrophys. J. 245, p. 960.
Weber, E.J., and Davis Jr., L.: 1967, Astrophys. J. 148, p. 217.

ATMOSPHERES AND WINDS OF T TAURI STARS

L. Hartmann
Center for Astrophysics
60 Garden St.
Cambridge, MA 02138

It is frequently necessary to construct models in order to interpret observations of complex astronomical phenomena. Even simple models can provide a great deal of insight in cases where naive or straightforward interpretations lead to confusion or contradictions.

Observations of T Tauri stars provide a good example of the need for model-building. Many of the emission lines observed pose complex radiative transfer problems, and the line profiles indicate the presence of poorly understood flow patterns. A few useful atmospheric models now exist which aid the analysis of T Tauri spectra, permitting more quantitative study of the problems involved in the activity of pre-main sequence stars (cf. the review by Calvet 1983). However, the models are still schematic and crude, and problems of non-uniqueness show that our understanding of the complicated phenomena involved is far from complete.

I. Infall or Outflow?

Before attempting to construct a model atmosphere, it is important to have at least a general idea of the mechanism(s) responsible for the observed emission. Explanations of T Tauri activity can be grouped into two general categories. In one group of theories, accretion of circumstellar material provides the energy for the observed emission (Lynden-Bell and Pringle 1974; Ulrich 1976; Appenzeller and Wolf 1977; Appenzeller et al. 1980; Uchida and Shibata 1983). The idea is appealing, in that accretion certainly must occur in early stellar evolution. Moreover, the energy release can probably be estimated more reliably than for the principal competing theories, based on analogy with solar-type activity (Herbig 1970; Dumont et al. 1973; Gershberg and Petrov 1976; Cram 1979; Calvet, Basri, and Kuhi 1983; de Campli 1981; Hartmann, Edwards, and Avrett 1982), in which the turbulent motions of star's convective envelope are ultimately responsible for generating the required mechanical energy fluxes (≈ 1 – 10% of the stellar luminosity; Cohen and Kuhi 1979; Brown, Ferraz, and Jordan 1983). In the latter group of theories, the presence of strong magnetic fields appears necessary in order to generate large fluxes of mechanical energy (Ulmschneider and Stein 1982), as well as to propagate the wave energy sufficiently far out into the stellar envelope (de Campli 1981). Since there is no reliable way of predicting the magnetic field strengths on T Tauri stars, and no method of measuring the existing fields, one can only say that very strong fields, covering large surface areas, are required (Calvet 1983).

If accretion provides the energy, the gas we are trying to model should be rotating about the star and falling in. The solar-type activity theories would probably tend to produce outflow (de Campli 1981; Hartmann, Edwards, and Avrett 1982). One would suppose that a clear-cut observational test can be posed; is the material around T Tauri stars falling in or flowing out?

Unfortunately, the answer seems to be, both. Infall is occasionally seen in certain stars (Walker 1972; Ulrich and Knapp 1979), and sometimes both infall and outflow are seen in the same star in different lines (Edwards 1980; Krautter and Bastian 1980). In addition, both infall and outflow can be seen in the same spectral line (Hartmann 1982; Fig. 1). It's enough to drive a theorist to observation.

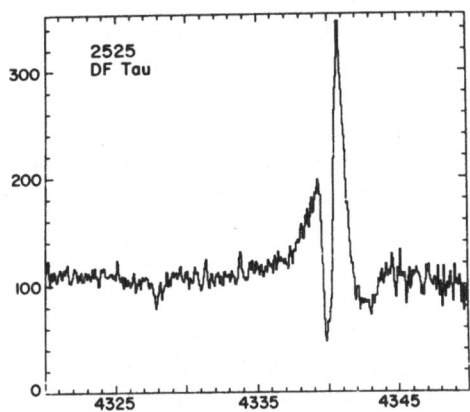

Fig. 1. The Hγ profile in DF Tau, exhibiting simultaneous redshifted and blueshifted absorption.

There are some regularities in the observations. P Cygni profiles are generally observed in the stronger lines, while the inverse P Cygni shapes suggesting infall are seen less frequently, and generally in weaker lines (Herbig 1977; Krautter and Bastian 1980; Ulrich and Knapp 1979; Hartmann 1982). The upshot is that the flow near the stellar surface can be quite complicated, but that expansion dominates at large distances.

We also know of many examples where some outflow from T Tauri stars has had a significant impact on the interstellar medium (cf. Schwartz 1983 and references therein). The mass loss rates inferred from Herbig-Haro objects and CO flows is often larger than can be comfortably accounted for by a steady T Tauri wind (Schwartz and Dopita 1980; Edwards and Snell 1982; Bally and Lada 1983; Bohm-Vitense et al. 1982; Mundt and Hartmann 1983; Calvet, Canto, and Rodriguez 1983). In any event, it seems likely that T Tauri stars eject enough material to blow away circumstellar matter in a very short time unless the material is confined to a thin disk.

Studies of Herbig-Haro objects and high-velocity motions in CO indicate that "bipolar" ejection is common (Herbig and Jones 1981; Snell and Edwards 1982; Bally and Lada 1983; Mundt, Stocke, and Stockman 1982; Mundt and Fried 1983). One likely way to achieve a bipolar flow is to channel a more or less symmetric wind by a surrounding disk (Canto and Rodriguez 1980; Konigl 1982).

Although accretion may still be occurring in some objects, it is not clear how the symmetric or normal P Cygni line profiles generally observed are naturally produced by accretion. Moreover, de Campli (1981) showed that the observed mass loss cannot be produced by thermal acceleration or radiation pressure, so that it is difficult to see how disk accretion can power the required outflow. Finally, it should be noted that "inverse P Cygni" profiles are observed in certain lines formed in the envelopes of evolved stars obviously losing mass, and not accreting from a protostellar disk (cf. Boesgaard 1979).

II. Magnetic Activity

The most popular explanation at present for the activity of T Tauri stars is that they exhibit extreme variants of solar-type activity. This means that magnetic fields must be important in producing the enormous mechanical energy fluxes required by observations.

Although the argument for magnetic fields on T Tauri stars is not conclusive, there are lines of argument which are suggestive:

1) It is thought that convective envelopes and differential rotation are necessary for magnetic fields to be generated by dynamo action. T Tauri stars have convective envelopes, and they are probably rotating more rapidly (Vogel and Kuhi 1981) than evolved stars in the same region of the HR diagram; the latter clearly exhibit solar-type chromospheric and coronal activity.

2) Periodic light variations have been observed for some stars, which are undoubtedly caused by starspot activity rotating into and out of view (Rydgren and Vrba 1983; Schaefer 1983). The observed color variations are generally consistent with the presence of cool spots (Herbst et al. 1982).

3) A class of objects called "post T Tauri stars" has been studied by Feigelson and Kriss (1981) and Mundt et al. (1983). The behavior of these stars is most easily explained by invoking scaled-up solar activity. The post-T Tauris help bridge the gap in mechanical energy fluxes between extreme T Tauri stars and the Sun.

III. Empirical Models for Emission Envelopes

a) Low-temperature (chromospheric) models

The notion of solar-type activity for T Tauri stars was first advanced by Herbig (1970), who suggested that a "deep" chromosphere (i.e., a high- density chromospheric region) was responsible for the extreme emission observed in the optical spectral region. Since then many investigators have taken up the problem (Dumont et al. 1973; Cram 1979; Calvet, Basri, and Kuhi 1983, hereinafter CBK).

The basic idea is to add a warm, dense layer of gas in hydrostatic equilibrium just outside the photosphere. This envelope is presumed to have a schematic temperature vs. height distribution similar to the solar chromosphere, but at significantly higher column densities. The model is qualitatively reasonable, in that higher fluxes of damped mechanical energy will move the temperature minimum inward, to higher densities. The structure is adjusted to match the observed emission, particularly the Balmer lines, Mg II h and k, Ca II H and K. Emission in the Na D lines and the Ca II infrared triplet are often seen (Hartmann 1982; Herbig and Soderblom 1980). Strong-emission stars exhibit Fe II emission and continuum "veiling".

Ideally, a model should incorporate a definite physical mechanism; but because our understanding of the heating processes is poor, empirical modelling is frequently necessary if any progress is to be made. One would like to construct a model atmosphere which accounts for all of the data with fewer adjustable parameters than observational constraints. The atmospheres of T Tauri stars are probably inhomogeneous, however, so that a single atmospheric model is unlikely to account for all the observed features of T Tauri activity.

The deep chromosphere model (Fig. 2) is able to account for the total energy fluxes observed in the chromospheric lines, and explains the photospheric veiling and uv excesses often observed (Table 1). Detailed modelling also makes clear what observations cannot be explained with the proposed mechanism. For example, the deep chromosphere models do not produce a large infrared excess, indicating that warm dust must surround many T Tauri stars (cf. CBK).

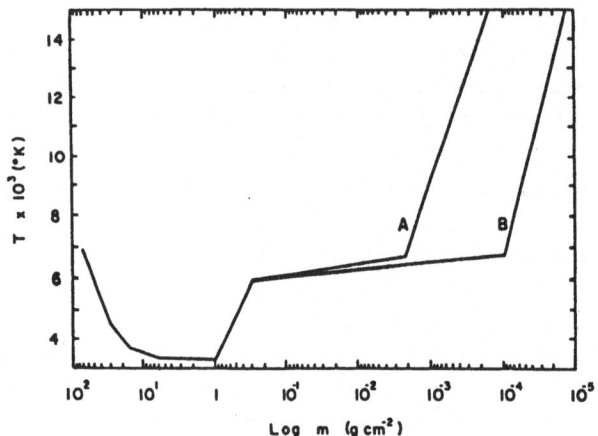

Fig. 2. Typical deep chromosphere models (from Calvet, Basri, and Kuhi 1983).

TABLE 1

EQUIVALENT WIDTH RATIOS AND LINE FLUXES

Star/Model	Hα/Hβ/Hγ[a]	F(Hα)[b]	F(Mg II)	F(Ca II)
T Tau	5.7/1.0/0.5	7.8×10^7	1.4×10^8	3.4×10^7
BP Tau	2.8/1.0/0.8	6.5×10^7	1.9×10^7	6.6×10^6
Deep chromosphere	1.1/1.0/0.85	2.6×10^7	1.9×10^7	6.6×10^6
Wind model	2.0/1.0/0.5	8.4×10^7	3.0×10^8	5.3×10^6
($\dot{M} = 7 \times 10^{-9}$ $M_\odot$ yr^{-1})				

[a] Equivalent widths, relative to Hβ.
[b] Line fluxes at the stellar surface in erg cm^{-2} s^{-1}.
  Data from Cram (1979), CBK, and Hartmann and Avrett (in preparation).

The parameters of the deep chromosphere model are not uniquely determined. As pointed out by Ulrich (1978), it is possible to produce a region with chromospheric temperatures in the context of an accretion model. One can vary the density and thickness of the chromospheric region together to provide a continuum of models with the correct total fluxes in many lines. In order to find out how deep the chromosphere really is, other diagnostic information is required. Photospheric veiling and continuum emission may be more easily produced by heating in dense layers. Unfortunately, it is clear that the usual assumption of LTE is not adequate to model the strong photospheric lines (cf. Cram 1979; CBK), as it overestimates the amount of emission produced. The matter is complicated by the fact that some of the continuum veiling can be produced at some considerable height in the chromosphere above the temperature minimum (cf. CBK).

When van der Waals broadening establishes the widths of the Mg II and Ca II resonance lines, measurements of the K_1 and k_1 widths can be used to set the column density of the temperature minimum (CBK). However, these determinations are probably not unique. Large turbulent motions are undoubtedly present in the atmospheres of T Tauri stars (cf. de Campli 1981); in addition, the Mg lI and Ca II lines often exhibit P Cygni profiles. Even if most of the photons are created in a deep chromosphere, scattering in an optically thick, expanding wind can obliterate the underlying chromospheric profile.

b) High-temperature atmospheres

The deep chromosphere models can naturally be bounded by a hot outer atmosphere which accounts for the observed coronal and transition-region emission (cf. Giampapa, Feigelson, this volume). With the lack of spectral information available in the X-ray region, it is difficult to construct a detailed model for very high temperatures. On the other hand, many transition-region lines are detectable in the uv with the IUE satellite, and these make it possible to construct detailed models of the material between 10^4K and 10^5K (Cram, Imhoff, and Giampapa 1980).

The most comprehensive model constructed to date is that of Brown, Ferraz, and Jordan (1983) for T Tau. The mean emission measure distribution derived by these authors is given in Table 2, and is similar to the results for other T Tauri stars (Cram, Imhoff, and Giampapa 1980). The emission measure distribution corresponds in shape roughly to a solar active region, but scaled upward by four orders of magnitude or more. There is evidence in T Tau for two or more atmospheric components; the permitted lines indicate $N_e \approx 10^{10} - 10^{11}$ cm^{-3}, while the intersystem lines like C III] λ1909 seem to require $N_e \approx 10^9$ cm^{-3}.

TABLE 2

MEAN EMISSION MEASURE DISTRIBUTION FOR T TAURI

log T_e(K)	3.9	4.0	4.2	4.4	4.6	4.8	5.0
log Em(T_e)	32.6	31.8	31.0	30.6	30.2	30.0	29.6

The X-ray emission observed (Gahm 1980; Feigelson and de Campli 1981) is typically not as enhanced over solar fluxes as is the transition-region (10^4K - 10^5K) emission. This effect has been attributed to the absorption of X-rays by the overlying cool wind (Walter and Kuhi 1981). This model predicts that soft X-ray absorption should be observed in the spectrum. Unfortunately, the spectral resolution of the IPC is so poor, and T Tauri stars are generally such weak sources, that it is difficult if not impossible to make this test. The best data is available for SU Aur (Sanders and Hartmann 1983, in preparation), indicating column densities $\leq 10^{23}$ cm^{-2}. This amount of material could hide soft X-rays efficiently, but will do little to the emission from the gas at T $\geq 10^7$K.

Another argument against the "smothered corona" model is that in some stars (notably RW Aur; Imhoff and Giampapa 1980, 1981), N V is substantially weaker than expected, indicating a turnover in the emission measure at $\approx$ 2 x 10^5K.

c) Wind models

The deep chromosphere models do not incorporate the large velocity fields required by the observations. Obviously, P Cygni profiles are not produced in a hydrostatic atmosphere. In addition, the observed line profiles indicate large "turbulent" line widths, $\approx$ 1/3 the escape velocity (cf. de Campli 1981; Hartmann 1982). Although it would be possible to insert these turbulent motions into a static chromosphere in an ad hoc manner, it seems likely that these motions would inflate the envelope considerably, or cause it to expand.

Deep chromosphere models do not reproduce the large Balmer decrements observed (Dumont et al. 1973; Cram 1979; Calvet et al. 1983). On the other hand, the simple model of Kuan (1975), in which an expanding isothermal envelope is treated using the Sobolev approximation, was able to produce larger decrements. These results indicate than an extended, expanding envelope is necessary to explain the Hα profiles of strong-emission T Tauri stars.

It is difficult to arrive at reasonable extimates of mass loss rates from the observations. Many emission lines, like the Balmer series, arise from highly excited electronic levels, and thus are very sensitive to the temperature structure, as well as radiative transfer effects. Resonance lines, such as Mg II and Ca II, can in principle be analyzed in a more straightforward fashion. However, these lines are often so optically thick that determinations of column densities are difficult.

Despite these uncertainties, a few simple considerations serve to constrain the amount of steady mass loss from T Tauri. Suppose we take $R = 6R_\odot$, $\dot{M} \approx 3 \times 10^{-8}$ $M_\odot$ yr^{-1} as indicated from radio observations (Simon et al. 1982), and suppose further that the flow accelerates rapidly near the surface to a constant velocity of 200 km s^{-1}. This produces a conservative volume emission measure of $\approx 6 \times 10^{56}$ cm^{-3}, or a column emission measure $\approx 6 \times 10^{32}$ cm^{-5}. Comparison with Table 2 shows why de Campli (1981) concluded that the winds of T Tauri stars cannot be thermally driven, unless the mass loss rates are very much lower than present estimates.

Table 2 shows that, if T Tauri does have a mass loss rate of 3×10^{-8} $M_\odot$ yr^{-1}, the bulk of the material must be at a temperature $\leq 1 \times 10^{-4}$K. The emission measure constraints make it difficult to produce a steady flow at the rates seemingly required by observations of the surrounding interstellar medium (although perhaps not in the specific case of T Tau; $\dot{M}v \approx 2 \times 10^{-6}$ $M_\odot$ yr^{-1} km s^{-1} is consistent with the CO data; Calvet, Canto, and Rodriguez 1983). For example, $\dot{M} \approx 3 \times 10^{-7}$ $M_\odot$ yr^{-1} from T Tauri can only be accomodated if the wind temperature is less than 8000K, so that hydrogen is only partially ionized. Although it is difficult to rule such a wind out completely, this temperature constraint appears to be unrealistically narrow, particularly in view of the evidence that Hα is formed in an expanding region (i.e., P Cygni profiles).

IV. Mechanism-specific models

Surprisingly, no one has computed shock-heated deep chromosphere models in analogy with efforts for the Sun and other stars. If one invokes a large enough flux of acoustic and slow-mode waves, which will damp rapidly near the temperature minimum (Ulmschneider and Stein 1982), a chromosphere probably can be produced which corresponds roughly with the empirical deep chromosphere models.

More consideration has been given to mass loss mechanisms for driving winds. Magnetic waves appear to be the most promising mechanism for producing the observed flows (de Campli 1981), since other mechanisms like thermal pressure cannot produce large mass loss rates. In order to get mass loss rates $\approx 10^{-8}$ $M_\odot$ yr^{-1}, magnetic fields of the order of a few hundred gauss are required. Although huge mechanical energy fluxes are required for such mass loss rates, it appears possible to generate such fluxes if strong magnetic fields are present (Calvet 1983).

Hartmann, Edwards, and Avrett (1982) discussed the first simple Alfven wave-driven wind models in which the radiative transfer and statistical equilibrium were solved together with the dynamics. Although these models are highly schematic, they are useful in providing an extreme contrast with the deep chromosphere model. There are a number of suggestive features predicted by the models.

It is assumed that a variety of wave modes are generated by convective motions in the photosphere, and all but the Alfven waves are dissipated very rapidly (forming the deep chromosphere). The wind model can be calculated if the initial magnetic field and wave flux, initial density, and damping rate are given. The wave flux is chosen to produce $\dot{M} \approx 10^{-8}$ $M_\odot$ yr^{-1}, and B is estimated so that the initial wave amplitude, δB, is $<<$ B.

a) Constant damping length model

The envelopes of T Tauri stars must be heated by mechanical energy dissipation. Initially, the dissipation of the Alfven waves is parametrized by a constant exponential damping length. Observed wind velocities are $\approx$ 100 - 300 km s^{-1} (Mundt 1983; Hartmann 1982) and there is evidence from the interstellar flows that wind velocities > 200 km s^{-1} are typical (cf. Schwartz and Dopita 1980; Hartmann and Raymond 1984). Experience suggests that a damping length $\approx$ 1 - $3R_*$ results in acceptable velocities. We choose a damping length of one stellar radius somewhat arbitrarily.

The results of such a calculation for a $1M_\odot$, $3R_\odot$ star with $B_0 =$ 300 gauss and $F_w = 9 \times 10^9$ erg cm^{-2} s^{-1} are shown in Fig. 3. The mass loss rate for this model is 7×10^{-9} $M_\odot$ yr^{-1}. (To compare with T Tauri, the mass loss rate should be scaled upward roughly as R^2, so this model corresponds to $\dot{M} \approx 2.8 \times 10^{-8}$ $M_\odot$ yr^{-1}.) The temperature rises outward at first because of the decrease in density, which makes radiative cooling less efficient. At large distances, the flow cools adiabatically and begins to recombine in various ions.

Gene Avrett and I have updated our results, using a more sophisticated model hydrogen model atom. Table 1 demonstrates that this model can account for the levels of emission in the Balmer lines and the Balmer decrements of T Tauri stars, in contrast to deep chromosphere models. The Mg II fluxes are much higher than the Ca II fluxes in the wind model, because the wind is sufficiently hot that Ca ionizes to Ca III (Fig. 3). Either a deep chromosphere is required for the Ca II fluxes, particularly for the infrared triplet lines, or the wind should be substantially cooler ($T \approx 1 \times 10^4$K). A combination of deep chromosphere and wind models would probably be most successful in accounting for the observed emission and line profiles. The distinction between "deep chromosphere" and "wind" is somewhat arbitrary, since even in the wind model most of the emission arises in the dense parts of the envelope close to the photosphere.

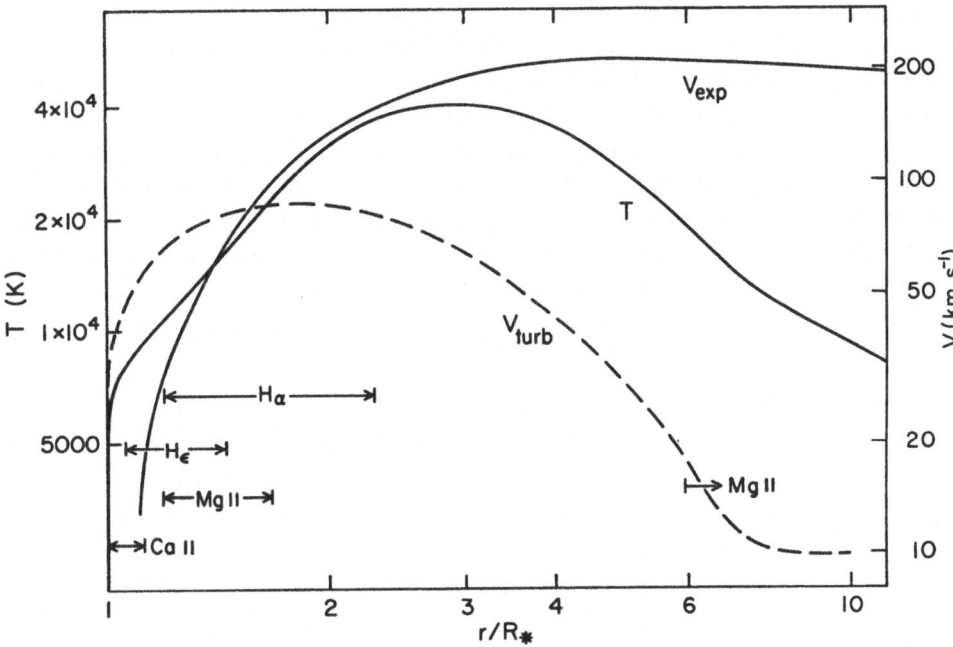

Fig. 3. Wind model structure for a damping length of $1R_*$ and $\dot{M} = 7 \times 10^{-9}$ $M_\odot$ yr^{-1}. The arrows span regions of line formation schematically. The left-most limit roughly indicates the site of maximum photon creation in the individual line, while the right-most limit indicates the place where the line becomes optically thin. The Mg II resonance lines become optically thin between 1.6 and $6R_*$; at larger distances, the cooling wind permits recombination and the formation of Mg II circumstellar shell absorption.

Another interesting feature of the model is the relative places of formation for several different lines (Fig. 3). The bulk of the emission in most lines occurs relatively near the star, so that the distinction between the wind and the deep chromosphere models is blurred. The high-order Balmer lines, $\gtrsim$ Hϵ, which often exhibit inverse P-Cygni profiles are formed significantly closer to the stellar surface than Hα. This demonstrates how different Balmer lines can exhibit qualitatively different line profiles, and indicates that the velocity field is very complicated within a few tenths of a stellar radius from the surface, but is more regular outflow for $r \gtrsim 2R_*$. The figure also indicates that the Mg II resonance lines are the most favorable for observing the terminal velocity of the wind.

In Fig. 4 we exhibit the line profiles of Hα in comparison with a typical T Tauri star. There are several interesting implications of the Hα profile. It is a so-called "type III" P Cygni profile, in which the absorption does not extend as blueward as the emission. It has been suggested that this type of profile is difficult to produce in a wind (Ulrich 1976). This is incorrect, if the Sobolev approximation cannot be used (Hartmann, Edwards, and Avrett 1982). All that a type III profile signifies is that there is a mechanism which broadens the underlying emission to higher velocities than the ejection velocity of the outer absorbing layers. In this model, the broadening velocities result from wave propagation (Fig. 3).

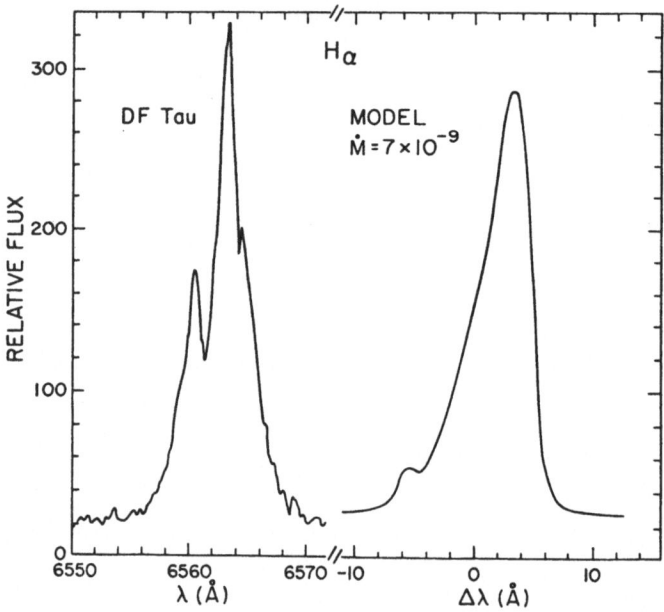

Fig. 4. The Hα profile produced by the wind model, contrasted with the profile of a typical T Tauri star.

Comparison of the calculated Hα and Mg II line profiles with observation suggests that while the broadening velocities in the model produce reasonable widths in the emission lines, the broadening at large distances from the star is too great.

The calculated Ca II resonance line shows only the barest hint of mass loss, in contrast to Mg II. This result is consistent with observations of LkHα332 by Penston and Lago (1982), but is not common in T Tauri stars. In T Tauri itself, both Ca II H and K and the Na I lines are observed to have P Cygni profiles, so that it is clear that at least some wind components have lower temperatures than in the present model.

One result of great practical importance for observers is the value of the Mg II resonance lines in determining the terminal velocity of the wind. The expansion produces a clearly visible asymmetry in the emission. However, the maximum velocity of the absorption is not clearly defined. One needs to obtain reasonable exposures in the faint line wings in order to detect the terminal velocity. Although such observations are difficult, they are extremely important to a proper understanding of T Tauri winds.

In contrast, Hα cannot be used to make a reasonable estimate of the wind terminal velocity, since it becomes optically thin long before Mg II does (cf. Fig. 3). One should beware of using Hα (or Na D for that matter) to determine wind velocities.

Note that the emission measure at 3×10^4 K in this model is an order of magnitude larger than that estimated by Brown et al. (1983) in their model for T Tau, which also suggests that the wind of that star is somewhat cooler than the present model, or that the mass loss rate has been overestimated. It is likely that the wind is not uniform over the surface, but depends upon the details of the (inhomogenous) magnetic field structure.

b) Shocks and heating

Estimates of mass loss rates depend critically upon the temperature structure assumed. In turn, the envelope temperatures depend upon the details of the heating mechanism. The presence of broadening velocities $\gtrsim 100$ km s^{-1} suggests that shocks may well be present, which might contribute to the ultraviolet emission as well as to the heating of chromospheric regions.

If this is the case, the temperature structure of the envelope could be quite inhomogeneous. John Raymond and I have computed some models in which the propagating waves shock, giving rise to temperatures in excess of 10^5K before cooling down. Such a model may produce quantitatively different Balmer emission than the wind model presented above, as well as contribute to the high-temperature emission. However, it is likely that two qualitative features of the constant damping length model will remain: an envelope at 2×10^4 K for Balmer emission, and cooling at large distances from adiabatic expansion.

It should be noted that the wind model is far too cold at large distances to produce detectable radio free-free emission. It may be that the winds of T Tauri stars are composed of streams at different temperatures. Hot flows probably need to be maintained by extended heating and/or conduction to provide the radio emission; such flows may not correspond to the material in which the low-excitation circumstellar lines are produced.

b) Time dependent effects and inhomogenous models

T Tauri stars are observed to vary in emission line strengths and profiles. This makes it likely that a realistic model must include both spatial and temporal inhomogeneities, which is beyond present capabilities. However, I would like to suggest a simple, qualitative picture in which magnetic waves drive the outflow from T Tauri stars in a rather time-dependent fashion. Observations of discrete, multiple circumstellar absorption lines strongly suggest time-variable mass loss (Mundt 1983). The mass loss is sporadic, and often material is ejected with only a fraction of escape velocity. Such gas will eventually fall back toward the star, accounting for the observations of infall in lines which probe the part of the emission envelope close to the stellar surface. The material which does escape dominates the envelope at large distances from the star; therefore, one sees mass outflow in the strong lines as a general rule.

This idea can be qualitatively tested by observations of variability. Since the velocity widths and shifts of many different lines are similar, one would expect the lines formed in a narrow region close to the stellar surface to vary much more rapidly than lines formed far away. For example, observations of the $H\alpha$ line profile of RW Aur (Hartmann 1982 and subsequent data) show that it is much more stable than the Na I D lines, which frequently indicate infall. It is my impression that this relationship between lines showing infall and those exhibiting mass loss is generally true. Clearly, more systematic work along these lines would be useful, and data on the timescales of line profile variations might help indicate the periods of the hypothesized wave impulses (cf. Mundt and Giampapa 1983).

I am grateful to Nuria Calvet and Reinhard Mundt for sending me manuscripts in advance of publication. This research was supported by NASA grant NAGW-100.

REFERENCES

Appenzeller, I., and Wolf, B. 1977, Astron. Ap., 54, 713.

_____, Chavarria, C., Krautter, J., Mundt, R., and Wolf, B. 1980, Astron. Ap., 90, 184.

Bally, J., and Lada, C.J. 1983, Ap. J., 265, 824.

Boesgaard, A.M. 1979, Ap. J., 232, 435.

Bohm-Vitense, E., Bohm, K.H., Cardelli, J.A., and Nemec, J.M. 1982, Ap. J., 262, 224.

Brown, A., Ferraz, M., and Jordan, C. 1983, M.N.R.A.S., press.

Calvet, N., 1983, Rev. Mexicana Astron. Ap., in press.

_____, Canto, J., and Rodriguez, L.F. 1983, Ap. J., 268, 739.

_____, Basri, G., and Kuhi, L.V. 1983, Ap. J., in press

Canto, J., and Rodriguez, L.F. 1980, Astron. Ap., 86, 327.

Cohen, M.H., and Kuhi, L.V. 1979, Ap. J. Suppl., 41, 743.

Cram, L.E. 1979, Ap. J., 234, 949.

_____, Giampapa, M.S., and Imhoff, C. L., 1980, Ap. J., 238, 905.

De Campli, W.M. 1981, Ap. J., 244, 124.

Dumont, S., Heidmann, N., Kuhi, L.V., and Thomas, R.N. 1973, Astron. Ap., 29, 199.

Edwards, S. 1980, Ph.D. Thesis, University of Hawaii.

_____, and Snell, R. 1982, Ap. J., 261, 151.

Feigelson, E.D., and de Campli, W.M. 1981, Ap. J. (Letters), 243, L89.

_____, and Kriss, G.A. 1981, Ap. J. (Letters), 248, L35.

Gahm, G.F. 1980, Ap. J. (Letters), 242, L163.

Gershberg, R.E., and Petrov, P.P. 1976, Sov. Astron. Lett., 2, 195.

Hartmann, L. 1982, Ap. J. (Suppl.), 48, 109.

_____, Edwards, S., and Avrett, E.H. 1982, Ap. J., 261, 279.

_____, and Raymond, J.C. 1984, Ap. J., in press.

Herbig, G. H. 1970, Mem. Roy. Soc. Liege, Ser. 5., 9, 13.

_____, 1977, Ap. J., 214, 747.

_____, and Jones, B.F. 1981, A.J., 86, 1232.

_____, and Soderblom, D.R. 1980, Ap. J., 242, 628.

Herbst, W., Holtzman, J.A., and Phelps, B.E. 1982, A. J., 87, 1710.

Imhoff, C.L., and Giampapa, M.S. 1980, Ap. J. (Letters), 239, L115.

_____, and Giampapa, M.S. 1981, in The First Two Years of IUE, ed. R.D. Chapman (NASA: Goddard Space Flight Center), p. 185.

Kuan, P. 1975, Ap. J.,202, 425.

Konigl, A. 1982, Ap. J., **261**, 115.

Krautter, J., and Bastian, U. 1980, Astron. Ap.,**88**, L6.

Lynden-Bell, D. and Pringle, J.E. 1974, M.N.R.A.S., **168**, 603.

Mundt, R. 1983, Ap. J., in press.

_____, and Fried, J. 1983, Ap. J. (Letters), in press.

_____, and Giampapa, M.S. 1983, Ap. J., **256**, 156.

_____, and Hartmann, L. 1983, Ap. J., **268**, 766.

_____, Stocke, J., and Stockman, H.S. 1983, Ap. J. (Letters), **265**, L71.

_____, Walter, F.M., Feigelson, E.D., Finkenzeller, E.D., Herbig, G.H., and Odell, A.P. 1983, Ap. J., **269**, 229.

Penston, M.V., and Lago, M.T.V.T. 1983, M.N.R.A.S.,**202**, 77.

Rydgren, A.E., and Vrba, F.J. 1983, Ap. J.,**267**, 191.

Schaefer, B. 1983, Ap. J. (Letters), **266**, L45.

Schwartz, R.D. 1983, Ann. Rev. Astron. Ap., **21**, 209.

_____, and Dopita, M. 1980, Ap. J., **236**, 543.

Simon, T., Schwartz, P.R., Dyck, H.M., and Zuckerman, B. 1983, in I.A.U. Colloquium 71, Activity in Red Dwarf Stars, ed. M. Rodono, in press.

Uchida, Y., and Shibata, K. 1983, preprint.

Ulmschneider, P. and Stein, R.F. 1982, Astron. Ap., **106**, 9.

Ulrich, R.K. 1976, Ap. J.,**210**, 377.

_____, 1978, in Protostars and Planets, ed. T. Gehrels (Tucson: University of Arizona Press), p. 716.

_____, and Knapp, G.R. 1979, Ap. J. (Letters), **230**, L99.

Vogel, S.N., and Kuhi, L.V. 1981, Ap. J., **245**, 960.

Walker, M. 1972, Ap. J., **175**, 546.

Walter, F.M., and Kuhi, L.V. 1981, Ap. J., **250**, 254.

POST-T TAURI STARS IN T ASSOCIATIONS

Frederick M. Walter
Joint Institute for Laboratory Astrophysics, University of Colorado and National
Bureau of Standards, and Laboratory for Atmospheric and Space Physics, University
of Colorado, Boulder, Colorado 80309

Recent _Einstein_ X-ray observations of T Tauri (TT) stars have detected a large
number of bright X-ray sources associated with pre-main sequence (PMS) stars. I
have investigated the serendipitously detected X-ray sources in 15 _Einstein_ IPC
images (~15 square degrees) using the IUE and using the 2.1-m telescope at Kitt
Peak. Twelve fields are in the Tau T association, two in Oph, and one in CrA.
These 15 fields contain 32 TT stars, 16 of which were detected as X-ray sources
(including two new optical identifications in Ophiuchus), as well as 16 other X-ray
sources with stellar counterparts, two of which are B stars. These stars appear to
be coeval with, yet exhibit little of the chromospheric activity of, the TT stars,
and appear to be a significant fraction of the PMS population of these T associations. The aim of this paper is to introduce the players and present an overview
of their chromospheric properties.

All 14 of these non-T Tauri, PMS stars are X-ray detected (and selected). Based
upon three deep follow-up fields, the sample is nearly complete, and not merely the
tip of a distribution of L_x. These stars fall in three groups by general spectral
morphology (and perhaps by stellar mass). All lie 1-2 magnitudes above the main
sequence. Chromospheric (Ca II and Mg II) and coronal (X-ray) surface fluxes are
comparable to those of rapidly rotating young main sequence G-M stars or of RS CVn
systems. A bestiary follows.

A. The sMe Stars (Super Me or Subgiant Me)

These five stars were first reported by Feigelson and Kriss (1981) and Walter
and Kuhi (1981). Mundt _et al._ (1983) report in some detail on their colors, and the
Hα and Li 6707 lines in their spectra. They are K7-M0 IV-V, with weak ($W_\lambda \sim 5$ Å) Hα
emission and strong ($W_\lambda \sim 750$ mÅ) Li I 6707 absorption. Rydgren and Vrba (1983) report no evidence for IR excesses in four stars, and find photometric periods of 2^d-
7^d for these stars. One is a close binary ($P = 4^d$, $q = 1$). Their blue spectra
(Fig. 1) resemble dMe stars (hence the sMe classification), with the Balmer series
in emission and the ratio of $F_{H\varepsilon}/F_{CaII\ H+K} \sim 0.2$. Mundt _et al._ estimated masses of
~0.6 M_θ, and radii of ~2 R_θ. X-ray luminosities are consistent with the rotation
periods. Mg II is in emission. Chromospheric and coronal surface fluxes are similar to those of the dMe stars.

B. The K (Not Yet) Dwarfs

Eight of the stars look very much like very active G8-K2 dwarfs or RS CVn systems (Fig. 2). Hα is in emission ($W_\lambda \lesssim 1$ Å) in four and in absorption in the rest.

Li 6707 is strongly in absorption (500-700 mÅ W_λ) in most. Ca II is in emission in all. The upper Balmer series (Hδ-H11) has been seen in emission in one on one occasion. Placing these stars on standard evolutionary tracks gives ages of ~5-10 × 10^6 yr and masses ~1 M_Θ.

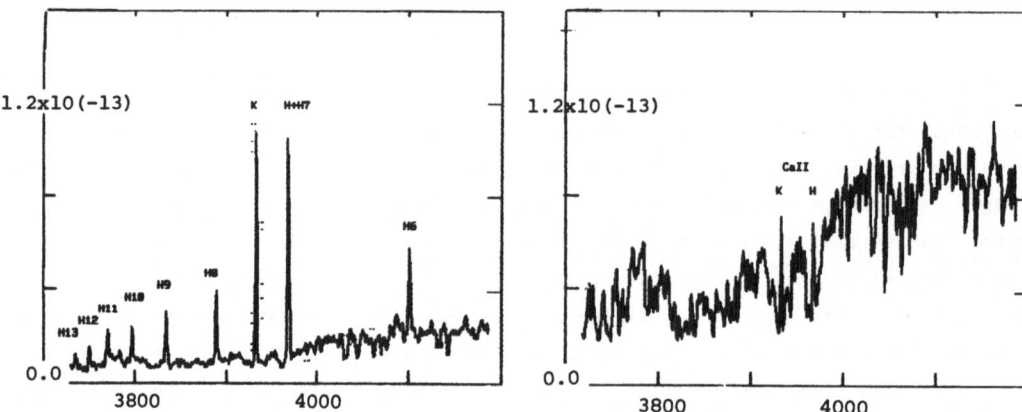

Fig. 1. Blue spectrum of P4 taken with the KPNO 2ᵐ1 IIDS system. The Ca II H & K flux is similar to the Mg II flux from IUE observations. The star resembles a dMe star.

Fig. 2. Blue spectrum of an anonymous 11th magnitude star in Taurus. Ca II H & K are prominent in emission. The photosphere is that of an early K dwarf. This star has been observed with the upper Balmer series (Hδ-H11) in emission.

C. The G Giant

BD+27°657 is a bright (m_V = 8.8) G0 III-IV near RY Tau. It exhibits all the characteristics of youth: rapid rotation (V sin i ~ 50 km s^{-1}), strong Li I 6707 absorption (~400 mÅ), filled in Hα absorption, strong Ca II H & K emission, an extremely rich UV emission line spectrum (Fig. 3), and, of course, strong X-ray flux. At 150 pc, the radius is ~3.5 R_Θ. This star bears some similarities to SU Aur, and has <u>stronger</u> Ca II H & K surface fluxes, but has no large IR excesses, no known optical variability, and lacks the Hα emission line. BD+27°657 may be a proto-A star, crossing the Hertzsprung gap on its way to the ZAMS.

Although the title refers to Post-T Tauri (PTT) stars, it is far from clear exactly what these stars really are. In an evolutionary sense, if a T Tauri star loses its characteristic emission, as it must before it reaches the MS, then it will likely pass through a state which looks like these stars. However, evolutionary calculations would indicate that PTT stars should be older than ~10^7 yr (cf. Herbig 1978). These PTT stars, while appearing to be in a more advanced evolutionary state (i.e., no dust), are no older than the T Tauri stars and are of comparable radii (based upon locations in the H-R diagram and conventional evolutionary tracks -- cf. Mundt <u>et al</u>. 1983). A few possible evolutionary scenarios follow.

First, they may indeed be old Herbig-type PTT stars. This is unlikely for the sMe stars (Mundt <u>et al</u>. 1983), but needs more analysis for the dK stars. Secondly,

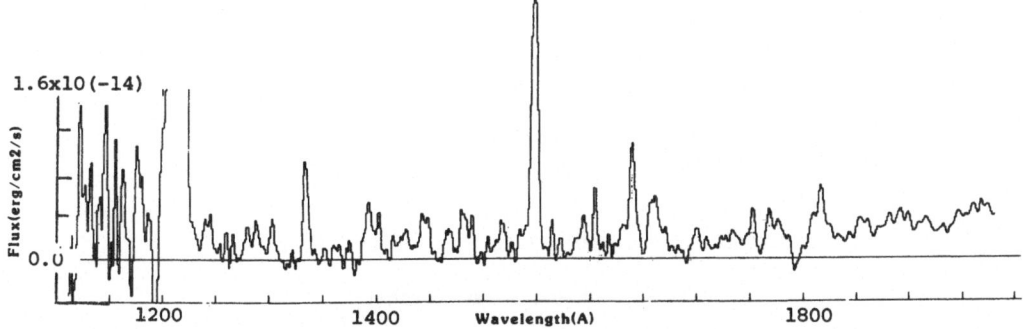

Fig. 3. The far UV spectrum of BD+27°657 taken with the IUE. All the normal chromospheric and transition region lines are in emission. The emission measure distribution (4 $\lesssim$ log T $\lesssim$ 7) is consistent with emission from solar-like loops. Note the many weak emission lines throughout the spectrum: These are low excitation lines (CI, SI, Si I, Si II, Fe II), and may be formed in a deep-lying chromosphere.

these PTT stars may be variants of TT stars which formed with different initial conditions. Perhaps they formed in a rather less dusty environment than the TT stars; they may have suffered less accretion and had less of a CS envelope to disperse, and may have dispersed it relatively early in their evolution. These PTT stars would then be unveiled TT stars, with the only major difference being in the absence of an extended atmosphere, and the attendant IR excess and Hα emission. This picture seems consistent with the data at hand. A third scenario is that these PTT stars are an inactive phase of TT stars. This would require that a TT star, having dispersed its CS shell to become PTT, later reform it (perhaps in discrete episodes of rapid mass loss) to return to the TT phase. This seems somewhat ad hoc, but cannot be ruled out.

The G giant poses another wrinkle, since we do not know which way it is going. Upper main sequence evolution is rapid, and it is conceivable (but not likely, given the ~5 × 10^6 yr age of the Taurus T asociation) that BD+27°657 is evolving off the main sequence. I consider it more likely that this star is similar to, but more evolved than (in the sense of far less CS material) a massive TT star like SU Aur.

Whatever their evolutionary status, these stars are ~1/3 of the known pre-main sequence by number in these T associations, and hence must be considered an important phase of PMS evolution. An understanding of their evolutionary status may prove important in elucidating how a low mass star evolves toward the main sequence, and how valid current evolutionary theories are.

REFERENCES

Feigelson, E. and Kriss, J. 1981, Ap. J. (Letters), 248, L35.
Herbig, G. 1978, in Problems of Physics and Evolution of the Universe (Yervan: Academy of Sciences of the Armenian SSR), p. 171.
Mundt, R., et al. 1983, Ap. J., 269, 229.
Rydgren, A. E. and Vrba, F. J. 1983, Ap. J., 264, 191.
Walter, F. M. and Kuhi, L. V. 1981, Ap. J., 250, 254.

THE FORMATION OF COOL STARS FROM CLOUD CORES

Frank H. Shu and Susan Terebey

Astronomy Department, University of California, Berkeley, CA 94720, USA

ABSTRACT

We consider models of the processes by which a molecular cloud acquires dense cores, a (magnetized) rotating core collapses to give a protostar plus nebular disk, and a powerful stellar wind sets in to reverse the accretion flow and reveal the central object as a pre-main-sequence star. At each stage, we rely on a combination of theory and observation to fix the basic parameters of the model. We show that core formation in a molecular cloud is an inevitable byproduct of ambipolar diffusion in a magnetized self-gravitating medium of low fractional ionization. We find that the gravitational collapse of a uniformly-rotating isothermal core, which possesses a $1/r^2$ density profile in its inner parts, has simple analytic properties. And we propose that strong stellar winds in T Tauri stars represent a phase of readjustment in the angular momentum distribution after deuterium burning drives convection throughout a strongly differentially-rotating protostar. We conclude that the major missing link in this picture is the evolutionary behavior of massive nebular disks that may accumulate around protostars. Otherwise, there seems to be a satisfying connection between the cloud cores observed by molecular-line radio astronomers and the active stellar atmospheres of young stars studied by optical and x-ray astronomers.

I. INTRODUCTION

Radio-line observations have established that the birth of stars in the Galaxy today occurs primarily in the dense cores of giant molecular cloud complexes (Evans 1978). These complexes contain on the order of 10^5 to 10^6 $M_\odot$ and are clumpy, with random motions for the clumps having magnitudes roughly appropriate for virial support of the complex (Solomon and Sanders 1980, Blitz 1980). Individual clumps typically have masses of several thousands of solar masses (e.g., Sargent 1977), and they are probably supported against their self-gravity by a combination of magnetic fields (Mestel and Spitzer 1956, Mouschovias 1976) and turbulence (Larson 1981). It is currently popular to think that this turbulence is driven by winds from embedded stars (see, however, Bally and Lada 1983).

When maps of carbon monoxide emission are made of molecular cloud clumps, they can often be separated into cores and envelopes (Rowan-Robinson 1979). In CO, the envelopes are seen as cool (T $\sim$ 10 K) extended regions, with molecular-hydrogen densities on the order of hundreds per cubic centimeter. The cores are seen as hot spots (T $\sim$ 50 K) if massive star formation has already taken place in them. Cooler cores can also be seen if they have H_2 densities high enough (n > 10^4 cm^{-3}) to excite emission or produce absorption in molecules such as ammonia, formaldehyde, etc. Some evidence exists that the cloud cores have density profiles which are proportional to r^{-2} or $r^{-3/2}$, where r is the radial distance from the center of the core (e.g., Loren, Sandqvist, and Wooten 1983).

Because of its many hyperfine splittings, ammonia is an especially useful probe for the small cores which are the likely progenitors of low-mass stars (Ho and Townes 1983). In particular, Myers and Benson (1983) have recently summarized the properties of a large number of NH_3 cores found near the sites of newly-born T Tauri stars. The narrow width associated with the ammonia lines in these cores indicate mean levels of turbulence which are quite subsonic. Moreover, the average densities, equivalent temperatures (including tubulence), and sizes of the cores are consistent with many of them being on the verge of gravitational collapse. Direct infrared evidence for associated stars in some of the cores is presented by Myers, Benson, and Wright (this conference). The coexistence of quiet dense regions and newly-born stars suggests that the birth of low-mass stars, at least, do not require external triggers. Rather, what seems needed is a quasi-static mechnism which will separate molecular clouds into envelopes and cores. Star formation then becomes inevitable as the process brings the cores ever closer to the brink of gravitational instability.

II. THE FORMATION OF MOLECULAR CLOUD CORES

A mechanism satisfying the requisite properties is ambipolar diffusion, by which charged particles and the magnetic field lines to which they are tied will drift relative to the neutral component of a lightly ionized gas (Mestel and Spitzer 1956, Nakano 1981, Mouschovias 1981, Black and Scott 1982). Magnetic fields, of course, exert forces only on charged particles; they can support the neutrals in a molecular cloud only indirectly, through the friction created as the neutrals, under the action of the self-gravity of the cloud, try to slip past the ions. The slip must be there, or there would be no force; thus, the very process by which the existence of magnetic flux can help to hold up a region of enhanced density, guarantees that the flux will gradually be lost from the same region. As the magnetic lines of force leak to the surrounding envelope, the neutral matter will pull itself more and more into a concentrated core (see, e.g., fig. 2b of Shu 1983). In a one-dimensional problem, where thermal pressure is the only other means of support for an isothermal gas, the density profile acquires in time the sech-squared distribution of Spitzer (1942). For contraction in three-dimensions by the core of an unbounded cloud, we suspect that the density asymptotically approaches a $1/r^2$ distribution. The latter problem is currently under study by Lizano and Shu.

The time required to produce cloud cores is of obvious interest. Let the mass fraction of ions be given by the law (see, e.g., Elmegreen 1979, or Umebayashi and Nakano 1980):

$$\rho_i/\rho = C\rho^{-1/2},$$

(1)

where ρ_i and ρ are, repectively, the densities of ions and neutrals, and C is a constant whose value is to be determined from theory or observations. Suppose further that the frictional force per unit volume exerted on the neutrals by the ions has the form (see, e.g., Draine, Roberge, and Dalgarno 1983):

$$\rho_i \rho \gamma w,$$

(2)

where γ is proportional to the collisional rate of ions and neutrals, and w is the relative drift velocity of the former with respect to the latter. The characteristic time that it then takes for a one-dimensional, self-gravitating, isothermal slab to lose an appreciable part of its initial magnetic flux, expressed as a fraction of the sound-crossing time in the final isothermal state, can be calculated to be

$$\tau_o \gamma C/2(2\pi G)^{1/2},$$

(3)

where τ_o is a dimensionless number of order unity that depends on the initial ratio of magnetic to gas pressure (see fig. 5 of Shu 1983). The number represented by equation (3) can be estimated to be 10–10^2 for conditions which apply in molecular clouds, whereas the sound-crossing times in the cores observed by Myers and Benson

(1983) are $\sim 10^5$ yr. Thus, if these cores form by the process just described, we can roughly estimate their ages to be 10^6-10^7 years, compatible with various chemical dating schemes (e.g., Allen and Robeinson 1976, Stahler 1983b). Since T Tauri stars also typically have such ages (Cohen and Kuhi 1979, Strom 1983), there should be roughly as many NH_3 cores as there are T Tauri stars in any molecular cloud clump.

It is tempting to speculate further that in a cloud clump which contains many Jeans masses (calculated with only thermal support to balance the self-gravity of the clump), many cores could eventually pull themselves into being from the turbulent magnetic background, until, perhaps, the process is interrupted by the disruptive formation of too many massive stars. The basic idea would be that the turbulence could be sustained only at sub-alfvenic velocities, and that as magnetic fields diffused from the regions of increasing density and the medium became increasingly inelastic, the level of turbulence would drop, until, finally, quiet cores of the variety observed by Myers and Benson are produced. In this picture, there would be no clean separation between the core and envelope of a molecular cloud. They would blend smoothly into one another, and it becomes better to regard the eventual star-forming region as an unbounded entity rather than as the pressure or volume bounded object common to most analyses of cloud equilibrium and collapse.

III. THE COLLAPSE OF SPHERICAL MOLECULAR CLOUD CORES

We have argued that there is reason for expecting the cores of molecular clouds to acquire $1/r^2$ density profiles, i.e., to try to become singular isothermal spheres (Chandrasekhar 1939, p. 157):

$$\rho = a^2/2\pi G r^2, \tag{4}$$

where a is the equivalent isothermal sound speed. In the absence of agents other than thermal pressure to counteract gravity, the singular isothermal sphere is the natural equilibrium state accessible to unbounded isothermal gaseous configurations because a sphere of any radius r then encloses about one Jeans mass. But since the resultant equilibrium state is unstable, it then remains unclear how closely the idealized $1/r^2$ distribution can actually be reached in a situation where other mechanisms for support, such as magnetic fields or turbulence, are slowly drained from the object's core.

If the singular configuration is approached closely, then the collapse process becomes extremely simple, for in the absence of rotation and magnetic fields, it has a similarity solution (Shu 1977, Hunter 1977, Boss and Black 1982). The inner-most densest parts collapse first, leaving behind a rarefied region into which the overlying material flows as the falling process progresses outward as an expansion wave travelling at the isothermal speed of sound. A condensed object, the protostar, builds up by accretion, with material flowing into the central regions at free-fall speeds, adding mass at a constant rate given by

$$\dot{M} = m_o(0) \ a^3/G, \tag{5}$$

where $m_o(0) = 0.975$ for the collapse of a singular isothermal sphere (Shu 1977). A numerical value of $\dot{M} \sim 10^{-5} \ M_\odot/yr$ might be typical for the unstable cloud cores in the Galaxy, requiring from equation (5), a = 0.35 km/s.

IV. THE EVOLUTION OF PROTOSTARS

The isothermal nature of the flow breaks down within $\sim 10^{14}$ cm of the protostar. During the main accretion phase, the radiating shock standing above the surface of the protostar generates appreciable luminosity (Stahler, Shu, Taam 1980a,b, 1981; hereafter SST). Dust suspended in the gas reprocesses the optical photons that try to emerge from the interior into infrared radiation, and this reprocessing heats up the inflowing material until the dust is destroyed at a radius of $\sim 10^{13}$ cm. The region within the dust photosphere at $\sim 10^{14}$ cm, where the accreting matter first becomes optically thick to the reprocessed infrared radiation, then comprises an "inner problem" for which the similarity collapse solution provides outer boundary conditions, namely, free-fall at the mass accretion rate $\dot{M}$. Since the time required to fall from the dust photosphere to the accretion shock and the the time needed for photons to diffuse across this region are both short in comparison to the accretion time, quasi-steady conditions apply in the accretion envelope, and the calculation of its structure simplifies considerably. Moreover, since the settling flow behind the radiative shock onto the protostar is highly subsonic, the evolution of the protostar itself can be followed using the familiar quasi-static equations of stellar structure theory. The only unusual element are the surface boundary conditions provided by the radiating accretion shock. The systematic exploitation of these considerations allow a more accurate and economical calculation of the structure and evolution of protostars than possible by direct finite-differencing of the equations of radiative hydrodynamics (e.g., Larson 1969, Winkler and Newman 1980a,b).

We may summarize the results for the accretion rate $\dot{M} = 1 \times 10^{-5} \ M_\odot/yr$ as follows. As long as the star is accreting, its gas photosphere, which is bounded by the accretion shock, suffers too much extinction by circumstellar dust to be optically visible. The protostar is observable only as an infrared source, whose color temperature is characteristic of the effective temperature of the dust photosphere, typically hundreds of K. Except for some initial transients, the radius of the protostar remains roughly 3×10^{11} cm throughout the entire calculated accretion phase, 10^5 yr. In the early stage of the protostar's evolution, its interior structure is radiative because the specific entropy of the shock-deposited material enters the star proper with values that increase in time, yielding an entropy distribution which increases outward. The star is, therefore, stable to convection. The outer convection zone which usually accompanies a low-mass star is absent because the outer layers are heated from above by the radiating accretion shock. Thus, the subphotospheric layers of high opacity -- the hydrogen and helium ionization zones -- do not need to transport outward fluxes of heat that would make them convectively unstable (Hayashi, Hoshi, and Sugimoto 1962). Indeed, appreciable convection does not arise in the accreting protostar until deuterium burning near the center begins to strongly heat the object from below. This happens in the calculations of SST when the protostar accumulates a mass of about 0.4 $M_\odot$, when the ratio of mass to radius has a value, $M/R = 0.1\ M_\odot/R_\odot$, that gives a temperature near the center of about 10^6 K. By the time the protostar has acquired 0.5 $M_\odot$, it is nearly completely convective except for negligibly small regions near the center and the surface. Thus, when the accretion is terminated (abruptly and completely artificially in the actual calculations), only the star's outer layers need to cool (along a locus of constant radius) before the star is ready to join a conventional pre-main-sequence track in the H-R diagram appropriate for its mass and radius.

If the actual accretion terminates because a stellar wind reverses the accretion flow and clears the surroundings of gas and dust, the star would become optically visible near this point. Since the youngest observed T Tauri stars seem to lie on completely convective pre-main-sequence tracks (Cohen and Kuhi 1979), this suggests that the onset of convection may have something to do with the triggering of a wind. We explore this possibility in greater detail in §VI. For now, we merely remark that because gas photospheres must lie to the left of the boundary to Hayashi's forbidden region (about 4,000 K for solar type stars), and because dust photospheres must lie (well) to the right of the destruction temperature of interstellar dust (1,500-2,000 K), there is a gap in the H-R diagram which protostars and pre-main-sequence stars do not occupy. Evolutionary tracks which show a continuous transition from infrared photospheres to optical photospheres (e.g., Appenzeller and Tscharnuter 1975, or Winkler and Newman 1980a) either overlook the fact that the opacity drops by many orders of magnitude when dust disappears, or misapply the concept of effective temperature to a situation when observers would speak of reddened optical photospheres rather than the temperature of an equivalent blackbody at optical depth 2/3 from the "cloud boundary."

Stahler (1983a, see also this volume) has used the results of SST to compute the "birthline" in the H-R diagram where T Tauri stars should first become optically visible. He adopts the assumption of spherical accretion at a uniform rate, $\dot{M} = 1 \times 10^{-5}\ M_\odot/yr$, and he takes the accretion to be terminated abruptly (by unspecified) mechanisms after a time $M/\dot{M}$ for various values of M. As is well known by now, Stahler's birthline forms a good upper envelope for the data points of Cohen and Kuhi (1979), and he argues that this suggests the adopted assumptions must be reasonably well fulfilled.

The fit is, indeed, so remarkable that it is hard not to take seriously the concept of a birthline, especially if additional observations support the sharpness of the cutoff of T Tauri stars above a well-defined curve. However, this sharpness is, theoretically, a double-edged sword. Mercer-Smith, Cameron, and Epstein (1983) have expressed scepticism that the properties of the collapse flow could be so narrowly constrained as to produce a clean birthline. In this regard, it is interesting to note that a reasonably sharp cutoff would always result, virtually independent of the geometry of the mass accretion or its rate as a function of time, providing (a) it is deuterium burning which triggers the reversal of the accretion flow, and (b) the inflow is stopped on a timescale that is short in comparison with $M/\dot{M}$. In particular, because of its sensitivity to temperature, deuterium burning would drive a low-mass star completely convective approximately at a point in the accretion process when the ratio of mass to radius, M/R, reaches the value $0.1 \, M_\odot/R_\odot$. We postpone for §VI the discussion why the onset of convection might be accompanied by the generation of a powerful stellar wind that could reverse the accretion flow. For now, we merely note that the locus connecting points of $M/R = 0.1 \, M_\odot/R_\odot$ along convective pre-main-sequence tracks of different masses produces nearly as acceptable an upper envelope for T Tauri stars as Stahler's construction. (The two loci intersect for a $0.5 \, M_\odot$ star.) In any case, in our interpretation, different accretion rates and geometries lead to different points at which a protostar will turn on a powerful enough wind to reverse the accretion flow. The point in the accretion history when this happens determines the star's ultimate mass, with higher effective rates tending to produce more massive stars. The fact that stars of $0.5 \, M_\odot$ are very common in the Galaxy suggests that an equivalent spherical accretion rate of $10^{-5} \, M_\odot/yr$ may well be typical of molecular cloud cores. It is then gratifying to find that such a value is also roughly suggested by the observations of Myers and Benson (1983).

V.   THE EFFECTS OF ROTATION IN THE COLLAPSE OF CLOUD CORES

Observations of molecular cloud envelopes show them to rotate relatively slowly if at all. Measured values in a few cases amount to velocity gradients of about a km/s-pc (e.g., Kutner et al. 1977, Ho et al. 1977, Schneps et al. 1978). A typical angular velocity might be $10^{-14} \, s^{-1}$. Cloud cores are generally too small to have their rotations measured reliably as yet, but if they form by the quasi-static process described in §II, then existing calculations of magnetic braking (e.g., Mestel 1965, Mouschovias and Paleologou 1980) lead us to expect that, as long as the core is not collapsing dynamically, it may be forced to rotate uniformly with its envelope.

Terebey, Shu, and Cassen (1983, hereafter TSC) have found an analytical equilibrium state which generalizes the singular isothermal sphere to include uniform rotation at a rate Ω (but no magnetic field). For r much smaller than a/Ω, the density behaves nearly as $a^2/2\pi G r^2$; for r much greater than a/Ω, the density is a constant, $\Omega^2/2\pi G$. The formal solution therefore looks reasonably like a molecular cloud core plus its envelope. However, the "envelope" of the model is centrifugally supported, an atypical state of affairs for real molecular clouds, and the use of this model should be restricted to regions well inside the turnover radius a/Ω.

For representative values, e.g., $a = 0.35$ km/s and $\Omega = 1 \times 10^{-14}$ s^{-1}, the radius, $R = GM/2a^2$, which contains $M = 1\ M_\odot$, equals 5.4×10^{16} cm, and is well within the turnover radius, $a/\Omega = 3.5 \times 10^{18}$ cm. Rotation is, in some sense, a small parameter for this part of the core if it is destined to collapse to form a protostar plus nebular disk. For example, the total angular momentum contained by this portion of the cloud,

$$\frac{2}{9}\ MR^2\Omega,\tag{6}$$

equals 1.3×10^{52} g cm^2/s, and would be not quite three times the value deduced for the augmented solar system, where the masses of the planets are scaled to reflect solar abundances (see Cassen and Moosman 1981). Thus, if magnetic braking does couple the rotation rate of a core of a molecular cloud to its envelope, then the so-called "angular momentum problem" seems largely solved insofar as the interstellar medium is concerned, because theory and observation would have agreed that sufficient braking has occurred to make it possible for the collapse to proceed to solar-system or binary-star dimensions (e.g., Mouschovias 1981). A problem would still remain in that the smooth distribution of mass with angular momentum in the infalling matter would generally require drastic readjustment before it would resemble the discrete spikes that characterizes multiple stars or planetary systems. We also emphasize that should molecular cloud cores turn out to rotate much faster than their envelopes, then the computations described by Bodenheimer (1981) may be more relevant than those described here.

To follow in detail how the collapse from interstellar dimensions to solar system dimensions proceeds in the case of low rotation, we (TSC) have performed a perturbational analysis of the axisymmetric collapse of a rotating singular isothermal sphere, using as the unperturbed state, the known similarity solution for the spherical case (Shu 1977). The resulting equations in radius, polar angle, and time for the perturbations can be reduced to a set of linear ordinary differential equations, whose solution we obtain by numerical means. The inner limit of the solution has simple analytic properties, and it can be matched by the outer limit of an inner solution of the type investigated by Ulrich (1976) and Cassen and Moosman (1981). In the inner problem, the gas and dust contain dynamically interesting amounts of angular momentum, but the pressure gradients have become negligible, so freely falling ballistic trajectories are a good approximation for the streamlines before the matter encounters the protostar or the nebular disk, and is arrested in a strong radiating shock. The thermodynamics of the flow, and the resultant evolution of the protostar, have to date not received as careful a treatment as the dynamical and radiative aspects, but the methods introduced by SST for the spherical inner problem should be generalizable to this case. The major uncertainty in carrying out definitive calculations concerns the rigorous treatment of the redistribution of angular momentum and mass in the disk (Cassen and Summers 1983).

In the absence of any redistribution, the infall of all the matter out to a radius, $R = GM/2a^2$, in the cloud core would eventually produce a disk of radius,

$$r_d \cong \Omega^2 R^4/GM = G^3 M^3 \Omega^2/16a^3. \tag{7}$$

If the protostar has a smaller radius, r_*, a fraction of mass (Cassen and Summers 1983)

$$M_*/M \cong 1.5 \ (r_*/r_d)^{1/3} \cong 3.8 \ \frac{r_*^{1/3} a^{8/3}}{GM\Omega^{2/3}} \tag{8}$$

will fall directly onto the protostar. The total mass M that falls in will depend on the mechanism which reverses the accretion flow; in particular, the above formula may be invalidated if a bipolar flow were to stop accretion onto the protostar, but not onto the disk. If we ignore such possibilities, for our fiducial values of a, Ω, and M, $r_d = 6.5 \times 10^{12}$ cm, and for $r_* = 3 \times 10^{11}$ cm, about half a solar mass will fall directly onto the protostar, and another half will flow into the disk. It remains to be seen whether the mechanisms of angular momentum and mass redistribution within the protostar and the disk would convert such a configuration into a star plus planets or into a binary star system.

VI. THE ONSET OF A WIND

One of the major observational discoveries concerning newly-born stars is that they all have strong winds; indeed, this is a recurrent theme in many of the papers presented at this conference. What is the source of this activity? For pre-main-sequence objects of low mass, it is natural to speculate that the activity is associated with the dynamo mechanism that accompanies convection in a differentially rotating star. Can we connect this idea to the mechanism by which protostars might be able to reverse their accretion flows? Yes, and in hindsight, the connection appears perfectly natural.

Consider the swirling matter which falls onto a protostar from the rotating core of a molecular cloud. This matter cannot know in advance that it is destined to become part of a star, so it is not likely to have a distribution of angular momentum with which the star will ultimately feel comfortable. In particular, the accreted material will generally accumulate in the protostar in a fashion as to make it strongly differentially-rotating. But as long as the star remains radiative and is not rotating close to breakup, there are no known instabilities that act faster than a Kelvin-Helmholtz timescale as long as the specific angular momentum of the matter increases outwards (Goldreich and Schubert 1967, Fricke 1967). Since the accretion timescale in low-mass protostars is always shorter than the Kelvin-Helmholtz timescale (SST), and since the low-angular momentum material in a rotating cloud core will always fall in first (TSC), low-mass protostars seem unlikely to become active as long as they remain radiative.

Imagine, however, what happens should this protostar want to become convective. Any seed magnetic field in the star would quickly be amplified and stretched as the convective elements carry the field lines across shearing layers. The strong magnetic fields generated by this dynamo action would rise buoyantly to the surface where they become available to create the violent disturbances that astronomers have long associated with the atmospheres of young stars (Herbig 1962, Kuhi 1964).

One might conjecture two stages to this process of converting kinetic energy of differential rotation into magnetic activity. The earlier phase is the tapping of the excess mechanical energy contained in the order-unity shear. When the shear is lessened, energy will be released because for a given total angular momentum, the uniformly rotating star has less energy than the differentially rotating one. The later phase involves the tapping of the kinetic energy of the nearly-uniform rotation as the star brakes with the help of a magnetic wind. The relatively slowly rotating T Tauri stars on convective pre-main-sequence tracks studied by Vogel and Kuhi (1981) may belong to the second phase. To get to their observed state, the stars might have released a total amount of rotational energy comparable to their gravitational binding energy. In this picture, the boisterous activity exhibited by young stars reflects the adjustment that they must make in reconciling the heritage they receive from the interstellar medium with the lifestyle they must pursue as mature stars.

In a low-mass protostar, we can envisage two distinct ways that convection can potentially be driven. The first is by being heated strongly from below; the second, by being strongly cooled from above. The first we have already discussed in connection with the ignition of deuterium. The second will occur if the accretion from a rotating cloud is not terminated before more and more of the matter falls into the nebular disk rather than directly onto the protostar. (There is always a cone near the rotation axis where matter will flow directly onto the protostar, but the opening angle of this cone shrinks with the passage of time.) As th accretion shock above the protostar weakens, its heating of the surface layers drops below a value which can prevent an outer convection zone from developing. An outer convection zone will appear because such a zone is natural for a pre-main-sequence star of low mass (i.e., for a non-accreting protostar), and it is conceivable that winds from some young stellar objects are triggered in this fashion. If accretion halts only after the star acquires a structurally important amount of angular momentum, it may contain enough rotational energy in some circumstances to become unstable to a barlike distortion and try to fission (Ostriker and Bodenheimer 1973). The calculation of the evolution of such rotating protostars and their nebular disks remains as a challenge for the future.

REFERENCES

Allen, M. A., and Robinson, G. W. 1976, Ap. J., 297, 745.

Appenzeller, I., and Tscharnuter, W. 1975, Astr. Ap., 40, 397.

Bally, J., and Lada, C. J. 1983, Ap. J., 265, 824.

Black, D. C., and Scott, E. H. 1982, Ap. J., 263, 696.

Blitz, L. 1980, in Giant Molecular Clouds in the Galaxy, ed. P. M. Solomon and M. G. Edwards (Oxford: Pergamon Press), p. 1.

Bodenheimer, P. 1981, in IAU Symp. No. 91, Fundamental Problems in the Theory of Stellar Evolution, ed. D. Sugimoto, D. Q. Lamb, and D. N. Schramm (Dordrecht: Reidel), p. 5.

Boss, A. P., and Black, D. C. 1982, Ap. J., 258, 270.

Cassen, P., and Moosman, A. 1981, Icarus, 48, 353.

Cassen, P., and Summers, A. 1983, Icarus, 53, 26.

Chandrasekhar, S. 1939, An Introduction to the Study of Stellar Structure (Univ. of Chicago Press).

Cohen, M., and Kuhi, L. V. 1979, Ap. J. Suppl., 41, 743.

Draine, B. T., Roberge, W. G., and Dalgarno, A. 1983, Ap. J., 264, 485.

Elmegreen, B. G. 1979, Ap. J., 232, 729.

Evans, N. J. 1978, in Protostars and Planets, ed. T. Gehrels (Tucson: Univ. Ariz. Press), p. 158.

Fricke, K. 1967, Z. f. Ap., 68, 317.

Goldreich, P., and Schubert, G. 1967, Ap. J., 150, 571.

Hayashi, C., Hoshi, R., and Sugimoto, D. 1962, Prog. Theor. Phys. Suppl., No. 23.

Herbig, G. 1962, Adv. Astr. Ap., 1, 47.

Ho, P. T. P., Martin, R. N., Myers, P. C., and Barrett, A. H. 1977, Ap. J. Lett., 215, L29.

Ho, P. T. P., and Townes, C. H. 1983, Ann. Rev. Astr. Ap., 21, 239.

Hunter, C. 1977, Ap. J., 213, 497.

Kuhi, L. V. 1964, Ap. J., 140, 409.

Kutner, M. L., Tucker, K. D., Chin, G., and Thaddeus, P. 1977, Ap. J., 215, 521.

Larson, R. B. 1969, MNRAS, 145, 271.

Larson, R. B. 1981, MNRAS, 194, 809.

Loren, R. B., Sandqvist, Aa., and Wooten, A. 1983, Ap. J., 270, 620.

Mercer-Smith, J. A., Cameron, A. G. W., and Epstein, R. I. 1983, preprint.

Mestel, L. 1965, Quart. J. R. A. S., 6, 265.

Mestel, L., and Spitzer, L. 1956, MNRAS, 116, 503.

Mouschovias, T. Ch. 1976, Ap. J., 207, 141.

Mouschovias, T. Ch. 1981, in IAU Symp. No. 91, Fundamental Problems in the Theory of Stellar Evolution, ed. D. Sugimoto, D. Q. Lamb, and D. N. Schramm (Dordrecht: Reidel), p. 27.

Mouschovias, T. Ch., and Paleologou, E. V. 1980, Ap. J., 237, 877.

Myers, P. C., and Benson, P. J. 1983, Ap. J., 266, 309.

Nakano, T. 1981, Prog. Theor. Phys. Suppl., No. 70, 54.

Ostriker, J. P., and Bodenheimer, P. 1973, Ap. J., 180, 171.

Rowan-Robinson, M. 1979, Ap. J., 234, 111.

Sargent, A. I. 1977, Ap. J., 218, 736.

Schneps, M. H., Martin, R. N., Ho, P. T. P., and Barrett, A. H. 1978, Ap. J., 221, 124.

Shu, F. H. 1977, Ap. J., 214, 488.

Shu, F. H. 1983, Ap. J., 273, 202.

Solomon, P. M., and Sanders, D. B. 1980, in Giant Molecular Clouds in the Galaxy, ed. P. M. Solomon and M. G. Edwards (Oxford: Pergamon Press), p. 41.

Spitzer, L. 1942, Ap. J., 95, 329.

Stahler, S. 1983a, preprint.

Stahler, S. 1983b, preprint.

Stahler, S. W., Shu, F. H., and Taam, R. E. 1980a, Ap. J., 241, 637; 1980b, Ap. J., 242, 226; 1981, Ap. J., 248, 727 (SST).

Strom, S. 1983, preprint.

Terebey, S., Shu, F. H., and Cassen, P. 1983, in preparation (TSC).

Ulrich, R. K. 1976, Ap. J., 210, 377.

Umebayashi, T., and Nakano, T. 1980, Pub. Astr, Soc. Japan, 32, 405.

Vogel, S. N., and Kuhi, L. V. 1981, Ap. J., 245, 960.

Winkler, K.-H., and Newman, M. J., 1980a, Ap. J., 236, 201; 1980b, Ap. J., 238, 311.

A DYNAMICAL ORIGIN FOR THE STELLAR BIRTHLINE

Steven W. Stahler

Harvard-Smithsonian Center for Astrophysics
60 Garden Street, Cambridge, MA 02138

1. The Observed Birthline

The location of T Tauri stars in the Hertzsprung-Russell diagram—above and to the right of the Main Sequence —indicates that they are low-mass stars undergoing quasi-static contraction prior to hydrogen ignition. Figure 1, taken from Figure 2 of Cohen and Kuhi (1979), is the diagram for T Tauri stars in the Taurus-Auriga molecular cloud complex. Superimposed are the theoretical Hayashi contraction tracks for several stars with the indicated masses (in solar units). It is apparent that the stars are not distributed randomly in this diagram, but are clustered below a well-defined boundary. Part of this boundary is shown as the heavy solid line cutting across the Hayashi tracks. The diagrams for other cloud complexes in Cohen and Kuhi's study exhibit a similar boundary. It is an attractive hypothesis that this boundary has a physical origin, specifically, that it represents a stellar "birthline" along which most T Tauri stars first appear as visible objects, following their optically veiled protostellar accretion phase. As this view has recently been presented in detail elsewhere (Stahler 1983), I limit myself in the next section to a brief summary of the main arguments.

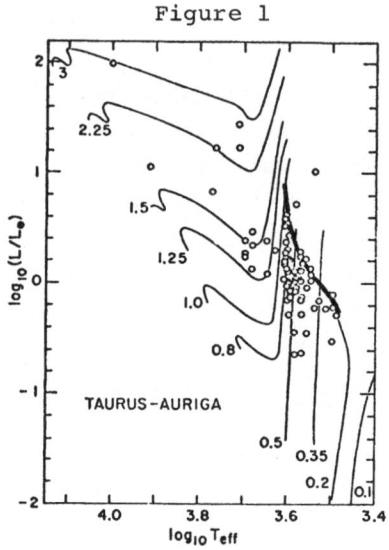

Figure 1

2. Protostars and the Birthline

Prior to their pre-Main-Sequence contraction phase, low-mass stars exist as accreting protostellar cores within a collapsing molecular cloud. These cores are optically invisible, hidden within their infalling dust envelopes (see Figure 1 of Stahler, Shu and Taam 1980). The process which ends the accretion phase is not understood, but it must be a rapid one, since few recognizable T Tauri stars seem to exhibit true circumstellar (and not just interstellar) obscuration

(one such star may be HL Tauri; see Cohen 1983). Following the end of
accretion, the newly visible core begins its slow contraction as a T
Tauri star toward the Main Sequence. Because the protostar—T Tauri
transition is rapid, the initial <u>radius</u> of a T Tauri star should be
the same as the radius of an accreting core of the same mass. If T
Tauri stars do indeed follow Hayashi tracks, their radii at each point
on the tracks are known. This information, combined with a protostellar
mass-radius relation, yields a theoretical prediction for the birthline. The birthline in Figure 1 was obtained in this manner, using
the protostar collapse calculation of Stahler, Shu and Taam (1980).

An immediate consequence of this picture of the evolution of T Tauri
stars is a revision of their ages. Traditionally, a T Tauri age is
taken to be the time for the star to contract from some very large
initial radius to its present value (see e.g., Iben 1965). With the
adoption of the stellar birthline, however, the contraction occurs
over a much smaller interval in radius. If we add to this contraction
time the period for the protostellar core to grow at a mass accretion
rate $\dot{M}$, the new age is related to the old by—

$$(1) \quad t_{new} = t_{old} - \frac{GM^2}{7\,R_o L_o} + \frac{M}{\dot{M}} \; ,$$

where R_o and L_o are the birthline values of radius and luminosity,
respectively, for a star of mass M. The factor of 1/7 on the right
side of this equation comes from approximating a T Tauri star's
internal structure as that of an n = 1.5 polytrope.

The new ages can be evaluated for observed stars by taking $R_o(M)$
from the aforementioned protostar collapse calculation, which adopted
a characteristic value for $\dot{M}$ of 10^{-5} $M_\odot$ yr^{-1}. For each star, t_{new} is
<u>shorter</u> than t_{old} by approximately 10^5 yr, since the second righthand
term in equation (1) is always larger than (but of the same order as)
the third term. Figure 2
shows the distribution of
T Tauri ages in Taurus-
Auriga. The star production
rate in this cloud complex
using the old ages (solid
curve) shows an irregular
but steady increase from the
past, but a marked drop in
the most recent 10^5 yr interval. This anomalous drop is
neatly filled in once the new

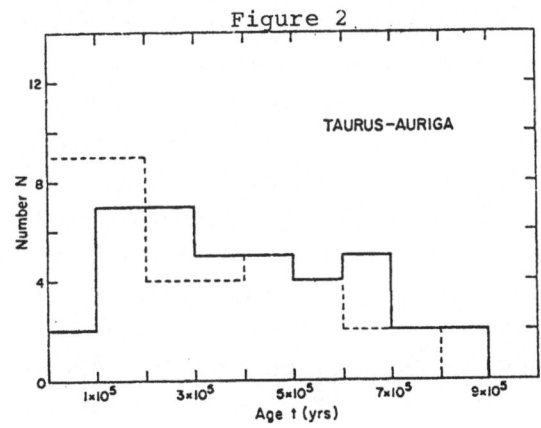

Figure 2

TAURUS-AURIGA

ages (dashed curve) are adopted. The results are similar with other cloud complexes containing T Tauri stars.

3. Molecular Clouds and the Birthline

The good agreement between the theoretical and observed birthlines indicates the basic correctness of the adopted protostellar mass-radius relation. The underlying collapse calculation was rather idealized, since it ignored any possible rotation, magnetic fields, or turbulent motion in the parent cloud. In actual molecular clouds, such effects must be present in varying degrees. Their alteration of the collapse dynamics will introduce a spread in birth radii at any stellar mass and thus a finite thickness to the theoretical birthline. Assuming a spherically symmetric collapse as a first approximation, the core mass-radius relation depends only on $\dot{M}$. It is important to note, however, that the protostellar radius—and hence, the birthline position—is insensitive to the exact value of $\dot{M}$, since R_o scales approximately as $(\dot{M})^{0.3}$ at fixed core mass (Stahler 1983).

If we consider the small dark clouds most likely to be sites of low-mass star formation ("cold cores" and "globules"), their actual degree of rotation and magnetization is not yet known. In the case of turbulence, however, the NH_3 radio measurements of Myers and Benson (1983) show that part of the observed line broadening in these clouds can be attributed to a Gaussian distribution of random velocities. This turbulence adds to the thermal pressure in supporting the clouds prior to collapse. We may define an effective sound speed by

$$a_{eff} \equiv \{(\frac{1}{\mu} - \frac{1}{17})(\frac{kT}{m_H}) + \frac{1}{8 \ln 2} (\Delta V)^2_{NH_3}\}^{1/2}$$

Here $(\Delta V)_{NH_3}$ is the FWHM of the line-of-sight component of the velocity distribution of NH_3 molecules, μ is the mean molecular weight of the cloud gas, and T its kinetic temperature.

Figure 3 shows the distribution of a_{eff} in 25 representative cores and globules. The values of T and $(\Delta V)_{NH_3}$ are taken from Myers and Benson (1983) and Benson (1983), and μ has been set at 2.33. The extreme narrowness of the distribution about the mean a_{eff} of 0.22 km s^{-1} can be attributed to the uniformity of cloud kinetic temperatures (all near 10 K) and the subsonic level of the turbulent speeds. Since $\dot{M}$ scales as a^3_{eff} in the collapse of a marginally stable cloud (Stahler, Shu and Taam 1980), the observed range in a_{eff} (all but four clouds in Figure 3 have a_{eff} from 0.20 to 0.25 km s^{-1}) will induce a factor of two variation in $\dot{M}$. Thus, the thickening of the birthline due to this

turbulent support is very
slight (Δlog L $\sim$ 0.2).

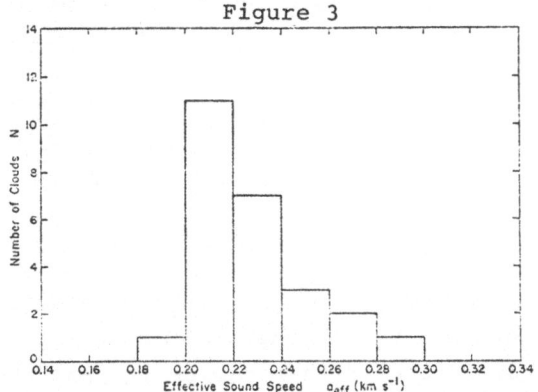

Figure 3

   More realistic collapse
calculations, using cloud
conditions taken from observation, are needed to
establish the true width of
the birthline. A small
proportion of clouds probably
exists with very different
conditions than the narrow
range considered here. These exceptional clouds, though statistically
insignificant, are important conceptually, since their presence would
explain the existance of the few overly luminous T Tauri stars situated
far above the birthline.

References

Benson, P.J. 1983, unpublished Ph.D. thesis, M.I.T.
Cohen, M. 1983, Ap.J. 270, L69.
Cohen, M. and Kuhi, L.V. 1979, Ap.J.Suppl. 41, 743.
Iben, I. 1965, Ap.J. 141, 993.
Myers, P.C. and Benson, P.J. 1983, Ap.J. 266, 309.
Stahler, S.W. 1983, Ap.J. in press.
Stahler, S.W., Shu, F.H., and Taam, R.E. 1980, Ap.J. 241, 637.

ROTATIONAL VELOCITIES OF LOW MASS STARS IN INTERMEDIATE AGE OPEN CLUSTERS

J. R. Stauffer
Harvard-Smithsonian Center for Astrophysics
60 Garden Street
Cambridge, MA 02138

Most previous observational studies of pre-main sequence (PMS) evolution have concentrated on very young, relatively high mass stars. The advantage of that choice is that the program stars are easily identifiable as young via various type of "activity". That characteristic is also a disadvantage, however, since the IR and UV excesses, emission lines, "blue-veiling", and variability make placement of those stars in an HR diagram difficult. In an effort to avoid those difficulties and to extend the observational study of PMS stars to a different portion of parameter space, I have begun a program of study of intermediate age, relatively low mass PMS stars. The primary goals of that program are:
(1) to determine the rotational velocity history of low mass stars;
(2) to determine the time spread of star formation in open clusters;
(3) to derive empirical PMS isochrones for low mass stars.
In this paper, I shall emphasize the rotational velocity portion of that program. My collaborators in that work are Lee Hartman, Dave Soderblom, and Neal Burnham.

I. Background

As is traditional, we have chosen to use open clusters as our source of young, low mass PMS stars (but see also Lindroos 1983, Duncan 1983 for an alternative). The ages and mass ranges of interest to us are $10^7 < \tau < 3 \times 10^8$ years and $0.2 < M/M_\odot < 1.0$. In order to be able to obtain optical photometry and high resolution spectra for those stars with the available instrumentation, the clusters must be located within about 700 parsecs of the sun. As a final selection criterion for the program, we need a proper motion survey for the cluster that extends to at least $M_V = 9^m$ (corresponding generally to an apparent magnitude fainter than 15^m). The age and distance requirements limit us to about 10 clusters. Until Fall 1983, the proper motion criterion was satisfied by only one of those clusters - the Pleiades ($\tau = 7 \times 10^7$ years, r = 125 parsecs). Consequently, the remainder of the paper will concentrate on our observational data for that cluster. Jones and Stauffer (1983) have now obtained a proper motion survey for the central portion of the α Perseus cluster ($\tau = 5 \times 10^7$ years, r = 160 parsecs) to V $\simeq 16^m$. Because of the low galactic latitude of the cluster, this was not an easy task - of approximately 4000 stars for which we determined proper motions, only about 40 appear to be cluster members. The first observational results for the proposed late type, α Perseus cluster members will also be reported.

II. Pleiades Photometry

To set the stage for the rotational velocity data, I will briefly review the photometric studies of the Pleiades cluster.

The first extensive photoelectric photometry for the Pleiades was obtained by Johnson and Mitchell (1958). Using the then current pre-main sequence and post-main sequence evolutionary theories, Herbig (1962) interpreted the photometry as indicating an age for the high mass stars (from the upper main sequence turn-off point) of $\tau = 6 \times 10^7$ years. Because the photometry showed no clear indication of a PMS turn-on point to V = 14^m, the low mass stars were assigned an age of $\tau > 2.2 \times 10^8$ years. The non-equality of those two ages led Herbig to postulate that stars were formed over a significant time period in the proto-cluster ("non-coeval star formation"), and that low mass stars apparently formed first.

Both the observational data and the theoretical models have evolved considerably since 1962, but Herbig's basic premise may remain viable. Figure 1 shows the current status of my Pleiades photometric program (Stauffer 1983), where the curve is a zero age main sequence derived from older clusters. The PMS turn-on point appears to occur at V $\simeq 14^m.5$. Using the now current theoretical models, the upper main sequence turn-off age is $\tau \simeq 7 \times 10^7$ years (Mermilliod 1981) and the PMS contraction age is $\tau_c \simeq 10^8$ years (VandenBerg et al. 1983). That is, the low mass stars still appear to be older than the high mass stars. It is impossible to tell with data from just one cluster whether that difference actually represents a time spread of star formation in the cluster, or is instead due to errors in the theoretical evolutionary models or errors in the interpretation of the data.

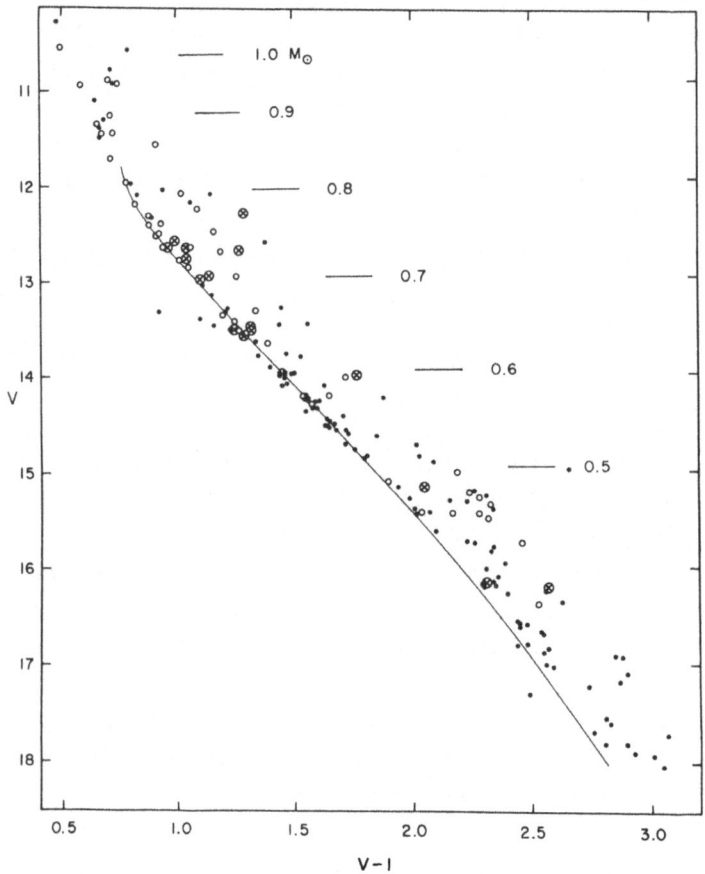

Figure 1: V versus V-I for Pleiades stars. Large symbols - vsini > 25 km-s^{-1}; open circles - vsini < 25 km-s^{-1}; filled dots - no spectra.

III.   Spotted, Rapidly Rotating K Dwarfs in the Pleiades

No rotational velocity estimates for Pleiades stars with M < 1 M$_\odot$ (V > 10^m) had been published prior to about 1981. I had not originally planned on obtaining such data, since I expected those rotational velocities to be small (probably below the detection limit of 10 km-s^{-1} for the MMT echelle). That expectation was based on a straightforward extrapolation of the existing information. The mean vsini of Pleiades stars, as well as field stars and other cluster stars, shows a broad peak near B5 and a sharp falloff for spectral types later than F0 (Kraft 1967). For stars slightly more massive than the sun, Kraft's data showed the Pleiades stars to have < vsini > ≈ 15 km-s^{-1}. Rotational velocities continue to decline toward later spectral type among field stars (Vaughan et al. 1981), so a mean rotational velocity less than 15 km-s^{-1} for Pleiades stars less massive than the sun seemed reasonable.

Evidence that the above reasoning was seriously in error was first reported by van Leeuwen and Alphenaar (1982 = VA). Those authors claimed that at least 12 of the Pleiades K dwarfs were short period photometric variables, with amplitudes $0.^{m}05$-$0.^{m}20$ and periods $0.24 < P < 1.25$ days. On the assumption that the variability is due to star-spots, the photometric period is equal to the rotational period. Given a period, it is only necessary to know approximate radii for the stars in order to determine rotational velocities. Since K dwarfs in the Pleiades are on the main sequence (see Figure 1), we adopt a standard, main sequence color-radius relation. Combining the periods and radii, we derive equatorial, rotational velocities for the VA variables of 35-170 km-s^{-1}, clearly much different from the previous estimate.

In order to attempt to understand the startling results presented by VA, we have now obtained high resolution echelle spectra for about 70 of the Pleiades stars in the $10^{m} < V < 16^{m}$ range. Our analysis of those data are presented in Stuaffer et al. (1983 = SHSB). In general, we confirm that the VA photometric variability is real and apparently due to spottedness. Seven of the VA stars have vsini's consistent with their photometric periods. The five smallest amplitude variables have vsini's much less than predicted by their published periods, and we believe the periods must be wrong. Van Leeuwen (1983) states that aliasing may have affected the results for those stars, and that periods of the order of 6 days are possible (corresponding to $v_{rot} < 10$ km-s^{-1}). A set of spectra showing the range in rotational velocity among the K and M dwarfs in the Pleiades is shown in Figure 2.

Our photometry also supports the star-spot model for the K dwarf variables. For the shortest period variable, HII 1883, we now have light curves measured through filters from 0.4μ to 2.2μ (BVRIJHK). The amplitude of the light curve decreases from $0.^{m}15$ at B to $0.^{m}06$ at K. By fitting a simple model to the variation of light curve amplitude with wavelength, D. Dorren estimates that a pair of spots covering about 3.5% of the total surface area of the star and having effective temperatures about 1700 K cooler than the surrounding photosphere can match the observations.

Photometric light curves like that shown by VA are sensitive only to the non-axisymmetric spot distribution. A star uniformly checker-boarded by spots would not vary appreciably. Since the integral colors of a spotted star differ from those for a "normal" star, a color-color

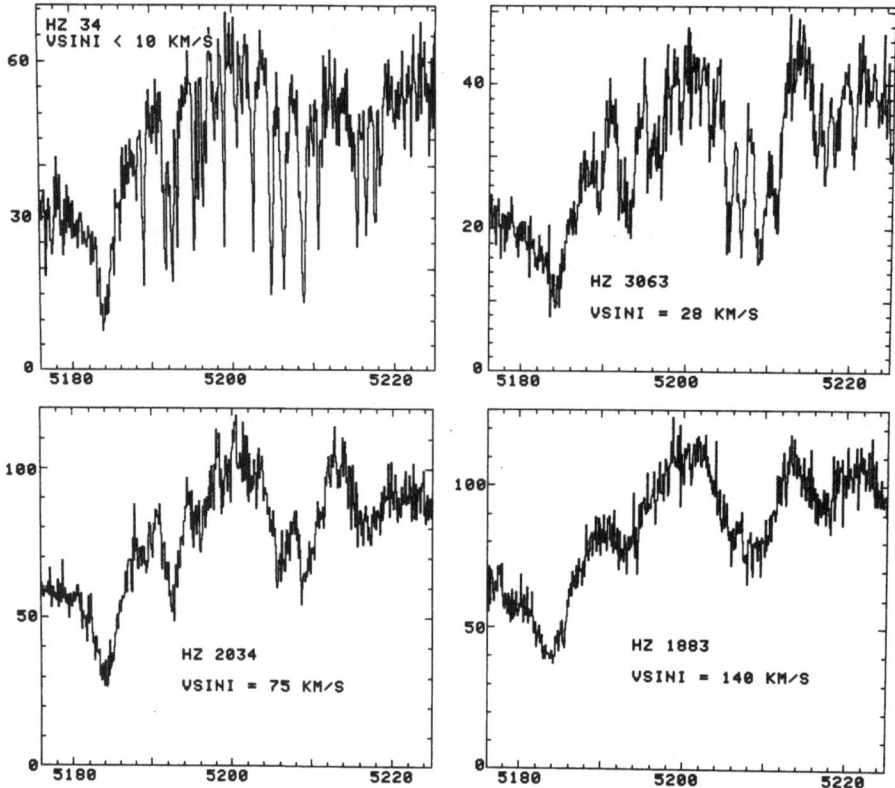

Figure 2: MMT echelle spectra of Pleiades K dwarfs.

diagram can be used to segregate heavily spotted stars from stars
with lower spot contributions. A B-V versus V-I diagram for all of the
late type Pleiades stars (Figure 6 of SHSB) shows the rapid rotators
to have B-V colors that average about $0^{m}04$ bluer than normal for their
observed V-I. That displacement is in the sense expected for spotted
stars, with the magnitude of the color anomaly indicating that spots
probably cover at least 10% of the stellar surface.

It is worth emphasizing at this time how useful color-color diagrams
can be for this type of investigation. Identifying rapid-rotators via
short period, photometric variability requires a quite lengthy observing
campaign. First, each star in the cluster must be observed several
times over a few day interval in order to identify possible variables.
Then, each of those stars must be monitored during another observing
run in order to determine a period. With a single observation per
star, the color-color diagram can identify heavily spotted stars
(probable rapid rotators) via their peculiar colors. Ideally, both

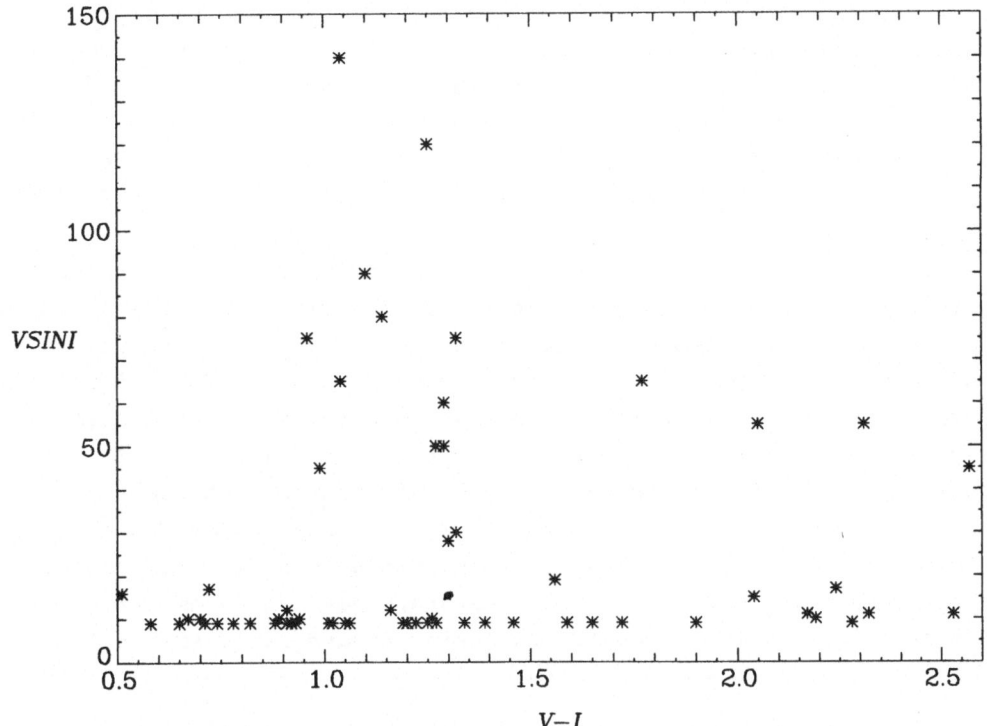

Figure 3: Spectroscopic rotational velocities for late type Pleiades stars. Spectral types G3, K2 and M2 correspond roughly to V-I = 0.5, 1.0 and 1.9, respectively.

techniques should be used since neither is infallible. The photometric variability technique will miss spotted stars seen "pole-on" or stars with nearly uniform spot coverage. The color-color method can in some cases misidentify a binary star as a spotted star candidate.

IV. The Rotational Velocity Distribution of Late Type Pleiades Stars

Of the 66 late type Pleiades stars we observed, 18 have rotational velocities greater than 25 km-s^{-1}. Figure 3 shows the distribution of vsini versus V-I. The location of the rapid rotators in the V versus V-I diagram is indicated in Figure 1. The salient features of those diagrams are that:

(1) There appears to be a turn-on point for rapid rotation, since none of the stars that we have observed with V-I < 0.95 (spectral type < K2) have vsini > 20 km-s^{-1}.

(2) The rapid rotators are main sequence stars, with approximately the same photometric binary frequency as slow rotators.

(3) In the range 0.95 < V-I < 1.35, perhaps as many as 50% of
    the Pleiades stars have vsini > 25 km-s^{-1}. But also note that,
    for that same color range, a similar fraction have
    vsini < 10 km-s^{-1}.

(4) The fraction of stars that are rapid rotators appears to be
    less among the M dwarfs (V-I > 1.75) than among the K dwarfs.

Two other bits of data support the identification of a "turn-on
point" for some phenomenon among the late type Pleiades stars. Figure 4
shows the relation between Hα equivalent width and (R-I) for the stars
observed spectroscopically. The transition from Hα being uniformly in
absorption to being uniformly in emission occurs at the same color as
the rapid rotation turn-on point. Near that color (R-I ≃ 0.48, corresponding to V-I ≃ 0.95), there is a good correlation between Hα and
rotational velocity - the rapid rotators with only one exception have
Hα in emission, while the slow rotators generally have Hα in absorption.
For later spectral types, all Pleiades stars have Hα in emission, with
the mean emission line strength increasing toward later spectral type.

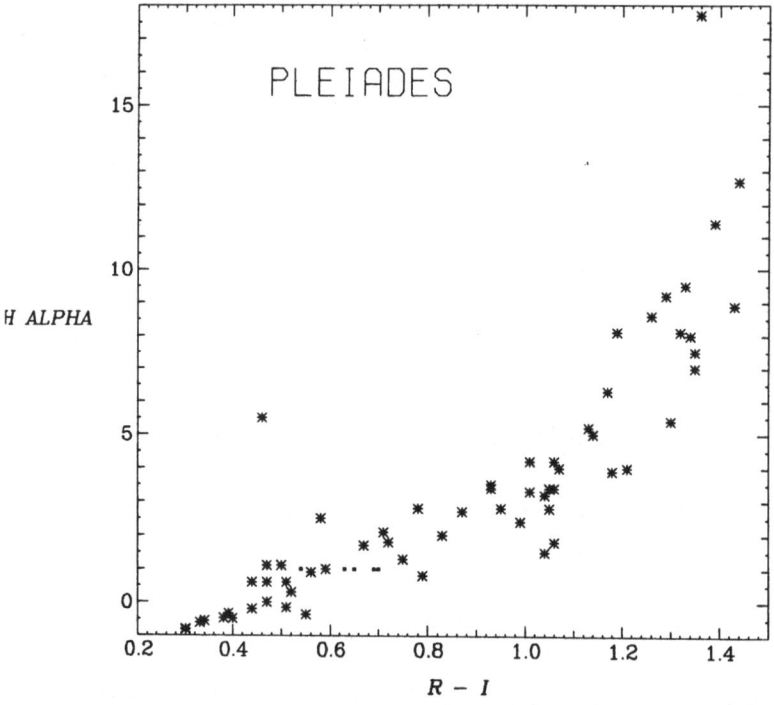

Figure 4: Hα equivalent width versus (R-I) for Pleiades dwarfs. Equivalent widths less than zero correspond to absorption.

An examination of flare star statistics for the Pleiades stars (Haro, Chavira and Gonzalez 1982) also shows a turn-on point. No flare stars have been detected with V-I < 0.95. HII 2034 (V-I = 0.96) is both the bluest star with a detected flare and the bluest rapid rotator in the Pleiades. The fraction of Pleiades stars with observed flares increases rapidly redward of V-I = 0.95, with probably all Pleiades M dwarfs being flare stars (see Figure 6 of Stauffer 1983).

One way in which late type stars can become rapid rotators is through synchronous rotation in a close binary. We believe that explanation is unreasonable for the Pleiades stars, since it would require that 50% of the K dwarfs be close binaries. Also, we see no evidence that the radial velocities of the rapid rotators deviate from the cluster mean by more than our expected measurement errors.

Is there a reasonable explanation for the high rotational velocities for some Pleiades late type stars? We believe the answer is yes, and in hindsight such a phenomenon could have been predicted.

Kraft (1967) has shown that the mean angular momentum per unit mass for high mass stars follows a $J/M \propto M^{2/3}$ relation. That relation is assumed to indicate the initial distribution of angular momentum among stars. High mass stars do not have outer convective envelopes, and so still follow the relation regardless of their age (at least while they are on the main sequence). Late type field stars have rotational velocities much below that predicted by the $M^{2/3}$ relation, presumably due to angular momentum loss from stellar winds. If the angular momentum loss occurs primarily on the main sequence, then by observing a cluster whose age is approximately equal to the PMS contraction time for a 0.8 M_O star, K dwarfs with rotational velocities of the order to that predicted by Kraft's relation might be detected. We believe this is the case for the Pleiades. VA have noted that the angular momentum of some of the Pleiades rapid rotators appears to be approximately consistent with the $M^{2/3}$ relation.

A theoretical evolutionary model predicting that solar mass stars could have rotational velocities up to about 100 km-s^{-1} upon arrival on the main sequence has been recently published by Endal and Sofia (1981). Those authors assume that the angular momentum per unit mass of PMS stars follows the Kraft relation. Angular momentum loss is through a Belcher-Macgregor type wind, where the angular momentum loss rate is roughly proportional to the surface rotational velocity. The predicted surface rotational velocities for three different assumed initial angular momenta are shown in Figure 4 of Endal and Sofia (1981). Because of their large radii while on convective tracks, the stars initially have low rotational velocities. During the radiative track phase of PMS evolution, the star contracts and also becomes more centrally condensed (non-homologous contraction). The rapid decrease in the moment of inertia of the star during the radiative track phase causes a corresponding increase in the surface rotational velocity, reaching about 60-70 $km-s^{-1}$ as the star arrives on the main sequence. The wind carries away some angular momentum during the radiative track phase, but that effect is small compared to the spin-up induced by contraction. The surface rotational velocity of the star begins to decrease once core nuclear fusion halts the contraction. The initial spin-down rate is quite rapid because the mass loss rate from the wind is at a maximum ($dJ/dt \propto$ the surface rotational velocity) and because the wind acts only to spin down the outer convective envelope. For the Endal and Sofia model B, the star drops from its maximum rotational velocity to about 15 $km-s^{-1}$ in 4×10^7 years.

The Pleiades data appear to conform reasonably well with the Endal and Sofia model. Theoretical evolutionary models predict that the early K dwarfs should have recently arrived on the main sequence. The G dwarfs in the Pleiades had shorter PMS contraction timescales. Therefore, while they presumably did arrive on the main sequence as rapid rotators, they have had enough time to spin down. The K dwarfs that are slow rotators could either be stars with lower than average initial angular momenta or, slightly older cluster members. If the time spread of star formation in the cluster were a few times 10^7 years (comparable to the main sequence spin down time), then we would predict a mix of slow and fast rotators at the turn-on point. Finally, rapid rotation is not common among the M dwarfs because stars in that mass range are convective all the way to the main sequence - they contract homologously and sedately, without any rapid changes in moment of inertia.

Evidence for a substantial wind in at least two of the rapid rotators is provided by their Hα profiles. For HII 686 and HII 1883, the width of the Hα emission line is much greater than can be accounted for by rotation. The most likely additional broadening mechanism is a wind. For HII 1883, the estimated wind velocity is about 350 $km-s^{-1}$.

Evidence that the K dwarf rapid rotators will spin down relatively
quickly, at least faster than a $t^{-1/2}$ law, is provided by spectra of
Hyades K dwarfs obtained by Latham and Stefanik (1983). None of the
Hyades K dwarfs so far observed have vsini > 10 km-s^{-1}. Assuming the
Hyades K dwarfs were rapid rotators when they reached the main sequence,
that implies that the spin down time for those stars must have been
less than 5 x 10^8 years (the age of the Hyades).

In summary, we believe that the Pleiades data show that low mass
stars arrive on the main sequence with rotational velocities up to at
least 150 km-s^{-1}. Strong stellar winds significantly decrease the
surface rotational velocities in a few times 10^7 years. Therefore,
other clusters in the 2 x 10^7 < τ < 2 x 10^8 year range should also
show rapid rotators among the late type dwarfs. Younger clusters should
have bluer turn-on points for rapid rotation.

V. The α Perseus Cluster

To test the veracity of our model to explain the Pleiades rapid
rotators, we need to observe other open clusters. As noted previously,
however, the proper motion studies necessary for that program do not
generally exist. Therefore, we have begun the task of obtaining those
proper motion studies ourselves. The first cluster for which we have
obtained a partial membership list is the α Perseus cluster (Jones and
Stauffer 1983). Since that cluster is about 20-30% younger than the
Pleiades, it should provide a good test of the model.

Spectra of about 15 of the newly identified α Per members were
obtained in September 1983 at the MMT. Five of those stars appear to
be rapid rotators, with rotational velocities ranging from about
30 km-s^{-1} to about 100 km-s^{-1}. Spectra of one of the rapid rotators
is shown in Figure 5. By the end of the 1983 observing season, we
hope to have obtained photometric and spectroscopic data comparable
to the Pleiades survey and thus have one check on our model. The mere
presence of rapid rotators among the late type α Per members at least
shows the Pleiades cluster is not unique.

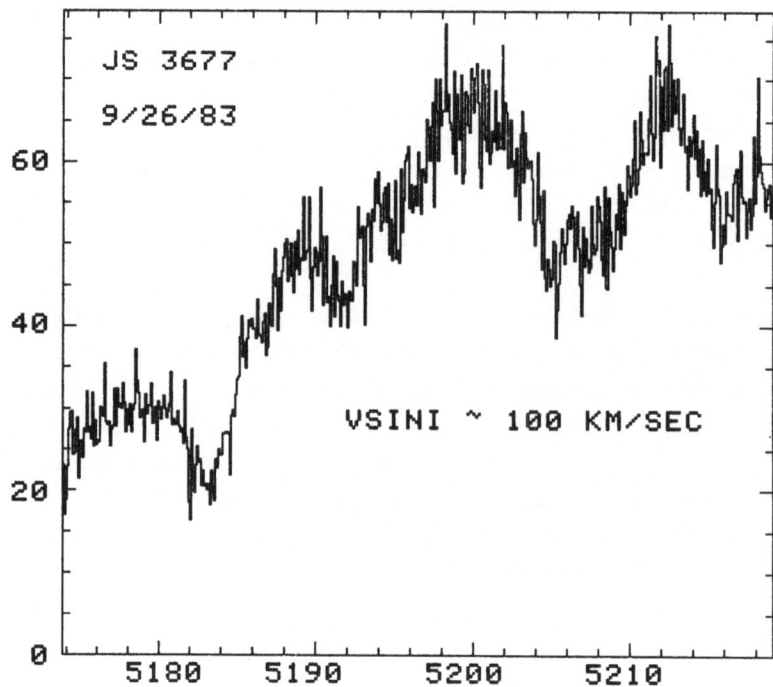

Figure 5: Echelle spectra for one of the late-type rapid rotators in the α Perseus cluster.

References

Duncan, D. 1983, preprint.

Endal, A. S., and Sofia, S. 1981, Ap. J., 243, 625.

Haro, G., Chavira, E., and Gonzalez, G. 1982, Boll. Obs. Tonantzintla and Tacubaya, 3, 3.

Herbig, G. 1962, Ap. J., 135, 736.

Johnson, H. and Mitchell, R. 1958, Ap. J., 128, 31.

Kraft, R. 1967, in Spectroscopic Astrophysics, ed. G. Herbig. (Berkeley: U. C. Berkeley Press), p. 385.

Latham, D., and Stefanik, R. 1983, in preparation.

Lindroos, K. 1983, A. A., 51, 161.

Mermilliod, J. C. 1981, A. A., 97, 235.

VandenBerg, D. A., Hartwick, F. D. A., Dawson, P., and Alexander, D. R. 1983, Ap. J., 266, 747.

Van Leeuwen, F. 1983, priv. comm.

Van Leeuwen, F., and Alphenaar, P. 1982, E. S. O. Messenger, No. 28, p. 15.

Vaughan, A. H., Baliunas, D. L., Middelkoop, F., Hartmann, L. W., Mihalas, D., Noyes, R. W., and Preston, G. W. 1981, Ap. J., 250, 276.

MEASUREMENTS OF ROTATIONAL VELOCITIES OF F AND G STARS

Giancarlo Noci - Istituto di Astronomia, Università di Padova
Sergio Ortolani - Osservatorio Astrofisico di Asiago
Pierluigi Rossi - Istituto di Astronomia, Università di Firenze
Leon Golub - Harvard-Smithsonian Center for Astrophysics

1. Introduction

As it is well known, stellar observations in the soft X-ray range
have shown that the presence of a corona, source of X-ray emission, is
a very common phenomenon. It is possible that stellar coronae are, like
the solar corona, inhomogeneous structures threaded by magnetic loops
and that the brighter regions are those of higher field. Since the amount of loop structures is, in the solar corona, a product of differe-
zial rotation, it is believed that a similar mechanism is at the base
of the coronal emission for solar type stars. Hence the finding of higher soft X-ray emission for faster rotating G stars is in agreement
with this line of thought. Similarly is interpreted the observed extension of this correlation to K and M stars. (Pallavicini et al., 1981;
Walter, 1982; and references therein.)

These results have promoted the measurement of the rotational velocity of stars for which the soft X-ray emission was known. Hence we
have undertaken a program at Asiago Astrophysical Observatory to make
such measurements. We report here on the preliminary results.

2. Determination of rotational velocities

The observations have been madeby means of the echelle spectrograph
attached to the 1.82 m reflector of Mt. Ekar station. In the setting
used for the observations the dispersion was 9 Å/mm and the instrumental profile had a width of 0.4 Å.

To date the projected rotational velocity, v.sini, has been determined for 7 stars of the spectral classes F0-G6, using the width of the
NaI doublet at the wavelengths 5890 and 5896 Å. For this category of
obseved stars the chosen lines appear to be strong enough without having

any trace of saturation; the line equivalent widths all fall in the interval 200-300 mÅ, so that we can expect similar radiative transfer effects for all the 7 observed stars. To get the rotational velocity for these stars we have constructed a calibration curve by a least square fit of a polynomial relation in a v.sini, observed half width plane to the points corresponding to 32 stars of known rotational velocity. The standard stars belong to the same spectral classes as the stars to be measured and the equivalent widths of the lines of the NaI doublet in their spectra fall in the interval quoted above. The value of v.sini for these stars was taken from the catalogue of Uesugi and Fukuda (1970).

The results of this determination are given in the following table, where the soft X-ray luminosity, observed by the Einstein satellite, is also reported.

Rotational velocity and X-ray luminosity

Star	Spectral type and luminosity class	v.sini (km/sec)	$\text{Log}(L_x)$ (erg/sec)
HD 80290	F3 V	<17	28.1
HD 133408N	F0 V	106±6	<29.0
HD 133408S	F0 V	<17	<29.0
HD 103928	A9 V F0 V	81±6	<28.8
HD 95638	F7 V	<17	29.1
HD 108944	F9 V	<20	29.0
HD 13994	G6 III	<25	30.3

The errors on the v.sini values are statistical errors defined by the 3 standard deviations confidence interval. At the low velocity end we set the measurement limit at twice the instrumental profile, except for some more noisy spectra where a larger measurement limit was assumed.

The rotational velocities and soft X-ray fluxes of the above table agree reasonably well with the correlation suggested by Pallavicini et al. (1981).

3. Acknowledgements

This work has been supported in part by Consiglio Nazionale delle Ricerche of Italy and by NASA through grant NAGW-112.

References

Pallavicini, R., Golub, L., Rosner, R., Vaiana, G. S., Ayres, T., and
    Linsky, J. L.: 1981, Astrophys. J. <u>248</u>, 279.

Uesugi, A., and Fukuda, I.: 1970, Contr. Inst. Astrophys. and Kwasan
    Observ. Univ. Kyoto, n. 189.

Walter, F.: 1982, Astrophys. J. <u>253</u>, 745.

HIGH-RESOLUTION OBSERVATIONS OF COOL STARS WITH THE ESO COUDE' ECHELLE SPECTROMETER

R. Pallavicini
Arcetri Observatory, Florence, Italy

M. Pakull
Technische Universitat, Berlin, Germany

I) INTRODUCTION

A new instrument capable of providing high spectral resolution observations of relatively bright objects became operational in 1982 at the European Southern Observatory (ESO) at La Silla (Chile). This is the Coudé Echelle Spectrometer (CES), which in its present configuration is normally fed from the 1.4 m Coudé Auxiliary Telescope (CAT). The optical configuration uses an echelle grating and a prism pre-disperser. The instrument can be operated either in a scanner or multi-channel mode. In the latter mode the detector used is an 1872 channel Reticon cooled to 140 K. The resolving power attainable in this mode is typically of the order of 10^5. The system appears to have a very pure instrumental profile and low scattered light. Further details on the instrument can be found in Enard (1982).

We have used the CES + Reticon detector for measuring rotational velocities and Ca II K fluxes for a number of Southern bright stars of spectral types F5 to K5 and luminosity classes III, IV and V. The observations were carried out in two runs (3-9 December 1982 and 18-30 July 1983). Most of these observations are under analysis at present. Some examples of the acquired data and preliminary results are presented.

II) ROTATIONAL VELOCITIES

Observations in the red part of the spectrum were carried out with the purpose of measuring rotational velocities for slowly-rotating late-type stars. We gave priority to stars for which transition-region and coronal UV and X-ray fluxes were available with the aim of improving the correlation between stellar rotation and transition region/coronal emission (Pallavicini et al. 1981).

Observations in the red were carried out at central wavelengths of 6020 A, 6250 A and 6450 A. In all cases the entrance slit was 200 µm, corresponding to a resolving power $\lambda/\Delta\lambda \approx 1.2 \times 10^5$. The spectral range covered was ≈ 50 A. Figs. 1-3 show examples of the data for each of the observed spectral regions. More than 50 stars have been observed at least at one wavelength.

The spectral regions chosen contain a number of unblended, intermediate strength lines which appear to be suitable for deriving projected rotational velocities v·sin i using Fourier transform methods (Smith and Gray 1976). These data are being analyzed in collaboration with D. Soderblom of the Harvard-Smithsonian Center for Astrophysics for main sequence stars, and in collaboration with D. Gray of the University of Western Ontario for giants. Preliminary inspection of the spectra shows that most stars in our sample have extremely narrow lines indicating rotational velocities not substantially different from that of the Sun. A few stars, however, have broadened lines which imply rotational velocities a factor 4 or 5 higher than for the Sun. The derived rotation rates will be used in conjunction with chromospheric and transition region/coronal fluxes to investigate processes of magnetic heating of stellar chromospheres and coronae.

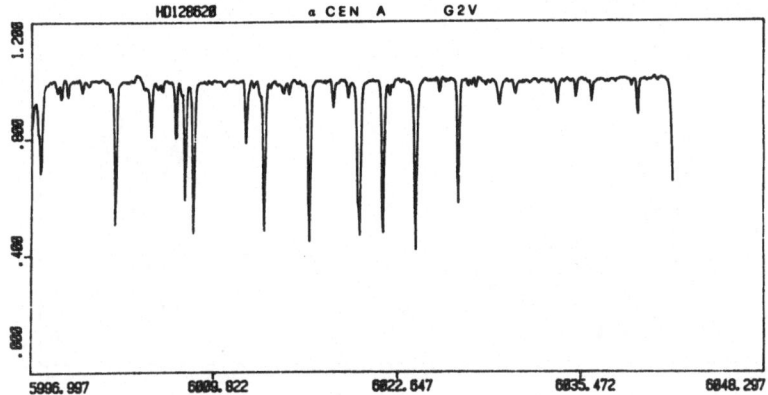

Fig. 1

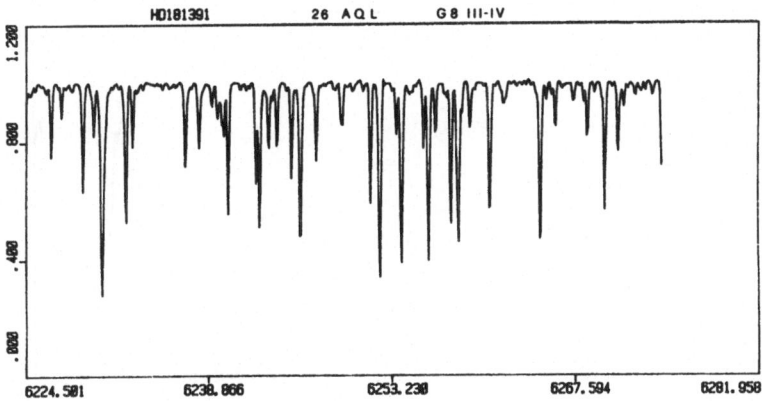

Fig. 2

Fig. 3

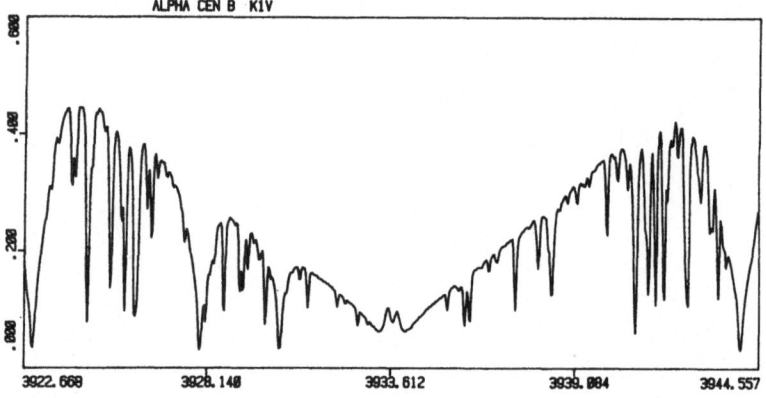

Fig. 4

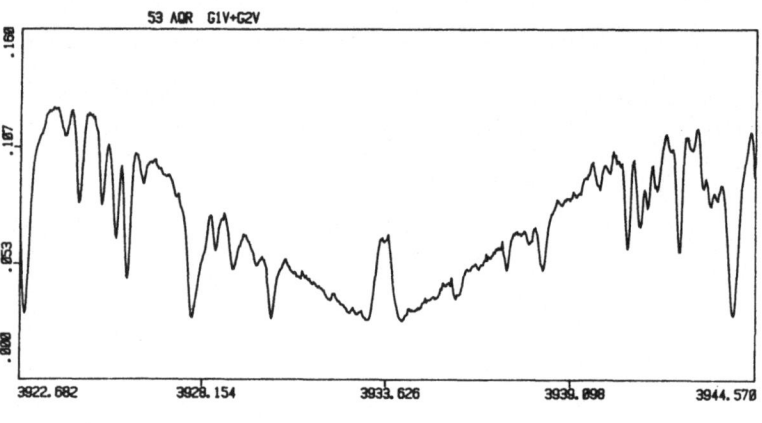

Fig. 5

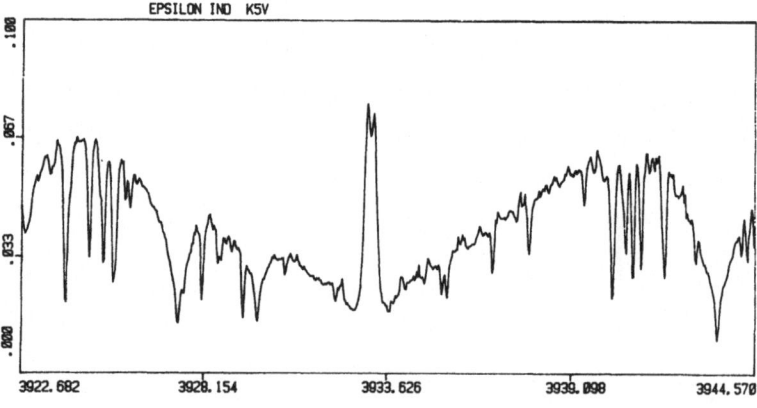

Fig. 6

III) CA II EMISSION FLUXES

Observations in the violet part of the spectrum were carried out in the K line of Ca II at 3933.7 A with the purpose of deriving chromospheric radiative fluxes from the central emission component. For a few program stars observations in the H line at 3968.5 A were also performed.

Observations in the violet with the CES + Reticon are much more difficult than in the red. The efficiency of the Reticon, which is about 70% at 6000 A, drops to 30% at the K line. In addition, in our program we were observing late-type stars which are intrinsically fainter in the U band than in the V band, and, more importantly, we were interested in the - usually small - emission component at the bottom of a very deep absorption line. For all these reasons, observations in the violet turned out to be practical only for quite bright objects, and even so at the expense of rather long integration times.

We have been able to obtain high resolution spectra ($\lambda/\Delta\lambda = 10^5$) of α Cen A and B in the H and K lines with reasonably short exposure times. For all other stars (≈ 40) we have used a somewhat reduced resolving power, ranging from 3×10^4 to 6×10^4, and typical exposure times from 1 to 4 hours. Even in the worst case, however, the spectral resolution is ≈ 120 mA, more than adequate to resolve structures in the central emission component. Figs. 4-6 give examples of the central portion of the observed spectra. The free spectral range observable in the K line is ≈ 30 A and the line was centered in such a way as to allow the simultaneous detection of the nearby pseudo-continuum at 3950 A.

Observed relative fluxes in the Ca II emission component are being converted to absolute fluxes using the calibration of Catalano (1979) and correcting for the underlying photospheric contribution (Linsky et al. 1979). The derived fluxes will be compared with UV and X-ray fluxes and correlated with rotation rates. As an example of the apparent dependence of Ca II emission on rotation, Fig. 5 shows the presence of a strong chromospheric emission component from the solar-type visual binary 53 Aqr A+B. Both components are known to be quite rapidly rotating stars for their spectral type (v·sin i ≈ 8 km/s, Soderblom 1982). Observations of this system at X-ray and UV wavelengths from EXOSAT and IUE are planned.

REFERENCES

Catalano, S. (1979) Astron. Ap. 80, 317.
Enard, D. (1982) Instrumentation in Astronomy IV, SPIE 331, 232.
Linsky, J.L. Worden, S.P., McClintock, W. and Robertson, R.M. (1979) Ap.J. Suppl. 41, 47.
Pallavicini, R., Golub, L., Rosner, R., Vaiana, G.S., Ayres, T. and Linsky, J.L. (1981) Ap.J. 248, 279.
Smith, M.A. and Gray, D.F. (1976) Publ. Astron. Soc. Pac. 88, 809.
Soderblom, D. (1982) Ap.J. 263, 239.

THE RAPIDLY-ROTATING FIELD M-DWARF GLIESE 890

Arthur Young and Andrew Skumanich
High Altitude Observatory
National Center for Atmospheric Research
Boulder, Colorado 80307
and
Clayton Heller and Scott Temple
Astronomy Department
San Diego State University
San Diego, California 92182

The causal connections between rapid rotation, dynamo generation, and a variety of magnetically driven surface phenomena on stars which have convective envelopes is now well established. At the High Altitude Observatory we are pursuing investigations which are designed to probe some of the details of the mechanisms which underlie such processes.

One line of investigation, concerned with the role of differential rotation in the aforementioned processes, involves a comparison being made between the surface activity of single, rapidly rotating stars and their counterparts of similar mass, radius, and luminosity which are components of strongly coupled binary systems. Our contention is that strong tidal coupling in a binary system affects the differential velocity field, and renders it different from an otherwise similar single star. Conventional radiative diagnostics such as the CaII H and K lines, the Ca II infrared triplet lines, and Hα are being used to model the chromospheric structures of such stars.

The RS CVn binaries provide a natural source for tidally coupled stars, but rapidly rotating single stars are more scarce, particularly at the lowest end of the main sequence. The so-called BY Dra stars provide both single and binary candidates in the latter domain, but, except for extremely young groups such as the Pleiades, the rotation periods of the known single photometric variables are generally longer than are those of the binaries with the shortest known periods.

Our survey of active dMe stars resulted in our discovery of GLS 890 which is an exception to that general behavior. The photospheric absorption features of GLS 890 appear to be broadened by considerable rotation. In a recent paper (Ap.J. in press) we demonstrate that GLS 890 is unlikely to be a binary star, and that the surface rotation ($v \sin i$) is of the order of 70 km/sec. Presuming normal dimensions for this dM 2.5 e star we predicted a rotation period of approximately 8 hours!

Broadband photometric observations of GLS 890 were made with the 1.0 m telescope at the Mount Laguna Observatory, and the light curve displayed here as Figure 1 was obtained. The observed variability is nearly sinusoidal with a peak-to-peak amplitude of ∼ 0.08 mag. in the V band. The currently available observations extend over a short time base-line, so our ephemeris is tentative and is given by

$$JD_{MAX} = 2445578.97 + 0.^{d}43103 \ E$$

with the stated epoch being a time of maximum brightness. The period of
∼ 10^h 20^m is in excellent agreement with our predicted value which we take to be
compelling evidence that GLS 890 is indeed a rapidly rotating single M-dwarf star.
Peculiar morphology of the light curve on the descending branch is probably an
artifact of the poorly determined ephemeris.

We observed the Hα line profile of GLS 890 using the 2.1 m telescope, coude
spectrograph, and CCD detector at the Kitt Peak National Observatory. Observations
were secured at phases 0.94, 0.18, and 0.35. Those observations are not yet presentable for publication, but they show that Hα is in emission over the observed
phases, and that it increases in equivalent width at phase 0.35 relative to the
other phases. A small blue-shift of the short wavelength portion of the line
core is also present at phase 0.18 relative to phase 0.94. We interpret the
current observations to mean that there is excess plage-region activity at the
leading edge of a large spot, or group of spots which are responsible for the
photometric variation.

From our experimental viewpoint, GLS 890 is a superb laboratory for observing
the effects of extremely rapid rotation in the absence of a tidal couple.
However, one should not overlook the enigma presented by its existence. Kinematically, GLS 890 is a member of the local disk population (U = −4 V= 0
W= −3 km/sec). If it is not an escapee from a very young group, then its
current rotation rate is not compatible with present views about the timescale
for rotational braking.

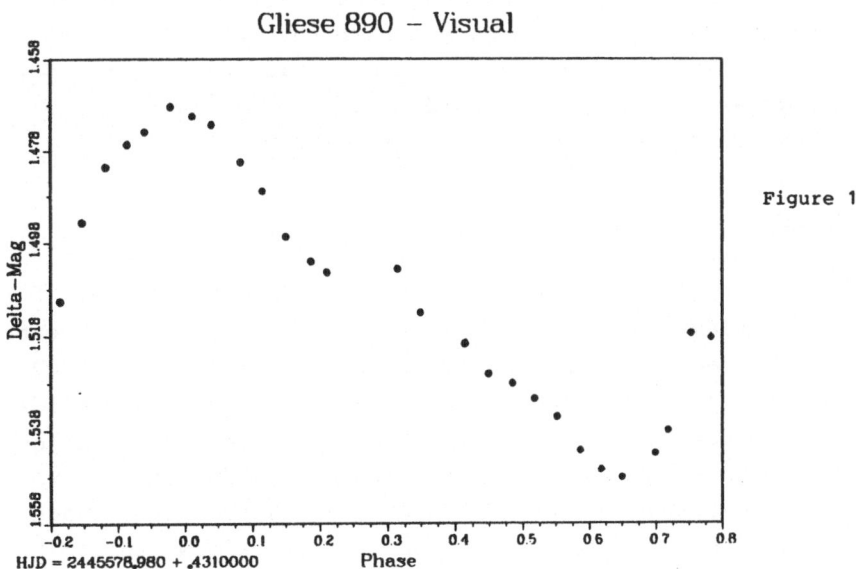

Gliese 890 − Visual

HJD = 2445578.980 + .4310000

Figure 1

EVOLUTION OF ROTATION IN MAIN-SEQUENCE STARS

Sallie L. Baliunas

Harvard-Smithsonian Center for Astrophysics

60 Garden Street, Cambridge, MA 02138

Abstract

A review of the classical and modern measurements of axial rotation and their implications for the evolution of rotation of stars on the lower main sequence is given. From stars of main-sequence spectral type A through early K, the dependence of rotation on mass and age is investigated, from results for stars in open clusters and in the field. The high-mass, single, normal dwarf stars of spectral type A display a common mean projected rotational velocity dependent on their masses and regardless of their ages. The angular momentum per unit mass in this range decreases slowly with decreasing mass.

At approximately mid-F spectral type, the rotation rate declines sharply then decreases again slowly with decreasing mass. The sudden decline in average rotation is probably precipitated by the effect of braking by magnetic torques in stellar winds. Along the lower main sequence, not only mass but also age determines rotation. Rotation slows with time at a given mass. Observational evidence is consistent with rotation slowing with the inverse square root of the main-sequence age of a star, $\langle vsini \rangle \propto t^{-1/2}$.

Among the youngest main-sequence stars, those in the Pleiades near spectral type K2, rapid rotation (up to 150 km s^{-1}) is common. These stars have gained angular momentum during the radiative-track phase of their pre-main-sequence evolution, and should shed their apparently excessive angular momentum on a rapid timescale, a few x 10^7 years, shorter than that predicted by the $t^{-1/2}$ relationship.

Introduction

The history of measuring stellar rotation, one of the fundamental physical attributes of a star, spans over fifty years. In early investigations, projected axial rotation that is rapid (vsini > 100 km s^{-1}) was obvious from the spectrum broadening of photospheric lines (cf. Struve 1930). Systematic investigations for modest rotation (100 < vsini < 20 km s^{-1}) were detailed nearly three decades ago with results at higher resolution (Huang 1953; Slettebak 1955; Herbig and Spalding 1955). Studies of field and cluster stars down to limits of about

10 km s^{-1} were achieved in the 1960's (Kraft 1967a), but lower velocities remained difficult to resolve until the last decade.

In main sequence stars earlier than about spectral type F5, rotational velocities are moderate to large and increase in single stars with earlier spectral type. The average angular momentum per unit mass increases slowly with mass $< J(M) > \propto M^{0.57}$, above about $1.5 M_O$ (Kraft 1970). Investigators with modest resolution noted that rotation and angular momentum decline rapidly, beneath detection limits, along the lower main sequence (Struve 1930). The slow rotation of the Sun, less than 2 km s^{-1}, indicated the expected velocities for advancing spectral type on the main sequence.

In addition to the measurement of rotation as a function of mass along the main sequence, rotation can be studied as a function of age. The comparison of field and cluster dwarf stars provides the evidence for the evolution of rotation with age during the main sequence lifetime. The ages of presumably coeval cluster stars can be combined with ages inferred from lithium abundances in field stars (Duncan 1981; Soderblom 1984). Along the lower main sequence, below the sharp decline in angular momentum at about $1.5 M_O$, the behavior of stellar rotation with age is radically different than for the higher-mass dwarf stars.

Rotation on the Upper Main Sequence

Among the brighter stars, for example spectral type A through early F, the average projected rotational velocity, <vsini>, as a function of mass, is different in each galactic cluster. The observational results summarized by Abt (1970) indicate that the cluster-to-cluster diferences are caused by variants in the mean velocities, rather than the axial inclinations. Thus, the brighter stars in clusters have significantly high or low average rotations compared to the field stars.

An informative example is a comparison of the Pleiades, α Per cluster and NGC 2516, which all share common space motions and similar color-magnitude diagrams. The distributions of <vsini> for the brighter stars are different in each cluster and in the field stars. At spectral type AOV, the α Per cluster stars rotate more rapidly than the Pleiades stars which in turn spin faster than the field stars (Kraft 1967b). The A-type stars in NGC 2516 rotate slower than those in the field (Abt and Morgan 1969; Abt 1970). Although both the Pleiades and α Per clusters are relatively poor in spectroscopic binaries among their bright stars compared to the field population, the α Per cluster is extremely poor in binaries. The exclusion of Ap stars, and binaries, including Am stars, from the velocity averages, produces similar distributions of rotational velocities in the Pleiades and α Per cluster (Kraft 1967b). Furthermore, the slow rotational velocities in NGC 2516 is accompanied by an unusually high proportion of Ap stars (Abt and Morgan 1969). Thus, mean rotation among the main-sequence A stars depends upon the frequency of binary stars and of the Ap stars, so that a preponderance of these stars lowers the mean rotation and a paucity of them raises the average projected rotational velocity. Tidal coupling in the close binaries (for example, Am stars) probably reduces mean rotation (Abt 1965) and magnetic braking likely slows velocities in Ap stars (Abt et al. 1967). Similar populations of A-type stars have remarkably similar average rotational velocities among cluster and field dwarf stars of dissimilar ages (Abt 1970).

For single, lower main sequence stars, however, the age of a star is an important determinant of its rotation rate.

Initial Results for Rotation on the Lower Main Sequence

Along the main sequence, Wilson (1966) noted the coincidence of the appearance of chromospheric Ca II H and K emission cores and the onset of predicted deep hydrogen convective zones. Below this spectral type also occurs the rapid decrease in angular momentum with mass. Magnetic braking may provide the strong drop in angular momentum along the main sequence (Schatzman 1962).

At a given mass along the lower main sequence, rapid rotation is associated with enhanced Ca II H and K emission. Wilson (1963) identified close binaries, with tidally-induced rapid rotation, and strong Ca II emission. Wilson (1963), and Wilson and Skumanich (1964) concluded that chromospheric emission decreases with time in single, main sequence stars. Kraft (1967a) noted that single stars of the same mass but stronger chromospheric emission rotate more rapidly than their weak-emission line counterparts. There is a progression of decreasing age, chromospheric emission strength and rotation in the solar-mass stars from the Pleiades (age about 8×10^7 yrs, Patenaude 1978), and the Hyades (age about 7×10^8 yrs), to the field stars (Kraft 1967a). From these measurements, Skumanich (1972) surmised that both Ca II and rotation decrease proportionately, and they both decay with the square root of the main-sequence age. This dependence of stellar axial rotation on age is colloquially described as the $t^{-1/2}$ law, with an e-folding time of about 1×10^9 years, slightly longer than the age of the Hyades.

Slowing a main sequence star with time is theoretically provided by magnetic braking caused by torques exerted by a stellar wind over time. The interplay of the motions of convection and rotation generates surface magnetic flux which corresponds to enhanced Ca II chromospheric emission, through the action of a magnetic dynamo. The magnetic fields permeate a stellar wind and provide an effective moment arm to brake rotation. This loss of angular momentum may appear in the solar wind (Brandt 1966; Weber and Davis 1967). The braking of rotation causes the dynamo to be age-dependent. Some theoretical justification is reviewed by Skumanich and Eddy (1981).

Recent Methods of Determining Lower Main Sequence Rotation

Over the last decade, two general techniques have been developed to measure either the slow rotation expected for main sequence stars or the velocities of faint members in clusters. First, Fourier analysis of photospheric lines provides Doppler broadening for small projected velocities of faint dwarf stars. Second, dedicated time-series measurements have traced rotation periods in cool dwarf stars.

The derivation of Doppler broadening from photospheric line profiles follows one of two methods, both consisting of velocity modeling in the Fourier frequency domain. For velocities below 5 km s^{-1}, a high spectrum resolution, high signal-to-noise ratio photospheric profile can be dissected for various broadening mechanisms (cf. Gray 1980). With an appropriate model atmosphere describing the thermal content of the line profile, additional broadening caused by macrovelocity and rotation can be identified. At levels below about 5 km s^{-1}, these nonthermal broadening mechanisms, while differing only subtlely in the line profile, become distinguishable in the Fourier transform. For example, Figure 1 shows the solar photospheric line of Fe I $\lambda 6430$ in the Sun and its power spectrum for different values of rotation and macrovelocity (Soderblom 1982).

Another Fourier analysis technique requires only moderate signal-to-noise ratio photospheric spectra at high resolution, and is suitable for faint stars. The method is differential; a spectrum of a narrow-lined, that is, slowly-rotating star, is used as a template against other stars with similar thermal properties but unknown rotation. The cross-correlation coefficients as a function of velocity are calculated between the two stellar spectra. The width of the cross-correlation peak centered at the radial velocity of the best superposition of the two spectra indicates the excess broadening, attributed to rotation. This technique derives from velocity-dispersion measurements of galaxy spectra (cf. Tonry and Davis 1980). At modest rotation (> 10 km s^{-1}) velocities can be surveyed rapidly with photon- counting detectors in many faint stars, for example, the K-type stars in the Pleiades discussed below (Stauffer et al. 1984; Stauffer, 1984).

The second general class of velocity determination directly measures the period traced out by atmospheric inhomogeneities producing a
large contrast in light monitored as a function of time. The advantages
of this method are that the period is measured directly, without the axial projection factor of sini, and that velocities as slow as just under
1 km s^{-1} have been inferred. The disadvantage is that dedicated blocks
of telescope time are required--several months in the case of slowly-
rotating stars. Such a project was begun in 1980 on the Mt. Wilson 60"
telescope with a spectrophotometer measuring the relative strength of
the chromospheric Ca II H and K emission cores(Vaughan et al. 1981;
Baliunas et al. 1983).

The two methods, spectroscopic and time-serial, are complementary.
The nightly accumulation of time-series measurements detects slow rotation directly but requires a substantial investment of telescope time
and fails for a rotation period shorter than about 2 days but measure
diretly the period. Although spectra are quickly measured, the Doppler
broadening method has the ambiguity of the rotation axis and potentially
large uncertainty under a few km s^{-1}.

Current Results for Rotation

a) Solar-Mass Stars

Soderblom (1982a,1983) reconsidered rotation as a function of age
in approximately solar-mass stars, in the spectrum range F8-G5V. Projected rotation velocities were inferred from Fourier analysis of photospheric line profiles and ages determined from Li I abundance analysis
in field stars, including the Sun, and Hyades and Ursa Major cluster
stars. Figure 2, from Soderblom (1983), shows the projected rotation
velocity as a function of age, and includes the information for the
Pleiades from Kraft (1967a). As Soderblom concludes, the line indicating the decay of rotation with the square root of time, <vsini> $t^{-1/2}$,
is consistent with these data. The Sun at the age of the Pleiades
should have been rotating at 12-15 km s^{-1}, the mean rotation of the
solar-mass Pleiades stars (Soderblom 1982b; Soderblom 1983).

b) Lower Main-Sequence Field Stars

For lower main-sequence field stars, periods have been determined
directly for over 50 stars from time-series measurements of chromospheric Ca II H and K. Figure 3 shows one season of data and its correspond-

118

ing autocorrelation from which the period is measured for six stars (Baliunas et al. 1983). This survey effort has produced rotations for over half the main sequence stars Wilson (1978) had monitored for long-term activity cycles. An initial result was the confirmation and extension of the trend of decreasing rotation with mass. The direct measurement of rotation has provided a fundamental, quantitative relation between chromospheric activity and rotation (Noyes et al. 1984). For these main sequence field stars, an average measure of the chromopsheric emission strength at a given stellar mass corresponds to the measured rotation, to within a surprisingly small scatter, about 20%. An interpretation of these findings (Noyes et al. 1984; Duncan 1984) is that the chromospheric emission is a monotonic decreasing function with increasing strength the dimensionless Rossby number, the ratio of the rotation period to the convective turnover time. The convective turnover time is a function of main-sequence mass, and relies on the calculation of the turnover time of theoretical models of convection zones. In generally accepted theories of the hydromagnetic dynamo producing magnetic fields and stellar activity, the Rossby number indicates the degree of magnetic field amplification. Thus for example, rapid rotation and long turnover times (late spectral type or large (B-V)) means the activity is stronger than for the same rotation but shallow convective zones (small (B-V)). A startling aspect of these empirical results is that for all field stars surveyed, the relationship has little scatter despite the wide range of ages, masses, and rotations. A byproduct of these results is that rotation can be predicted from two stellar parameters--the chromospheric radiative loss at Ca II H and K and the mass, or empirically, the S-strength and color (B-V).

c) Results for the Pleiades Early K-Type Dwarf Stars

For masses smaller than solar, results for K dwarf stars have become available in the field (Vaughan et al. 1981; Baliunas et al. 1983) and in the Pleiades (Stauffer et al. 1984), and so far indirectly in the Hyades (Duncan et al. 1984).

The early K-dwarf stars (near K2V) in the Pleiades have added an interesting wrinkle to the picture of angular momentum loss with age both before and during main sequence lifetimes. A survey of rotational velocities in K-type stars in the Pleiades (Stauffer et al. 1984) reveals the following: (1) The stars near K2V are main sequence stars although some have just arrived there; (2) about half of these are rotating with surprisingly large velocities, in excess of 25 km s^{-1}, with a few as rapid as 130-150 km s^{-1}; (3) about half of these have projected rotational velocities smaller than 10 km s^{-1}, the resolution limit of the survey; (4) no rapid rotators were detected at 1 M_O; and (5) no rapid rotators were detected (among 35 stars) near spectral type K2 in the older Hyades (Latham and Stefanik, private communication, 1983).

The explanation of these empirical facts has interesting implications for angular momentum loss discussed by Stauffer et al. (1984). First, the Hyades K stars have lost considerable angular momentum between the ages of the clusters, on a timescale shorter than a few x 10^8 years, which requires much quicker braking than predicted by the t$^{-1/2}$ relationship. Second, the rapidly-rotating K stars gained angular momentum, during radiative-track contraction prior to ZAMS, a process described by Endal and Sofia (1981). The age of the completion of radiative-track contraction phase for K-type stars is about 7 x 10^7 years Van den Berg et al 1983), which is also the nuclear age of the Pleiades determined from the turnoff of the upper main sequence. The Pleiades G-stars (1 M_O) complete radiative track contraction at 3 x 10^7 years (Iben 1965) and hence must shed angular momentum in only 4 x 10^7 years, again a shorter timescale than that predicted for main sequence rotation braking t$^{-1/2}$. Models of the pre-main sequence Sun (Endal and Sofia 1981) suggest that the core and convective envelope of the just-ZAMS G-stars may be decoupled and the envelope could be rapidly braked, for example by magnetic torques in young stellar winds, on relatively short timescales, shorter than 3 x 10^7 years. The bifurcation of rapid and slowly-rotating Pleiades dwarf stars further indicates the possibility of a rather wide spread in initial angular momenta in the K-dwarfs or two bursts of star formation, with an age spread of at most a few x 10^7 years.

Summary

The rotation of lower main-sequence stars is schematically presented in Figure 4. It should be emphasized that the observational data are drawn from a wide variety of sources, with inaccuracies both in the rotations and spectral types. It is nevertheless instructive to examine the trends represented by the numerous measurements represented here.

Above 1 M_O, earlier than approximately G0, the average vsini's for clusters are from Bernacca and Perinotto (1974) and Fukuda (1981), who consolidated many contemporary catalogues. Included are the stars observed in the Pleiades, Coma, Praesepe and Hyades. Also included are the dispersions for the Hyades and Pleiades. The length of the error bars is two standard deviations where the dispersion is calculated from the number of stars in each spectral-type bin. Generally the dispersions for each cluster are similar, so that it is puzzling why Coma, Praesepe and Hyades are discrepant even with similar ages. The distinction, however, may be attributable to uncertainty between the inhomogeneous sources of rotational velocities.

The weak-emission line field stars earlier than G0 are the averages of Kraft (1967a). These field stars are presumably of solar age, and their statistical dispersions are quite small, about 20%.

At spectral type G2, $1M_O$, the Pleiades average is 12-15 km s^{-1} (Soderblom 1982b). Later than G2 the measurements of, or upper limits to, vsini in the Pleiades are shown individually with the exception of the upper limit at spectral type K2 (Stauffer et al. 1984). Below the resolution limit of 10 km s^{-1} at K2 the upper limit drawn represents as many stars slowly rotating as are drawn individually above the limit.

For the Hyades, the measurements of chromospheric Ca II H and K, the S-values (Duncan et al. 1984), have been used to estimate the rotation from the prediction of Noyes et al. (1984) discussed above. Inferred velocities near 1 M_O are in agreement with photometrically-determined periods (Radick et al. 1984). Finally, the strong emission-line field stars, about as young as the Hyades, and the weak emission-line field stars, near the age of the Sun, have periods determined by chromospheric rotation modulation (Baliunas et al. 1983). The ages of the Pleiades and Hyades (Paternaude 1979) along with that of the field stars (Soderblom 1983) are labels for the schematic trends of the decay of rotation with age of the Pleiades, Hyades and weak emission line field stars.

Both Figure 4 and the results presented here can be summarized for the behavior of main-sequence rotation velocities.

At masses larger than those of early F-type stars, exemplified by the results from the A-type stars, the age of a galactic cluster is not significant in determining rotation. Differences between clusters in the average rotation of A-stars can be predominantly explained by differing populations of, for example, closely-spaced binaries or Ap stars. Above early F, the angular momentum for unit mass declines slowly with mass.

At smaller masses, corresponding to spectral types later than F0-F5, rotation decreases with decreasing mass along isochrones corresponding to chromospherically active (relatively young) and inactive (relatively old) main sequence stars, and at a given mass, rotation decreases with increasing age. In solar mass stars, around spectral type F8-G2V, the decay of rotation is consistent with the form $t^{-1/2}$. Considering the early K-stars, especially in the Pleiades, the $t^{-1/2}$ decay would be consistent with the measured rotation in Pleiades, Hyades and field stars _if_: there is a rapid loss in angular momentum in rapidly-rotating K-stars on timescales shorter than a few x 10^7 years, and even shorter timescales for the G-type stars, none of which are spinning rapidly, and there is either an age spread on the order of a few x 10^7 years or a wide distribution in initial angular momenta in the Pleiades.

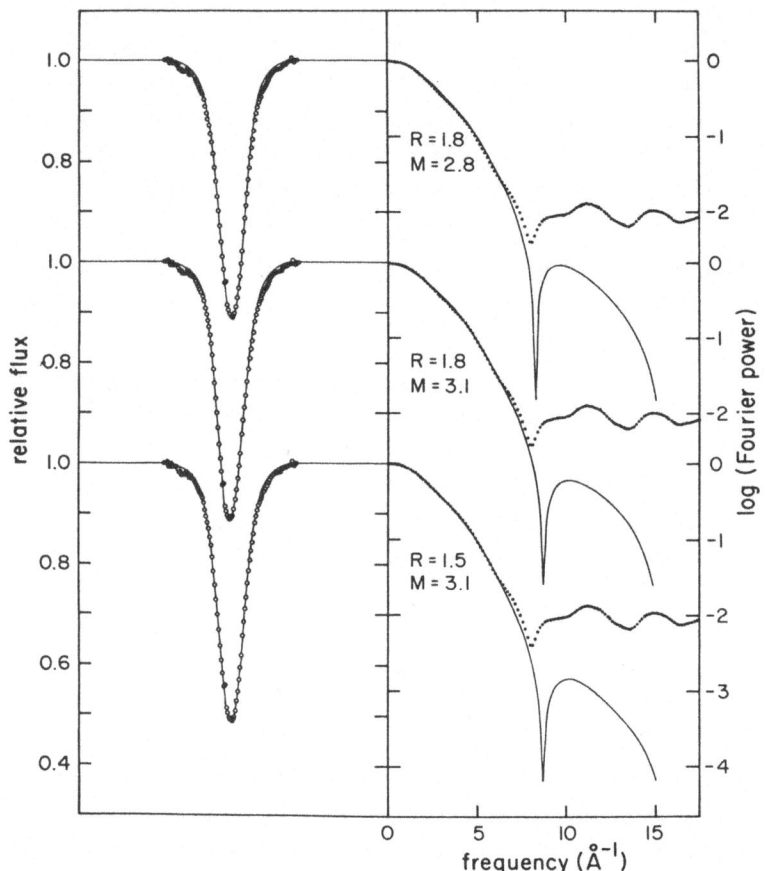

Figure 1 -- (left panel) The observed photospheric line profile of Fe I λ6430 (points) in the Sun and computed profiles for three different combinations of rotational velocity (R) and macroturbulent velocity (M). The power spectra of the observed and computed profiles are shown in the right panel. The best fit to the data for both the profile and its power spectrum is R = 1.8, M = 2.8 km s^{-1} (Soderblom 1982a, by permission of the author).

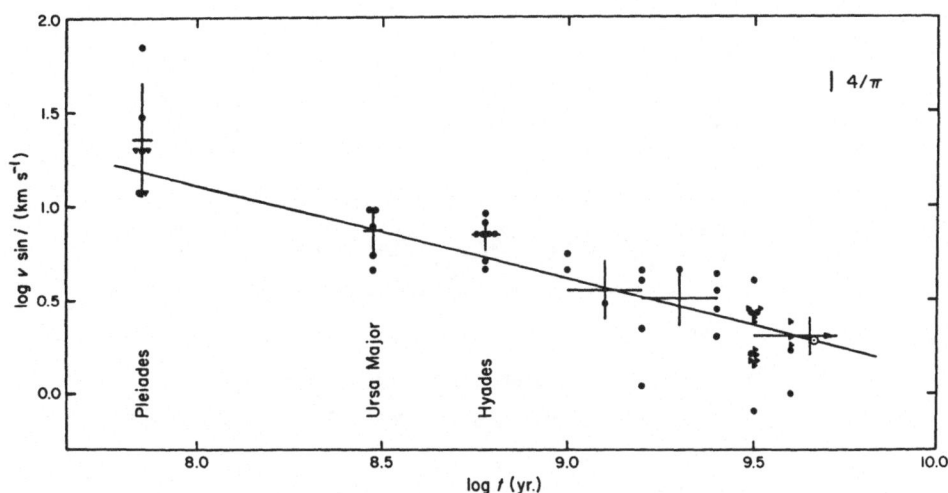

Figure 2 -- The measured projected rotational velocity of approximately solar mass stars is plotted as a function of their ages. The line shown is the mean projected rotational velocity proportional to $t^{-1/2}$ (Soderblom 1983, by permission of the author).

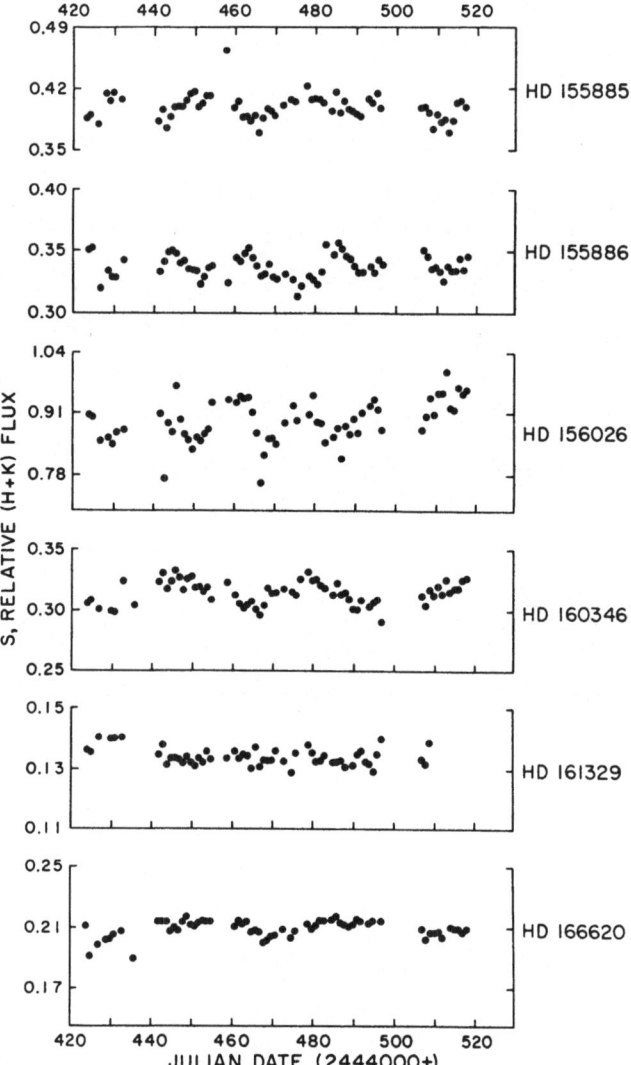

Figure 3 -- (a)

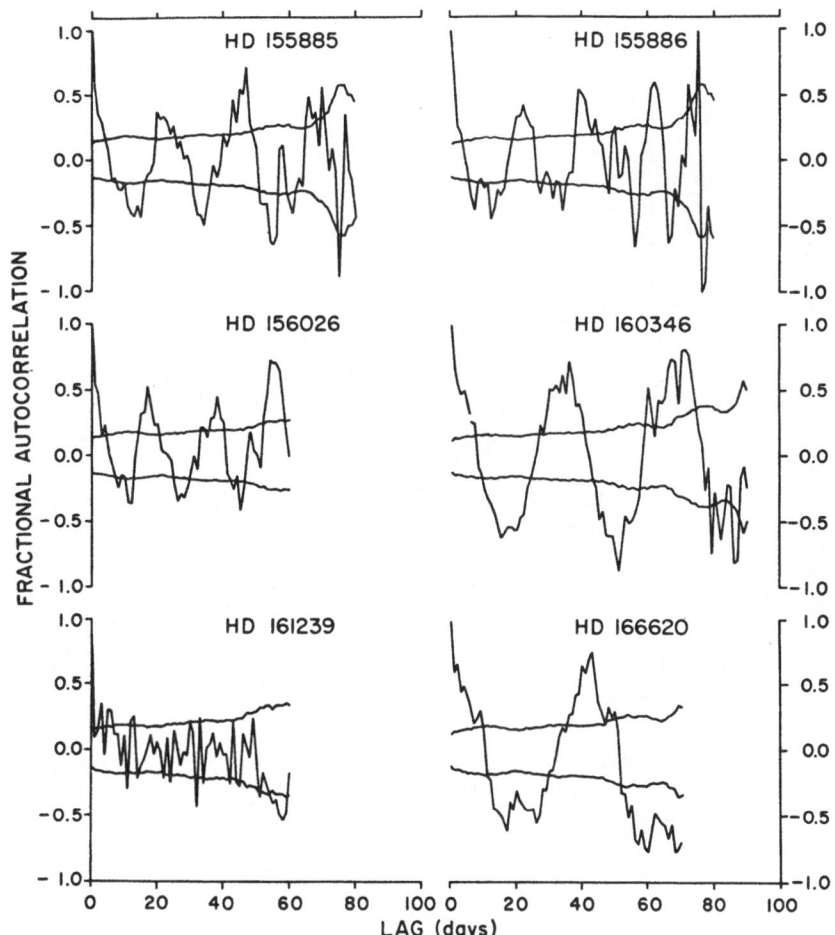

Figure 3 -- (a) For six lower main sequence stars, the nightly means of the relative chromospheric emission strength, S, plotted as a function of time observed in 1980.
(b) The autocorrelation coefficients calculated from these data. Peaks in the autocorrelations mark successive timing of the rotation period. For example, HD 160346 has a rotation period of about 34 days, corresponding to an equatorial velocity of about 1 km s[-1] (Baliunas et al. 1983, by permission of the authors).

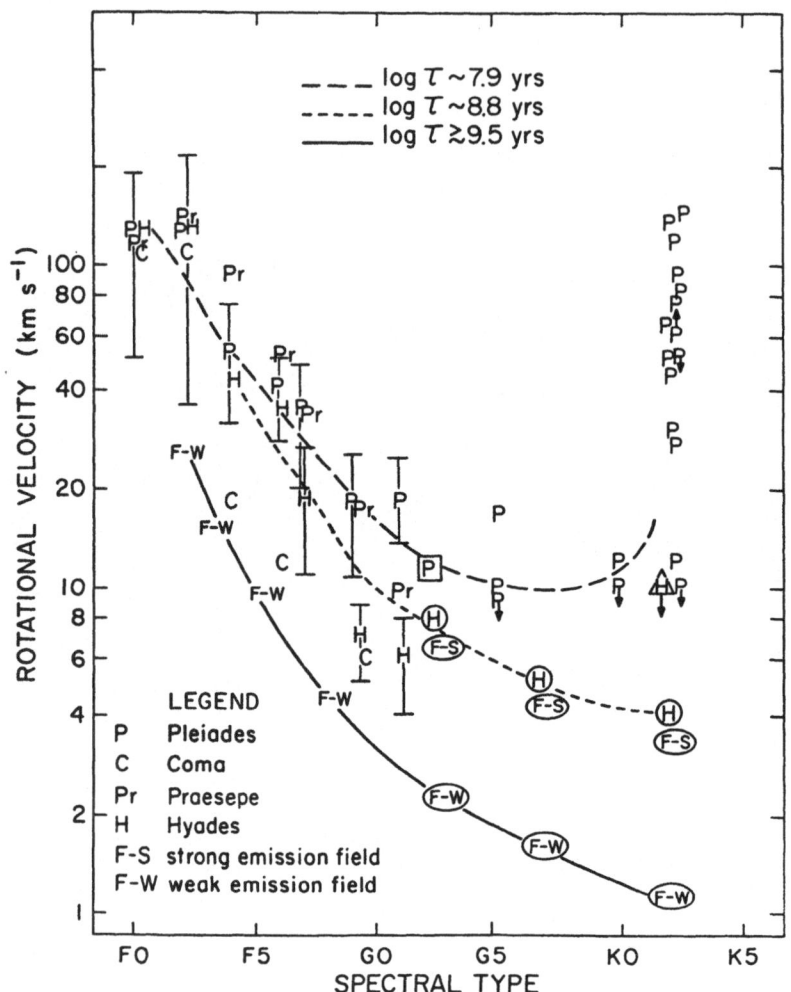

Figure 4 -- A schematic plot of mean rotational velocities as a function of main-sequence spectral types. Spectral types have been grouped and average velocities calculated in each bin. Earlier than spectral type G0, averages for cluster dwarfs are from Bernacca and Perinotto (1974). The length of the bars for the cluster stars indicate two standard deviations calculated from the number of stars in each bin. The weak-emission field dwarf data earlier than G0 are from Kraft (1967b). At spectral type G2, the Pleiades' average (box) are from Soderblom (1982b). Individual Pleiades' vsini measurements later than G2 are from Stauffer et al. (1984). The upper limit at K2 represents about one dozen velocities slower than the limits of detection at 10 km s⁻¹. The Hyades' measurement at K2 (triangle) is an upper limit of about 35 stars from Latham and Stefanik (1983). The encircled field stars later than G0 are measurements of the velocity inferred from the rotation periods (Baliunas et al. 1983). The Hyades values, with circles, were predicted by the measurement of chromospheric emission (Duncan et al. 1984). Schematic lines at the ages of the Pleiades, Hyades and weak-emission, presumably old, field stars follow the decline of rotation with decreasing mass.

REFERENCES

Abt, H.A. 1965, Ap. J. Suppl., **11**, 429.
Abt, H.A. 1970, in Stellar Rotation, ed. A. Slettebak (Dordrecht, Holland: D. Reidel), p. 193.
Abt, H.A., Chaffee, F.H. and Suffolk, G. 1967, Astron. J., **72**, 783.
Abt, H.A. and Morgan, W.W. 1969, Astron. J., **74**, 813.
Baliunas, S.L., Vaughan, A.H., Hartmann, L., Middelkoop, F., Mihalas, D., Noyes, R.W., Preston, G.W., Frazer, J. and Lanning, H. 1983, Ap. J., in press.
Bernacca, P.O. and Perinotto, M. 1974, Astron. Ap., **33**, 443.
Brandt, J. 1966, Ap. J., **144**, 1221.
Duncan, D.K. 1981, Ap. J., **248**, 651.
Duncan, D.K. 1984, these proceedings.
Duncan, D.K., Baliunas, S.L., Noyes, R.W., and Vaughan, A.H. 1984, in preparation.
Endal, A.S. and Sofia, S. 1981, Ap. J., **243**, 625.
Fukuda, I. 1982, Pub. A.S.P., **94**, 271.
Gray, D.F. 1980, In Stellar Turbulence, ed. D.F. Gray and J.L. Linsky (New York: Springer), p. 75.
Herbig, G. and Spalding, J. 1955, Ap. J., **121**, 118.
Huang, S.-S. 1953, Ap. J., **118**, 285.
Iben, I., Jr. 1965, Ap. J., **141**, 993.
Kraft, R.P. 1967a, Ap. J., **150**, 551.
Kraft, R.P. 1967b, Ap. J., **148**, 129.
Kraft, R.P. 1970, in Spectroscopic Astrophysics, ed. G. Herbig (Berkeley: Univ. Calif. Press), p. 385.
Lockwood, G.W., Thompson, D.T., Radick, R.R., Osborn, W.H., Baggett, W.E., Duncan, D.K. and Hartmann, L.W. 1984, Pub. A.S.P., in press.
Noyes, R.W., Hartmann, L.W., Baliunas, S.L., Duncan, D.K. and Vaughan, A.H. 1984, Ap. J., in press.
Patenaude, M. 1978, Astr. Ap., **66**, 225.
Schatzman, E. 1962, Ann. Astr., **25**, 18.
Skumanich, A. 1972, Ap. J., **171**, 565.
Skumanich, A. and Eddy, J.A. 1981, in Solar Phenomena in Stars and Stellar Systems, ed. R. Bonnet and A.K. Dupree (Dordrecht: Reidel), p. 349.
Slettebak, A. 1955, Ap. J., **121**, 653.
Soderblom, D.R. 1982a, Ap. J., **263**, 239.
Soderblom, D.R. 1982b, Carnegie Institution of Washington Yearbook, **81**, 662.
Soderblom, D.R. 1983, Ap. J. Suppl, **53**, 1.
Soderblom, D.R. 1984, these proceedings.
Stauffer, J.R. 1984, these proceedings.
Stauffer, J.R., Hartmann, L., Soderblom, D.R. and Burnham, N. 1984, Ap. J., in press.
Struve, O. 1930, Ap. J., **72**, 1.
Tonry, J. and Davis, M. 1979, A. J., **84**, 1511.
Van den Berg, D.A., Hartwick, F.D.A., Dawson, P. and Alexander, D.R. 1983, Ap. J., **266**, 747.
Vaughan, A.H., Baliunas, S.L., Middelkoop, F., Hartmann, L.W., Mihalas, D., Noyes, R.W. and Preston, G.W. 1981, Ap. J., **250**, 276.
Weber, E.J. and Davis, L., Jr. 1967, Ap. J., **148**, 217.

Wilson, O.C. 1963, Ap. J., **138**, 932.
Wilson, O.C. 1966, Ap. J., **144**, 795.
Wilson, O.C. 1978, Ap. J., **226**, 379.
Wilson, O.C., and Skumanich, A. 1964, Ap. J., **140**, 1401.

ON THE ORIGINS OF CHROMOSPHERIC ACTIVITY IN LATE-TYPE STARS

Douglas K. Duncan
Mount Wilson and Las Campanas Observatories
Carnegie Institution of Washington
Pasadena, California 91101

INTRODUCTION

The correlation between stellar rotation rate and overall chromospheric
activity levels in late-type stars (which I take to mean stars with subsurface
convection zones) has been known for some time (Kraft 1967; Skumanich 1972). As
such stars age they slow down, and their chromospheric emission decreases with their
surface rotation rate. The details of this picture and the work leading up to it
have been reviewed by Skumanich and Eddy (1981). Similar relations even more
strongly dependent on rotation have been found for coronal X-ray emission
(Pallavicini et al. 1981; Walter 1981,1982).

The causal nature of these relationships - the details of the physical
mechanisms responsible - have been more difficult to elucidate. However, the last
few years have seen a tremendous increase in the available data on chromospheric
emission and rotation rates of late-type stars, as well as much theoretical work on
dynamos (see, for example, the recent reviews by Schlussler 1983, and Gilman 1983)
and chromospheric heating mechanisms (e.g. Ulmschneider and Stein 1982). I will
take as my purpose in this short review the discussion of the new data and the
constraints they place on theories· of the origin of chromospheric activity, and the
suggestion of further observations which would be especially useful.

MEASUREMENTS OF ROTATION AND CHROMOSPHERIC EMISSION LEVELS

Rotational velocities of late-type stars are small, typically less than 10
km/s, and very difficult to measure with conventional techniques. The Fourier
transform method of line profile analysis pioneered by Gray (1973;1982) and used by
Smith (1979) and Soderblom (1982) represents a significant advance since the
signatures of other broadening mechanisms (e.g. macroturbulence) are more

distinguishable from that of rotation in the Fourier domain. Judging from intercomparison of the results of these three investigators, the method can achieve an accuracy of about 1 km/s when carefully applied to very high quality spectra.

Another step forward came with the demonstration by Vaughan et al. (1981; also Baliunas et al. 1983) that simply by monitoring the Ca II H and K line core emission from the chromospheres of late-type stars one could often detect a periodic modulation due to stellar rotation. This method has proven to be extremely powerful. To date over 40 rotation periods have been determined (Noyes et al. 1984), generally with a precision of a few percent, even for stars rotating as slowly as 1 km/s. The observed HK modulation periods are combined with standard relations for stellar radii to yield rotational velocities. Velocities so found have the significant advantage that they are free of the sin i projection factor inherent in Doppler (line profile) measurements. Strictly speaking it is the rotation period of the latitude on a star where the active region resides that is determined by the HK modulation technique, and indeed the method may be sensitive enough to detect differential rotation if more than one active region can be observed (see below).

Complete specification of chromospheric activity levels is also difficult. Principal energy loss mechanisms in the solar chromosphere as estimated by Vernazza, Avrett, and Loeser (1981) include the Ca II H and K lines (approximately 20% of the total chromospheric losses), the Mg II h and k lines (20%), and the Ca IR triplet (40%). Linsky and Ayres (1978) find the contribution of Mg II II h and k to be somewhat higher and that of Ca II H and K to be somewhat lower.

Mg II fluxes can be obtained with the International Ultraviolet Explorer (IUE) satellite, and Hartmann et al. (1984) present results for approximately 30 G and K dwarfs. One expects that to first order the loss mechanisms would vary together and Figure 1 (from Hartmann et al.) indicates that they do: stars which appear active in the Mg II h and k lines are also strong Ca II H and K emitters.

If we therefore proceed under the assumption that the strengths of the Ca II H and K line core emission are good indicators of overall chromospheric activity we may make use of two very useful samples:

(1) The 91 late-type dwarfs Olin Wilson (1978) began to study for long-term cyclical variations in 1966. These stars continue to be followed at Mt. Wilson and approximately 40 have had rotation periods determined from the study of periodic modulation of their HK emission.

(2) The so-called "Solar Neighborhood Survey" (Vaughan and Preston 1980): A sample of almost 400 late-type stars chosen from the Woolley et al. (1970) catalogue of stars within 25 pc of the sun. Stars in this survey were typically observed just a few times each.

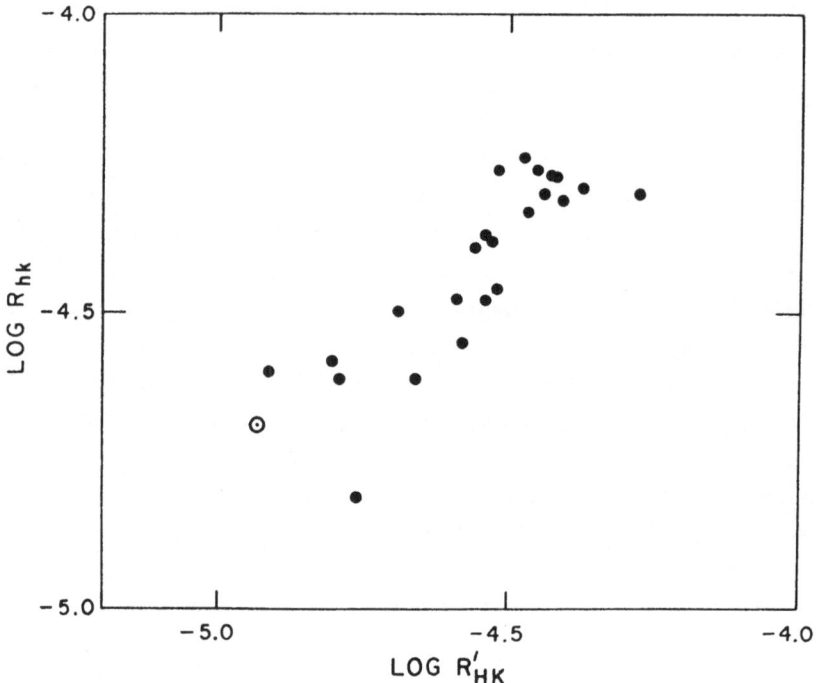

Figure 1. The correlation between CaII H and K emission and Mg II h and k emission (Hartmann et al. 1984).

THE NATURE OF THE ROTATION-ACTIVITY CORRELATION

Noyes et al. (1984) present a detailed study of the relation between rotation and chromospheric activity. Fig. 2 is their plot of chromospheric HK emission for many of the stars in Wilson's sample. Numbers give known rotation periods, in days. The ordinate, R'_{HK}, is the fraction of a star's bolometric luminosity that emerges as chromospheric emission in the H and K lines. [Previously Mt. Wilson HK measurements have been published in terms of an index, "S", which is essentially the equivalent width of the H and K lines. Since the continuum near the H and K lines falls rapidly with advancing spectral type, the H and K emission of cooler stars is unduly emphasized by the S index. Calibration of the continuum in absolute flux units as a function of B-V allows transformation to F_{HK}, the average flux per cm² at the stellar surface, or $R_{HK} = F_{HK}/\sigma T_e^{4}$, the fraction of a star's luminosity which emerges in the H and K lines. A second correction should be made for the fact that

130

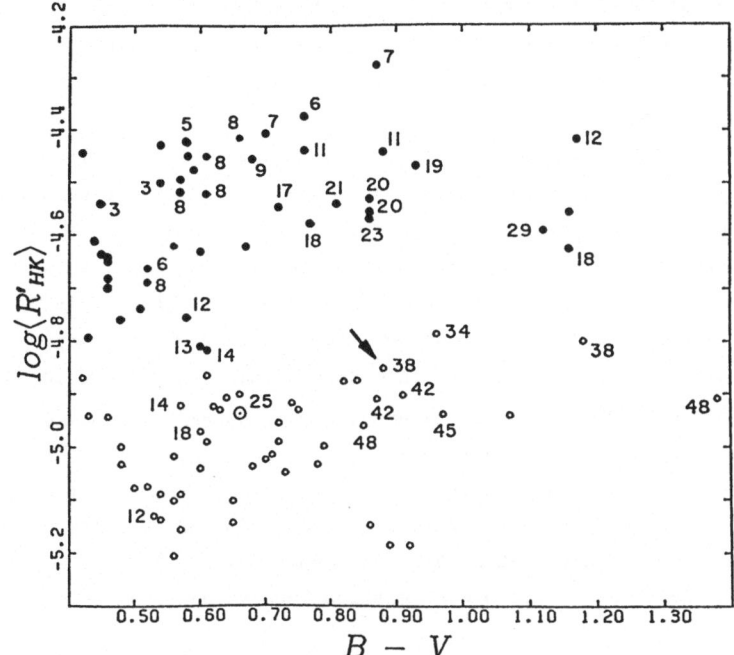

Figure 2 Chromospheric CaII H and K emission (expressed as a fraction of bolometric luminosity) for stars of various spectral types. Numbers next to each point give the stellar rotation period in days. Filled circles represent stars above the so-called "Vaughan-Preston Gap", open circles stars below the Gap. (Noyes et al. 1984).

even in the absence of chromospheric emission some photospheric light will be detected in the H and K bands (which are 1A wide in the Mt. Wilson measurements). This photospheric background correction is especially important in weak emission stars, amounting to about 50% in the case of the sun. Indices corrected for this background will be denoted F'_{HK} and R'_{HK}.]

It is clear from Figure 2 that, in general, more rapid rotators have stronger emission. However, it is also clear that there must be second, color-dependent parameter in the problem. Thus, for stars with a given B-V faster rotators have stronger emission, but for stars with a given rotation period later spectral types have more emission.

Until recently, correlations drawn from this data have been entirely empirical. A successful example of the predictive capability of Figure 2 is provided by the star HD 4628, indicated by an arrow. In the summer of 1980, this star appeared to have a rotation period of 19.0 days. It was the only star which violated the regular distribution described above and Vaughan et al. (1981) speculated that it might be showing an alias of the true period; that perhaps there were two active regions on opposite hemispheres of the star which made the modulation period 19 days while the rotation period was 38 days. Indeed, the next year the star was observed to have a 38.0 day period.

Noyes et al. (1984) sought a physical understanding of the rotation-activity correlation. They make the assumption that Ca II H and K fluxes represent the amount of non-thermal chromospheric heating, which in turn is associated with surface magnetic fields. This assumption is supported by many observations in the case of the sun (e.g. Leighton 1959; Skumanich, Smythe, and Frazier 1975).

Figure 2 shows that for a given rotational period stars of later type convert a much greater fraction of their luminosity into chromospheric heating than do earlier type stars. It may be argued that R'_{HK} is not the most appropriate measure of chromospheric activity. On the basis of more limited data than that of Noyes et al., Catalano and Marilli (1983) suggest a correlation of total K line luminosity with rotation, and Middlekoop (1982) suggests a correlation of HK surface flux with rotation, both relations independent of spectral type. Figure 3, from Noyes et al. confirms that there is a reasonably close relation between chromospheric HK flux and rotation period.

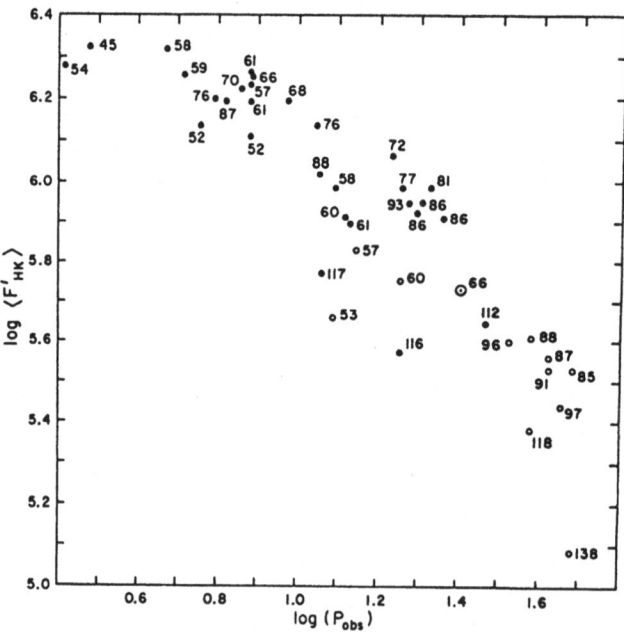

Figure 3. The relation between stellar suface chromospheric flux and rotation period. Numbers near each point give the B-V color x 100. As in Figure 2 filled circles represent stars above the Vaughan-Preston Gap and open circles stars below the Gap. (Noyes et al. 1984).

However, in most dynamo models convection zone properties would be expected to influence the generation of magnetic fields. The nature of that influence of course depends on the location of dynamo activity. Considerations of magnetic buoyancy require that the site not be too near the top of the convection zone, and many models suggest the primary dynamo activity is distributed near the base of the convection zone, where the largest scale convective motions occur.

Dynamo action can be caused by the combined effects of differential rotation (the ω effect) and the helicity introduced into the convective velocity field by rotation (the α effect), as is postulated in standard kinematic $\alpha\omega$ dynamo models, or by the α effect alone (cf. Schlussler 1983). The fully nonlinear 3-D hydrodynamic models of Gilman (1983), which self-consistently solve for the velocity fields rather than assuming them, also show the α effect to be important.

Such considerations led Noyes et al. (1984) to investigate how chromospheric activity depends on the Rossby number, Ro. This dimensionless number is the ratio of rotation period P to the convective turnover time τ_c in that part of the convection zone where dynamo activity is situated. It is a measure of the importance of coriolis forces in introducing helicity into convective motions. The Rossby number is related to the commonly used dynamo number approximately as $N_D \propto Ro^{-2}$. (cf. Durney and Latour 1978). Noyes et al. take their estimates of turnover times near the base of the convection zone from Gilman (1980), who calculated them as a function of spectral type using mixing length theory. A more-or-less free parameter in mixing length theory is the ratio of mixing length to pressure scale height. (This ratio is usually denoted α.) Larger values of α correspond to deeper convection zones. Stellar evolution calculations generally require values of α from 1.5 to 2 to reproduce the age, luminosity, and composition of the sun (Gehren 1982). Abundances of light elements in the sun and studies of solar p-mode oscillations also suggest that the sun's convection zone is best modelled with α greater than 1. Noyes et al. used results for α = 1,2, and 3, and plotted R'$_{HK}$ against the Rossby number (P/τ_c). The plot using α = 2 mixing length models is shown in Figure 4.

The scatter in Figure 4 is remarkably small - just over half that of Figure 3. This good fit is only achieved for mixing length ratios near 2. The data of Figure 4 are consistant with the mean level of chromospheric activity in lower-main-sequence stars being primarily governed by the single parameter P/τ_c, where τ_c is an empirical function of B-V very similar to theoretically calculated convective overturn times. Figure 4 supports the notion that the Rossby number is in fact a major determinant of magnetic field amplification and that the site of dynamo activity is near the base of the convection zone.

The small scatter of Figure 4 is surprising, even if there is a unique relation between field generation and Rossby number. What is observed is not magnetic fields deep in the convection zone but Ca II H and K emission resulting from chromospheric heating. The value of surface magnetic fields produced and the details of their

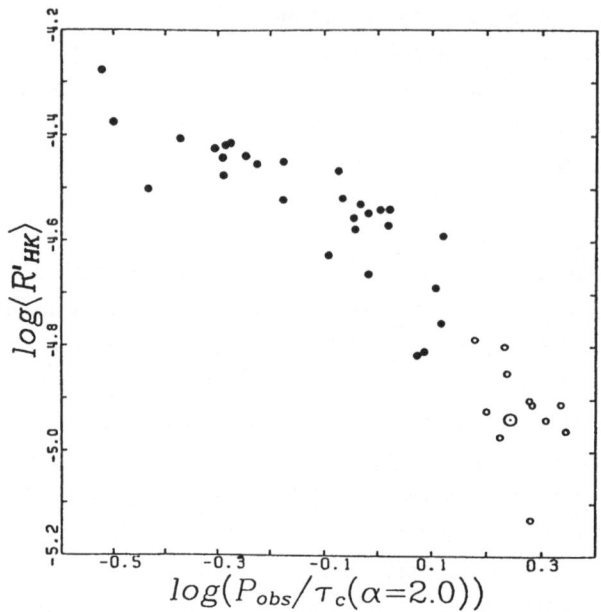

Figure 4. Chromospheric Ca II H and K line emission (expressed as a fraction of bolometric luminosity) vs. Rossby number. Symbols are the same as in Figures 2 and 3. (Noyes et al. 1984).

heating of the chromosphere are not explicitly taken into account. That this complicated chain of events should be spectral type independent is not expected. In fact, the most likely chromospheric heating mechanisms show efficiencies which depend on temperature approximately as $T^2 - T^3$ (e.g. Ulmschneider and Stein 1982). Such a strong additional temperature dependence cannot be introduced into the data of Figure 4 without destroying the good fit, even if other values of the mixing length ratio besides 2 are allowed. This is primarily because the shape of the convective turnover time curves given by Gilman (1980) are not well fit by power laws. Further investigation is called for as well as an extension of the observations to include more cool stars so that a wider temperature range is spanned.

It of course could be that the goodness of fit in Figure 4 is fortuitous. Series of models exist in which dynamo activity is not situated near the base of the convection zone but confined to an overshoot layer beneath it: so-called "shell dynamos" (Schlussler 1983; Rosner 1983). Even for shell dynamos, spectral-type dependent variations in the state of the material in the overshoot region may be expected to affect dynamo activity, but such variations are more difficult to predict than those of τ_c. Realistic predictions may not even be possible, since present models of convective overshoot are not reliable. They fail to account for the observed abundances of light elements, principally Li, which is destroyed in or near the overshoot layer (Strauss, Blake, and Schramm 1976; Cayrel et al. 1984). A critical test of the shell dynamo hypothesis would be observations of chromospheric activity in stars sufficiently cool that they are entirely convective and hence have no overshoot layer. Giampapa (1983; also private communication) has made some such observations but the results are not conclusive.

If the relation of Figure 4 is not fortuitous it provides additional information on dynamo behavior. Decreasing Rossby number causes increasing activity. The observed rate of increase is large for large Rossby number and it becomes less for small Rossby number. This could be due to negative feedback produced by large magnetic fields in the convection zone acting to reduce the α-effect dynamo and differential rotation (Robinson and Durney 1982; Gilman 1983). However it could also simply be due to surface saturation effects – a limit to the HK emission in regions of high magnetic flux. Certainly the data of Figure 4 deserve further explanation and investigation.

If the smooth relationship of Figure 4 is causal in nature it indicates that the same mechanism is present in all late-type dwarfs, active and inactive, old and young, regardless of spectral type. It then presents a puzzle: what is the nature of the Vaughan-Preston Gap? Previous explanations postulated a fundamental difference between stars above and below the Gap. Durney, Mihalas, and Robinson (1981) suggested that a critical rotation rate separates a regime of higher rotation where several dynamo modes are excited (the young stars above the Gap), from one of lower rotation where only the fundamental mode is excited (the old stars below the Gap). On the other hand Knobloch, Rosner, and Weiss (1981) suggested that above a critical rotation rate convection occurs not in random eddies but in longitudinal rolls, and that in the latter case surface fields might be amplified to higher values than in the former. In either case the relation between chromospheric emission and rotation would change discontinuously at a critical (possibly spectral-type dependent) rotation rate, and an offset of approximately 0.2 dex (the width of the Gap; cf. figure 2) should be seen between the young and old stars in Figure 4. It is not.

Three other possible explanations of the Gap include:
(1) A lack of intermediate age stars. Even if rotation rate and as a result chromospheric activity decrease smoothly with age, there could be in the solar neighborhood a lack of stars of that age (about 1 b.y.), and hence of that rotation rate which would produce emission levels corresponding to the gap. Independent evidence seems to rule this out, however. Twarog's (1980) extensive study of the ages of solar neighborhood F stars, based on the interpretation of Stromgren photometry, shows no such lack of 1 b.y. old stars.
(2) The gap results from an epoch of rapid spindown, perhaps due to an enhanced stellar wind (cf. Durney et al. 1981). The same relationship between HK emission

and rotation would hold for stars above and below the gap, but stars would pass quickly through intermediate rotation rates. There is at present no independent evidence to support this hypothesis.

(3) The Gap is not as significant as it seems - it might be the result of a chance fluctuation in the chromospheric activity distribution. Hartmann et al. (1983) have made Monte Carlo simulations of the Gap. They emphasize that the photospheric background provides a "floor" to the distribution when dealing with S or R_{HK} measurements (rather than R'_{HK}), and an absence of very young stars in the solar neighborhood could explain the top of the distribution. A simulation requires knowledge of (1) the age distribution of solar neighborhood stars, and (2) the way in which chromospheric activity declines with age. For (1), Hartmann et al. assume a constant rate of star formation over the last 10 b.y. (cf. Scalo and Miller 1980) and the scale height corrections of Twarog (1980). For (2) they try various possibilities. Figure 5 compares the solar neighborhood survey data (5a) to the simulations. (5b) assumes that chromospheric activity declines as $t^{-1/2}$, the original Skumanich (1972) relation, and that there are no solar neighborhood stars less than 2×10^8 yrs. old (since their emission would be higher than that of any observed stars). An alternate simulation (5c) presumes an exponential decay of chromospheric activity with age. This avoids the problem of stars with too intense emission, but makes a poor fit. The most successful simulation (5d) assumes an exponential decay for 3 b.y., followed by a $t^{-1/2}$ decay.

I interpret these simulation results as interesting and useful in pointing out that the Gap need not be as significant as is often assumed, but certainly not conclusive. A principal problem is that the actual form of the chromospheric activity decay law is not well known.

In fact, detailed determinations of the decline of chromospheric HK emission with age in late-type stars have been published by Barry et al. (1981;1983). They studied Ca II H and K emission in a number of stars in each of seven open clusters ranging from 10^7 to 6×10^9 years in age and in the sun, as well as in some field stars. These results have not been widely used, perhaps in part due to the relatively low resolution of the spectral scans employed (approximately 3 A). At such resolution HK emission cores are not seen except in the youngest clusters, but filling in of the bottom of the H and K absorption profiles is measurable in all clusters. [For unclear reasons Barry et al. state that in the few cases where emission cores were seen they were ignored.] From a star by star intercomparison of field star observations I have found that the Barry et al. measurements may be readily transformed to the Mt. Wilson scale, and that they are about one third as accurate as Mt. Wilson measurements.

Observation of a number of field stars of independently known ages is now underway at Mt. Wilson for the express purpose of better defining the decline of HK emission with age. These F and G dwarfs are secondaries in wide visual binaries in which the primaries are earlier-type stars whose ages have been estimated from

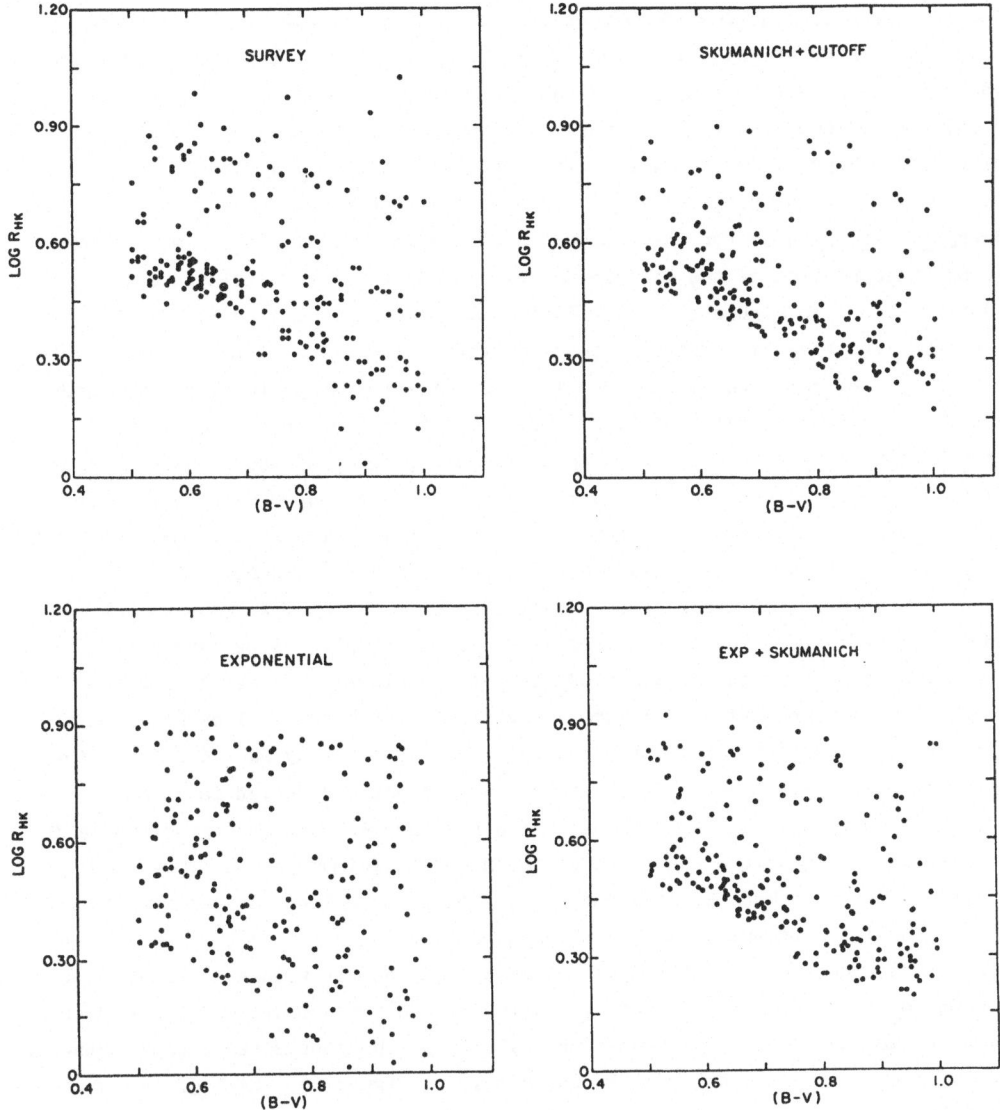

Figure 5. The "Solar Neighborhood Survey" (a) and various Monte Carlo simulations of it (b,c,and d. Hartmann et al. 1984). Each simulation assumes the same distribution of stellar ages but a different description of the decline of Ca II H and K emission with age (see text).

Stromgren photometry (Duncan 1984). In some of the systems the primaries have not evolved off the main sequence so only upper limits to their ages may be deduced. Only about one third of the sample has as yet been observed, but the results to date are shown in Figure 6, along with the transformed measurements of Barry et al. (crosses), the assumed decay law of Hartmann et al. from Figure 5d (dashed line), the range of a large number of Hyades stars observed at Mt. Wilson (cf. the final section below), and the range of a few Pleiades stars observed at Mt. Wilson. That the Barry et al. results for the youngest clusters lie relatively low may be the result of their neglect of the visible HK emission cores in those clusters.

Several stars in Figure 6 deserve individual comment. HD 138268B and HD 196310B have significantly higher HK emission than other stars of their age. It is well known that HK emission is enhanced in close binaries of period less than about 20 days (e.g. Young and Koniges 1977). These two stars need to be checked for possible multiplicity before definite conclusions can be drawn about their enhanced emission. The results will be important in indicating the appropriateness of the Hartmann et al. fit repeated in Figure 6a or a power law fit (Figure 6b). HD 23630 C and D, and HD 27638B are very weak emitters for their age. There is little doubt that they are young. The former two are companions to the star Alcyone in the Pleiades and the latter the companion of a B9V star. It will be extremely interesting to determine the rotation of these three stars. If they follow the rotation-activity correlation of Figure 4 their rotation rates will be relatively low, and they will demonstrate that there is no unique activity-age relation which holds for all late-type stars. On the other hand if their rotation is high one might be tempted to speculate whether active stars can show a temporary decrease in chromospheric emission along the lines of the Maunder minimum period of reduced solar activity. In either case the results may need to be incorporated into future statistical analyses of the Solar Neighborhood Survey.

The stars other than the five just discussed may be fit with a straight line (not shown in Figure 6a for the sake of clarity), with a slope of -1/3 rather than the -1/2 originally proposed by Skumanich (1972). If the photospheric background is subtracted from the stars in Figure 6a the result is Figure 6b, which gives a more accurate idea of the decline of purely chromospheric HK flux with age. The straight line in Figure 6b has a slope of -1/2. More measurements such as those of Figure 6 should allow an empirical determination of the age-chromospheric activity relation which could be used to make simulations such as those of Hartmann et al. more definitive, and thus make more certain the interpretation of the Vaughan-Preston Gap.

USE OF THE ROTATION-ACTIVITY CORRELATION AS A PREDICTOR OF ROTATION

Regardless of the physical basis of the "Rossby number" relation of Figure 4 it is a

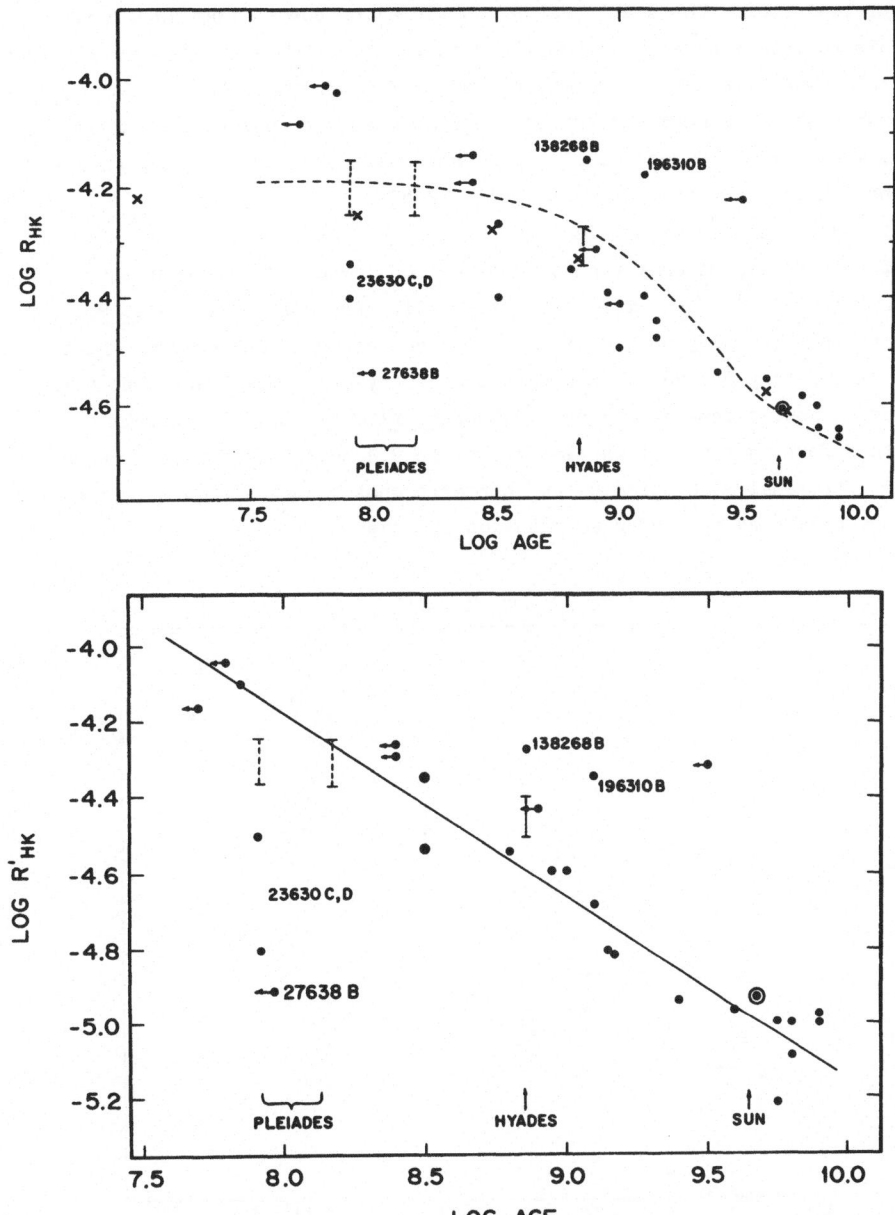

Figure 6. (a) Ca II H and K emission in objects of known age (see text). (b) The same as (a) except with the photospheric background subtracted to leave the purely chromospheric component. The decay law assumed in Figure 5(d) is given by the dashed line in Figure 6(a).

useful empirical tool. The small scatter should allow one to take just a few measurements to determine the mean activity level of a star, use its B-V color to estimate the convective turnover time, and solve for (predict) the rotation period. This of course would be much easier than making dozens of observations to detect rotational modulation. Duncan et al. 1984 have made such predictions and compared them to observation. During the winter of 1982-83 extensive observations of HK emission in Hyades dwarfs were made at Mt. Wilson. These observations were not sufficient to directly determine rotational periods from modulation of the HK emission but they defined the mean activity levels very well. Duncan et al. used these means and the Rossby number relation to predict rotation periods. During the same time period very accurate broad band photometry of Hyades dwarfs was being carried out at Lowell Observatory by Lockwood et al. (1984). The Lowell investigators found periodic light variations in 9 Hyades stars, which presumably are due to rotational modulation of photospheric spot groups. Figure 7 shows the agreement of predicted and measured rotation periods.

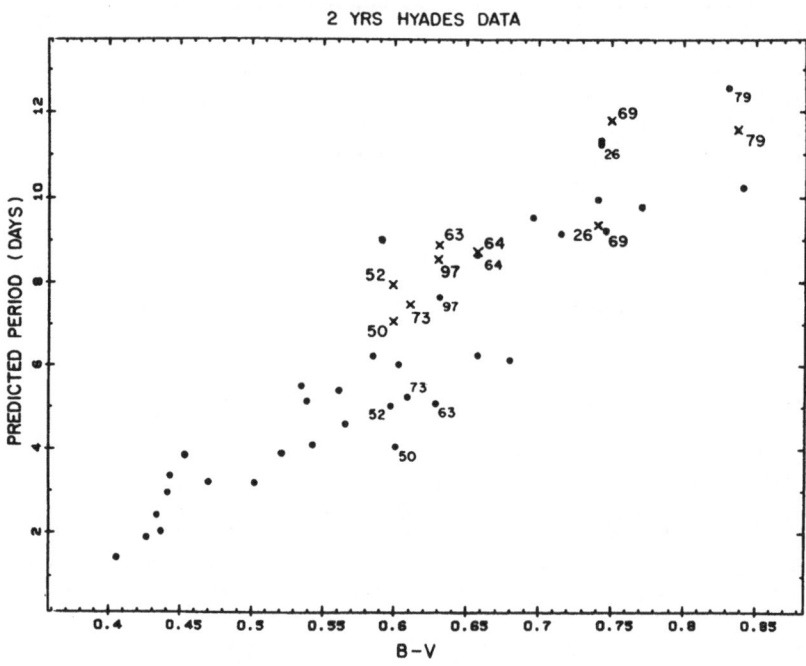

Figure 7. Hyades rotational velocities predicted from mean chromospheric activity levels and the activity-rotation ("Rossby number") correlation of Figure 4 (dots) compared with periods found from broad-band photometric variations by Lockwood et al. 1984 (crosses). Numbers are Van Bueren star identifications.

REFERENCES

Baliunas, S.L., Vaughan, A.H., Hartmann, L., Middlekoop, F., Mihalas, D., Noyes, R.W., Preston, G.W., Frazer,J., and Lanning, H. 1983, _Ap. J._, in press (Dec. 1).

Barry, D.C., Cromwell, R.H., Hege, K., and Schoolman, S.A. 1981, _Ap. J._, **247**, 210.

Barry, D.C., Hege, K., and Cromwell, R.H. 1984, _Ap. J._, in press.

Catalano, S. and Marilli, E. 1983, _Astr. Ap._, **121**, 190.

Cayrel, R., Cayrel de Strobel, G., Campbell, B., and Dappen, W., 1984, _Ap. J. (Letters)_, in press.

Duncan, D.K. 1984, _A. J._, in press.

Duncan, D.K., Baliunas, S.L., Noyes, R.W., and Vaughan, A.H. 1984, _P.A.S.P._, in preparation.

Durney, B.R., and Latour, J. 1978, _Geophys. Ap. Fluid Dyn._, **9**, 241.

Durney, B.R., Mihalas, D., and Robinson, R. 1981, _P.A.S.P._, **93**, 537.

Gehren, T. 1982, _Astr. Ap._, **109**, 187.

Giampapa, M. 1983, in _Solar and Stellar Magnetic Fields: Origins and Coronal Effects_, Stenflo, J.O., ed., Reidel. 187.

Gilman, P. 1980, in IAU Collq. 51, _Stellar Turbulence_, ed. D. Gray, and J. Linsky, Springer, 19.

Gilman P. 1983, in _Solar and Stellar Magnetic Fields: Origins and Coronal Effects_, Stenflo, J.O., ed., Reidel, 247.

Gray, D. 1973, _Ap. J._, **184**, 461.

Gray, D. 1982, _Ap. J._, **261**, 259.

Hartmann, L., Baliunas, S.L., Duncan, D.K., and Noyes, R.W. 1984, _Ap. J._, in press (Apr. 15).

Hartmann, L., Soderblom, D., Noyes, R.W., Burnham, N., and Vaughan, A. H. 1983, _Ap. J._, in press.

Knobloch, E., Rosner, R., and Weiss, N.O. 1981, _M.N.R.A.S._, **197**, 45P.

Kraft, R.P. 1967, _Ap. J._, **150**, 551.

Leighton, R.B. 1959, _Ap. J._, **130**, 366.

Lockwood, G.W., Thompson, D.T., Radick, R.R., Osborn, W.H., Baggett, W.E., Duncan, D.K., and Hartmann, L. 1984, _P.A.S.P._, in preparation.

Marcy, G.W. 1984, _Ap. J._, in press (Feb. 15).

Middlekoop, F. 1982, _Astr Ap._, **107**, 31.

Noyes, R.W., Hartmann, L., Baliunas, S.L., Duncan, D.K., and Vaughan, A.H. 1984, _Ap. J._, in press (Apr. 15).

Pallavicini, R., Golub, L., Rosner, R., Vaiana, G., Ayres, T., and Linsky, J. 1981, _Ap. J._, **248**, 279.

Robinson, R., and Durney, B.R. 1982, *Astr. Ap.*, 108, 322.

Rosner , R. 1983, in *Solar and Stellar Magnetic Fields: Origins and Coronal Effects*, Stenflo, J.O., ed., Reidel, 279.

Scalo, J.M., and Miller, G.E 1980, *Ap. J.*, **239**, 953.

Schussler, M. 1983, in *Solar and Stellar Magnetic Fields: Origins and Coronal Effects*, Stenflo, J.O., ed., Reidel, 213.

Smith, M. 1979, *P.A.S.P.*, **91**, 737.

Skumanich, A. 1972, *Ap. J.*, **171**, 565.

Skumanich, A. and Eddy, J. 1981, in *Solar Phenomena in Stars and Stellar Systems*, R.M. Bonnet and A.K. Dupree, ed., Reidel: Dordrecht, 349.

Skumanich, A., Smythe, C., and Frazier, E.N. 1975, *Ap. J.*, **200**, 747.

Soderblom, D. 1982, *Ap. J.*, **263**, 239.

Strauss, J.M., Blake, J.B., and Schramm, D.N. 1976, *Ap. J.*, **204**, 481.

Twarog, B. 1980, *Ap. J.*, **242**, 242.

Ulmschneider, P. and Stein, R.F. 1982, *Astr. Ap.*, **106**, 9.

Vaughan, A.H., Baliunas, S.L., Middlekoop, F., Hartmann, L., Mihalas, D., Noyes, R.W., and Preston, G.W. 1981, *Ap. J.*, **250**, 276.

Vaughan, A.H. and Preston, G.W. 1980, *P.A.S.P.*, **92**, 385.

Vernazza, J.E., Avrett, E.H., and Loeser, R. 1981, *Ap. J. (Suppl.)*, **45**, 635.

Walter, F.M. 1981, *Ap. J.*, **245**, 677.

Walter, F.M., 1982, *Ap. J.*, **253**, 745.

Wilson, O.C. 1978, *Ap. J.*, **226**, 379.

Woolley, Sir R.v.d.R., Epps, E.A., Penston, M.J., and Pocock, S.B. 1970, *A Catalogue of Stars Within 25 Parsecs of the Sun*, Roy. Obs. Annals, No. 5.

TIME SERIES MEASUREMENTS OF CHROMOSPHERIC
EMISSION AND POSSIBLE EVIDENCE FOR DIFFERENTIAL ROTATION

J.H. Horne, S.L. Baliunas and R.W. Noyes
Harvard-Smithsonian Center for Astrophysics
and
D.K. Duncan and A.H. Vaughan
Mt. Wilson and Las Campanas Observatories

Measurements of the long-term chromospheric Ca II H and K emission variations in nearly 100 lower main-sequence stars begun in 1966 by Wilson (1978) have continued for several seasons. The ongoing nightly monitoring at the Mt. Wilson 60" telescope begun in 1980 provides a data base unique in the study of rotation and activity cycles in cool main-sequence stars. Rotation periods have been determined for many of these stars from the periodic modulation of chromospheric emission as spatial inhomogeneities pass through our line of sight (Vaughan et al. 1981; Baliunas et al. 1983; Noyes et al. 1984). We have analyzed the periods determined from the data in each observing season with the goal of detecting surface differential rotation (Baliunas et al. 1984). In the pattern of variation of chromospheric emission with time there may be an instantaneous indication of surface differential rotation which results from two active areas marking two different periods in a season. Other explanations are also possible, for example, the growth and decay of emission-producing regions in the stellar atmosphere.

The data are the sum of the fluxes in 1A passbands centered on H (λ 3965) and K (λ 3934) divided by the sum of the fluxes in 20A passbands in the nearby stellar continuum, at λ 3900 and λ 4000, the S-index, as a function of time. The chromospheric S-index and the equipment used to collect the data are described in detail by Vaughan et al. (1981) and references therein.

We analyzed the chromospheric emission strength, S, for each star during each observing season. In the best stretches of data, both most complete and longest, we calculated power spectra as described by Scargle (1982). This method treats without bias these measurements which are unequally spaced with time.

Multiple significant peaks occasionally appear in the power spectra (Figure 1). To test for aliases of a period produced by the window function, we filter out a particular frequency and recompute the power spectrum (Ferraz-Mello 1981). This method is equivalent to removing a least-squares sine curve of the filtered frequency from the data (Scargle 1982). Some of the power spectra initially show double peaks that clearly contain alias frequencies. For those stars that showed two significant peaks that were not aliases, further tests were made. We fit, using least-squares techniques, the sum of two sine curves to the data for each season for each star. This represents two periods simultaneously present in the data. We also fit a sine curve to the data that had the same period throughout the season, but at a certain time during the season, the function was allowed to change phase with various continuity conditions. These fits represent the evolution of active areas. The reduced x^2 values were calculated for all of these fits to the data.

In ten cool stars, two distinct frequencies near that of the rotation period are present in the variation of chromospheric Ca II H and K emission with time (Baliunas et al. 1984). In most of these stars, the data are insufficient to distinguish unambiguously between differential rotation and evolution of active areas. Two stars, however, probably show differential rotation (see Table 1). In HD 190406, there is a secular increase in rotation period over three seasons, and the third season shows a second period corresponding to that of the first season. The reduced x^2 is best for the sum of two sines compared to the other single-period sine curves fit to the S-data. This star has a 2.6 year long-term activity cycle, so the periods determined from its modulations correspond well to the model where active regions appear at one latitude at the beginning of an activity cycle, gradually move throughout the cycle to a different latitude with a different rotation velocity, and then reappear at the original latitude at the beginning of a new cycle. When the chromospheric emission is strong, the period is short, similar to the behavior of rotation tracers on the Sun. The star HD 206860 has one dominant period during all three seasons analyzed and a second period during the first and second seasons (Figure 1). In this case, the constancy of the primary frequency and its phase over all three seasons analyzed together weakens the possibility of active area evolution and phase change as an explanation of the behavior of S with time. The fractional differential rotation in these two stars is 11% and 5%, respectively. From these data, it is impossible to measure the latitude and hence differential rotation rate. The values are, however, well within the predicted maximum fractional differential rotation of 40% (Gilman 1976).

References

Baliunas, S.L., Horne, J.H., Noyes, R.W., Duncan, D.K. and Vaughan, A.H. 1984, Ap.J., submitted.

Ferraz-Mello, S. 1981, Astron. J., 86, 619.

Noyes, R.W., Hartmann, L., Baliunas, S.L., Duncan, D.K. and Vaughan, A.H. 1984, Ap.J., in press.

Scargle, J.D. 1982, Ap.J., 263, 835.

Vaughan, A.H., Baliunas, S.L., Middelkoop, F., Hartmann, L., Wilson, O.C. 1978, Ap.J., 226, 379.

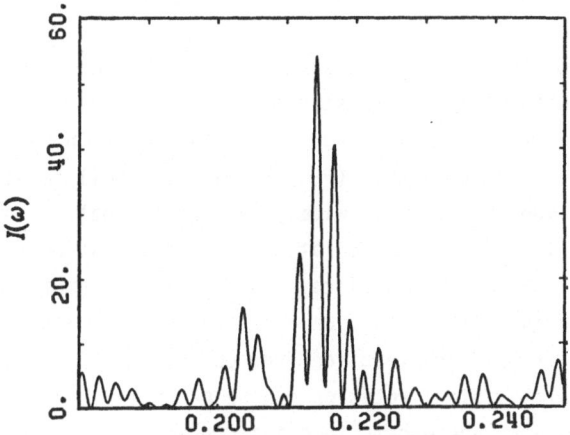

Frequency (cycles day⁻¹)

Figure 1 -- The power spectrum of HD 206860 in a limited frequency range for the 1980 and 1981 seasons together. The two main frequencies 4.67 and 4.90 days (0.204 and 0.214 cycles day⁻¹) are not aliases. The other frequencies clustered near these peaks are aliases caused by the inter-season gap in data. The ordinate $I(\omega)$ is the power divided by the variance. These same periods, 4.67 and 4.90 days, appear in the power spectra of individual seasons.

TABLE 1. ANALYSIS OF MULTIPLE PEAKS IN THE POWER SPECTRA OF TIME SERIES OF CHROMOSPHERIC EMISSION

STAR Name	HD	Spectral type	P_1[a] (days)	Q_1[b] (%)	P_2[a] (days)	Q_2[b] (%)	Date range (JD 2444000+)	N[c]
15 Sge	190406	G1 V	13.7	$<10^{-7}$	-	-	420 - 510	57
			14.5	0.2	-	-	770 - 880	52
			15.3	$<10^{-7}$	13.6	$<10^{-7}$	1100 - 1265	74
	206860	G0 V	4.7	$<10^{-7}$	5.0	0.33	420 - 570	80
			4.7	$<10^{-7}$	4.9	3.0×10^{-5}	770 - 925	84
			4.7	$<10^{-7}$	-	-	1100 - 1265	85

[a] P_1 is the period corresponding to the most significant peak in the power spectrum and P_2 the next most significant. P_2 is determined after filtering out P_1.

[b] Q_1 and Q_2 are the probabilities that a peak as high as or higher than the one given would appear if the data were Gaussian noise with the same variance.

[c] Number of nights with data in the specified range.

DETERMINISTIC TIME SERIES MODELING OF Ca II H & K STELLAR INDEX DATA

Ronald L. Gilliland
High Altitude Observatory
P.O. Box 3000
Boulder, CO 80307

Time series of chromospheric Ca II H & K variations have been used by O.C. Wilson (1978) to demonstrate unambiguously the existence of stellar activity cycles. Similar information obtained at nightly intervals over an observing season has been used by a group from Mt. Wilson, CFA and elsewhere (Vaughan, et al. 1981) to derive rotation periods for some 50 late type stars. It is widely recognized that if some of the objects have strong differential rotation and latitude variation of activity centers through a magnetic cycle, then differential rotation is, in principle, detectable as a variation of rotation period with time. Such a detection of period change would not distinguish equatorward drift and acceleration from poleward drift with acceleration. Since the general form of latitudinal rotation can be predicted with rotating hydrodynamic convection zone models, and dynamo theories predict the direction of activity migration, observational results for these quantities are of considerable importance for testing and directing further theoretical developments.

It should be possible to derive the sign of equator to pole differential rotation and cyclic migration through modeling the time series to derive the stellar inclination (i) and activity region latitude (ϕ). I have attempted such modeling for 19 stars with well determined periods from the 1980 Mt. Wilson H-K flux data. Reasonable parameter estimation resulting in good least-squares fits to the data can be done to yield period, phase, amplitude, background level, i and ϕ for most of these stars. However, non-uniqueness problems between sets of i-ϕ points nearly always arise. It is not possible to derive these parameters uniquely from one season's data. For successful modeling two requirements can be stated; 1) an improved signal-to-noise for the chromospheric variation index, and 2) availability of data over most of an activity cycle -- this could break the degeneracy between sets of i and ϕ, since i will be constant while ϕ varies. Chances for success seem good, but are likely to be realized only slowly.

I. Model Summary and Assumptions

The light curve model used in this study was developed by Budding (1977). The light curve variation is assumed to be dominated by one activity region on the stellar surface. Parameters of the model include: phase (T), period (P), stellar inclination (i), plage latitude (ϕ), activity region radius (γ), background intensity (U), contrast (KW), and linear limb-darkening coefficient (L). For the analyses below I will assume $KW = 3.0$, $L = 0.0$ (no limb-darkening), which leaves the model with 6 free parameters.

Analysis of one star from the Mt. Wilson 1980 Ca II H & K data set and studies of artificial light curves as sensitivity tests will be presented below.

II. Analysis of 1980 Mt. Wilson data -- HD 1835

HD 1835 is a G2 dwarf with a well determined rotation period of 7.9 days (Vaughan, et al. 1981). In attempting to derive light curve parameters the following steps are taken: 1) Estimate period with power spectrum analysis for unevenly-spaced data (Scargle, 1982) -- this yields an unambiguous determination for HD 1835 of P = 7.87 days. 2) Determine phase of variation via least-squares fit of cosine to data. Assume reasonable values for activity region size, background intensity, contrast, and limb-darkening. 3) Generate light curves using above estimated parameters over a full i, ϕ grid. Search for maximum of $1/\chi^2$ from linear least-squares fits to the data over the i, ϕ grid (see Fig. 1). 4) Using estimated i, ϕ from grid search perform a full nonlinear least-squares fit to estimate the 6 free parameters. Solutions with P = 7.875 days, $\gamma = 18.0°$, $U = 0.323$ and (i, ϕ) = (16,20) and (70,74) are identically equivalent fits (see Fig. 2). 5) For this case a vsin (i) measure exists (Soderblom, 1982) and from an estimate of the stellar radius (Vaughan, et al. 1981) it follows that sin (i) ~ 1, which favors the high inclination solution. 6) An active region latitude of 74° suggests (Mihalas, 1983) that an oblique rotator model, in which the active region resides at a magnetic pole slightly inclined from the rotation axis, might possibly be appropriate for HD 1835.

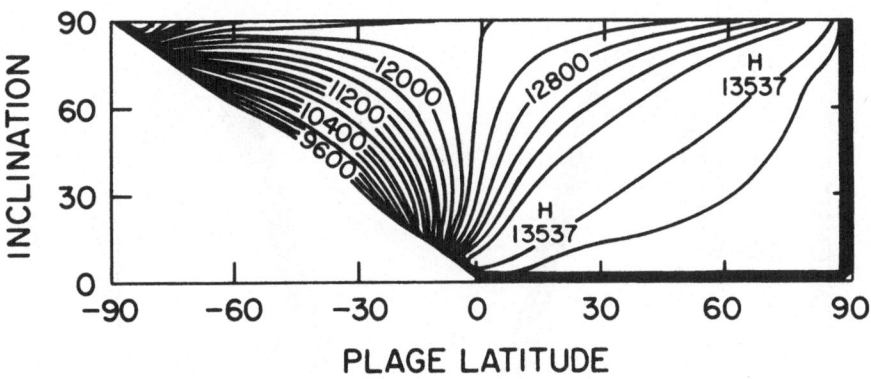

Fig. 1. Plot of $1/\chi^2$ (arbitrary units) as function of i, ϕ for HD 1835. Note the presence of two identically preferred solutions.

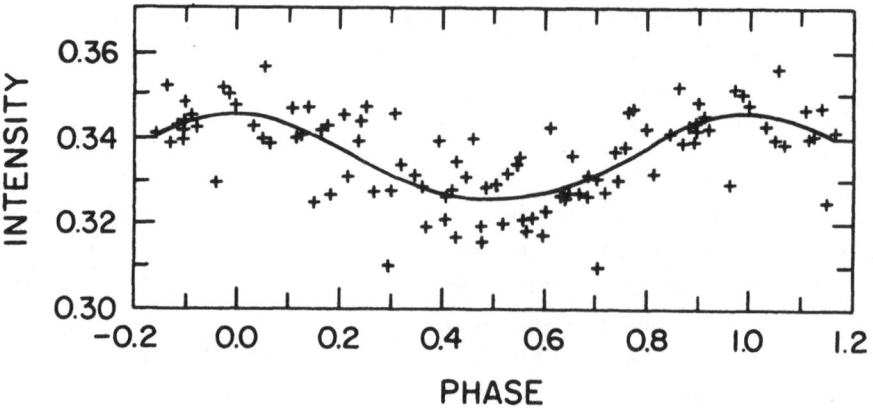

Fig. 2. S index data for HD 1835 plotted versus phase of derived period with parameters for model (smooth curve) as given in text.

III. Sensitivity of Light Curve to i, ϕ Variations.

I next consider what observational constraints must in general be met to derive reliable estimates of i,ϕ from detailed light curve modeling. To illustrate the problem a series of light curves generated along a diagonal $(i,\phi) = (45,21)$ to $(69,45)$ show no real variation -- curves differ only in amplitude. Therefore in practice an infinity of degenerate solutions may exist. However, a series of light curves (see Fig. 3) at constant i and variable ϕ do show discernable differences.

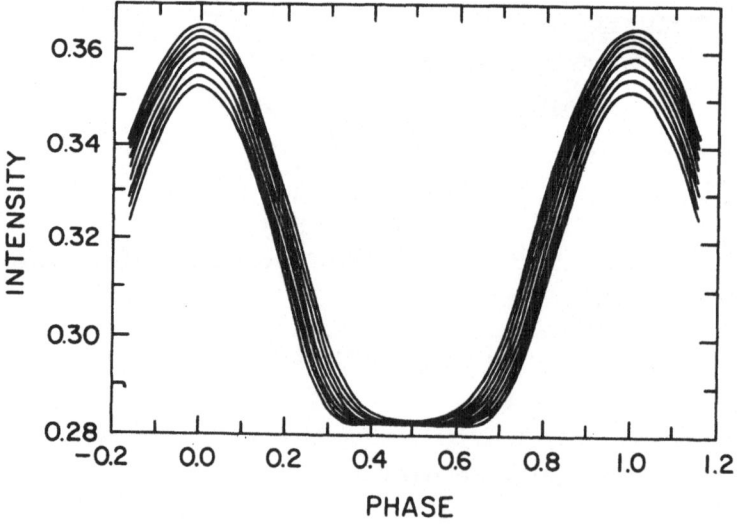

Fig. 3. Light curves generated with $i = 45°$, $\phi = 9-33°$.

The problem is now to investigate the signal-to-noise (S/N) required to successfully determine ϕ, if i is assumed known. I assume that the multiple degeneracy problem can be circumvented if data exists over several observing seasons (all at same i, different ϕ). Artificial light curves (P = 8.8 days) at various noise levels have been generated on observation times for HD 1835 in the 1980 Mt. Wilson data set (# data points = 72). The light curves are generated on an i,ϕ test grid; i = 15,30,..., 75, ϕ = -35, -20, 55 and then a general analysis (assuming only that i is known) is performed. The S/N and success rates are as follows (S/N $\sim$ 50 in Mt. Wilson data set): S/N = 50 yields ϕ to < 10^0 error for only 50% of the cases. S/N = 100 yields ϕ to < 7.5^0 error 75% of the time, while S/N = 200 has ϕ < 5^0 for 85% of the test points.

IV. Conclusion

Requirements for successful modeling of i and ϕ to allow direct determination of differential rotation are: 1) S/N $\gtrsim$ 100. 2) Data must exist over most of an activity cycle -- this breaks i, ϕ degeneracies and should allow for i determination nearly independent of ϕ for individual observing seasons.

I thank Douglas K. Duncan and Sallie L. Baliunas for making the Mt. Wilson - CFA, S index data set available to me.

References
Budding, E. 1977, Astrop. and Space Sci., 48, 207.
Mihalas, D. 1983, private communication.
Scargle, J.D. 1982, Ap. J., 263, 835.
Soderblom, D. 1982, Ap. J., 263, 239.
Vaughan, A.H., Baliunas, S.L., Middelkoop, F., Hartmann, L.W., Mihalas, D., Noyes, R.W. and Preston, G.W. 1981, Ap. J., 250, 276.
Wilson, O.C. 1978, Ap. J., 226, 379.

THE EVOLUTION OF CORONAL ACTIVITY IN MAIN SEQUENCE COOL STARS

Robert A. Stern
Lockheed Palo Alto Research Laboratory
3170 Porter Drive
Palo Alto, CA 94304

Abstract

Stars spend most of their lifetime and show the least amount of nuclear evolution on the main sequence. However, the x-ray luminosities of cool star coronas change by orders of magnitude as a function of main sequence age. Such coronal evolution will be discussed in relation to our knowledge of the solar corona, solar and stellar flares, stellar rotation and binarity. The relevance of X-ray observations to current speculations on stellar dynamos will also be considered.

Introduction

The mass and chemical composition of a star is sufficient to determine its location on the main sequence. This statement, the Russell-Vogt theorem, is basic to the understanding of the nuclear evolution of stars. The specification of a cool star's location on the main sequence on an H-R diagram is not, however, sufficient to predict the properties of its outer atmosphere: that is, its chromosphere, transition region, and corona.

The pioneering work of Wilson (1966), Kraft(1967), and Skumanich (1972), showed that both stellar rotation decreased and chromospheric activity, as measured by the Ca II emission reversal, decayed as solar type stars grew older on the main sequence. This work has been confirmed and extended by new observations of stellar coronas with the Einstein Observatory, and observations of stellar chromospheres and transition regions with IUE (for reviews of these observations, see Stern 1983 and Hartmann 1983).

Einstein results have provided evidence for the evolution of the level of quiescent coronal activity in main sequence cool stars. The most active stars detected by Einstein are invariably young single stars or rapid rotators in binary systems. In both cases, the X-ray emission is thought to be the result of enhanced magnetic activity produced by stellar dynamos.

It is generally accepted that magnetic flux in cool stars is generated by a dynamo mechanism in which the differential rotation couples with subphotospheric convective motions and thus provides continual regeneration of the stellar field (see, e.g. papers in Bonnet and Dupree 1981). Unfortunately, the details of the dynamo mechanism are too uncertain at present to yield reliable predictions of magnetic flux generation, although order of magnitude estimates have been made (e.g. Durney, Mihalas and Robinson 1981; Belvedere, Chiuderi, and Paterno 1981)

The precise way in which solar or stellar coronas are heated is thus unknown at present. Yet the observations support an intimate connection between magnetic flux generation and enhanced coronal x-ray emission. A number of mechanisms involving coupling of subphotospheric turbulence through magnetic waves or, alternatively, twisting of flux tubes have been proposed to explain coronal heating, though none is universally accepted (see, e.g. Leibacher and Stein 1981, Sturrock and Uchida 1981, Rosner et al. 1983).

Although we would like to use observational evidence as a guide to constrain such theories of magnetic flux generation and coronal heating, the stellar parameters such theories require are not easily measured using current instrumentation (see Figure 1). In particular, the influence of differential rotation must be estimated by observing the projected rotational velocity of a star, or the period of rotational modulation of Ca II plage regions across the stellar disk (see Baliunas et al. 1983 and references therein). Observations of apparent differential rotation in spot groups have been reported at this conference (Dorren and Guinan 1983), but such observations will require extensive long-term monitoring. Our only hope in obtaining information about the radial variation of differential rotation within the convective zone may be from observations of stellar oscillations; the first such observations have only recently been made (Fossat 1981, Noyes et al., 1983).

As a first step, therefore, observers have attempted to correlate X-ray emission with various observable stellar parameters. The fact that such correlations between X-ray luminosity and rotation or age are successful tells us that we are at least on the right track.

The first Einstein surveys clearly demonstrated that main sequence stars of virtually all spectral types emitted x-rays with the possible exception of single A stars (Golub et al . 1983, Golub, private communication; see review by Stern 1983). Most main sequence stars of spectral type (F8 - G5) detected by Einstein are brighter than the sun in x-rays. The x-ray luminosities of these stars range from 10^{27} - 10^{30} erg s^{-1} or more. Since a third parameter at least, besides L and T$_{eff}$, is required to explain these observations, rotation was the logical suggestion (Vaiana et al 1981), especially in light of the work of Skumanich (1972) and hints from HEAO-1 data (Ayres and Linsky 1980).

The work of Walter (1981) and Pallavicini et al. (1981) attempted to quantize such relations via power law scaling relations between the ratio of x-ray to bolometric luminosity (L_x/L_{bol}) and rotation period or x-ray luminosity and projected rotational velocity. In general, the scatter and sampling biases in the observational data allowed for either formulation of this rotation-activity connection, as pointed out by Stern and Skumanich (1983).

Observational and Theoretical Stellar Parameters

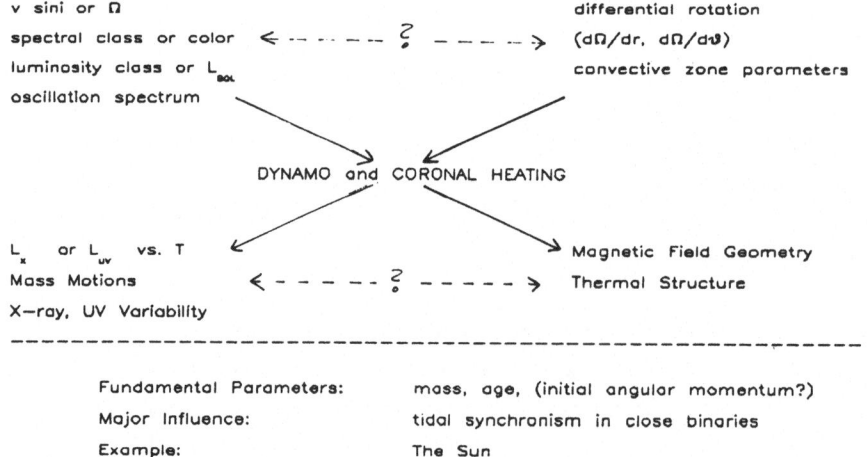

Figure 1. Relation between Observational and Theoretical Stellar Parameters

Recent observational results have, however, called into question the use of simple power-law scaling relations to predict both chromospheric and coronal activity. Hartmann et al (1983) give evidence that the Skumanich t $^{-1/2}$ relation is not obeyed for Ca II emission in Pleiades stars, despite the fact that the rotational velocities of the Pleiades F-G stars are what one would expect from the Skumanich relation. In addition, Caillault and Helfand (1981) and, more recently, Micela et al. (1983) have reported Einstein observations of the Pleiades which also suggest that their x-ray emission is not as high as one would expect from their rotational velocities and the x-ray activity-rotation scaling relations cited above.

Hartmann et al. (1983) have suggested that their results are evidence for a possible saturation in the chromospheric emission of stars younger than the Pleiades (7×10^7 yr). Such a saturation effect might be a clue to the ultimate efficiency of the stellar dynamo in its ability to generate magnetic flux. However, observations of cool stars with higher rotational velocities than the Pleiades, but possibly different subphotospheric structures, may also shed light on this problem: namely, the recent observations of W UMa stars by Cruddace and Dupree (1983), and those of the Pre- Main Sequence (but post T Tauri phase) G stars in the Orion star formation region by Smith, Pravdo, and Ku (1983).

A summary of the most recent published and unpublished results on x-ray activity and rotation in solar type main sequence stars (F8-G5) is shown in Figure 2, with unpublished x-ray data provided by L.Golub (field stars and Pleiades), J.P. Caillault (Pleiades), and unpublished rotational velocity data from D. Soderblom (field stars and Pleiades) and S. Baliunas (Hyades stars). The data are presented in the L_x vs. v sini formulation of Pallavicini et al. (1981). It should be noted that plotting L_x/L_{bol} vs v sini or estimated period produces a similar correlation plot, with no obvious advantage of one formulation over another.

Examination of Figure 2 confirms the correlation between x-ray luminosity and rotation over 4 orders of magnitude in L_x and roughly 2 orders of magnitude in v sini: the quadratic relation proposed by Pallavicini et al. (1981) indicates the general trend. Several important points should be remembered, however, while interpreting the data in Figure 2.

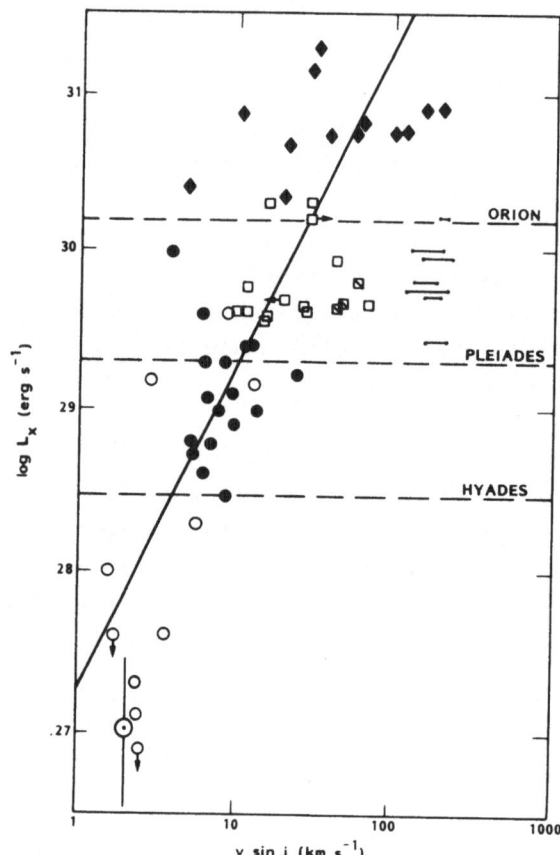

Figure 2. Correlation between X-ray Luminosity and Projected Rotational Velocity. Symbols are as follows (O - field stars, ● - Hyades, □ - Pleiades G stars, ◧ - Pleiades G-K Stars, ◆ - Orion PMS G stars, ●—● - W UMa systems). Solid line is relation of Pallavicini et al. (1981). Dashed lines indicate detection thresholds for each cluster.

(1) the velocities are all projected, hence the average correction factor (towards increasing v sini's) will be about $4/\pi$, but individual velocities may have to be corrected by several times this,

(2) the approximate Einstein detection thresholds are indicated by the horizontal dashed lines for the cluster data: for the Hyades, about 80% of the solar-type stars were detected, but for the Pleiades, the figure is <50% (Micela et al., 1983, Caillault and Helfand, 1981) and for the Orion PMS stars, the detection rate is not given, but may be significantly less. Upper limits are not indicated for purposes of clarity.

(3) a number of slowly rotating solar-type stars with v sini < 2 km s-1 and nearby enough to have a low detection threshold were simply not observed by Einstein due to the limited lifetime of the observatory,

(4) Several Pleiades stars with nominal spectral type around G–K
are described in Stauffer   et al. (1983a) as rapidly rotating K
stars which have recently (~few x 10^7 years) arrived on the main
sequence. Two other Pleiades stars which have apparently anomalous rotational velocities for their x-ray emission (at   least if
the Pallavicini et al. relation is used as a reference) are ~1
M $\odot$ stars observed by Kraft (1967).

(5) The W UMa systems plotted are those from Cruddace and
Dupree (1983) with solar-type spectral classifications.

(6) Although x-ray error bars are not indicated, they range
from probably < 10–20% for the nearby field stars to up to
50% for the weakest Hyades stars (Stern et al. 1981) and up to
a factor of 2 or more for the Orion PMS stars (Smith, Pravdo,
and Ku 1983).

The most obvious exception to the general trend is the W UMa
group. This was noted in a similar comparison by Cruddace and Dupree
(1983), who suggested a saturation effect in the x-ray luminosity when
a stellar surface is completely "covered" by x-ray emitting magnetic
loops, with the plasma pressure limited in some way by tidal forces. I
suspect that the pressure limitation is more fundamental: in at least
one Hyades G dwarf, for example (BD +14° 693), even full surface
coverage at solar-like pressures is not sufficient to produce the
observed x-ray luminosity (Stern 1983). Indeed, combining IUE and
Einstein data on this star in a simple static loop model predicts
coronal base pressures of more than 1000 times solar with very small
surface coverage factors (<1%; Zolcinski et al. 1982). Furthermore,
the magnetic field coupling between the stars in W UMa contact binary
systems may yield field geometries so different from the sun's that
using standard solar loop "building blocks" may not even be appropriate here. The answers will likely come when we can measure
temperature distributions and plasma densities in such stars, possibly
with the spectroscopic instruments on AXAF or future missions.

The W UMa's also exhibit a property characteristic of other
clusters or groups of stars such as the RS CVn systems when plotted
separately on a L_x vs v sini plot: the rotation-activity correlation
is generally much weaker or even non-existent within groups of stars
than it is among groups as a general trend. I think this is primarily

due to the distribution in sin i's, the uncertainties in the x-ray measurements (and possibility of cyclic behavior), and possible sampling biases. I make this remark to caution against comparing the slope of power law relations of individual samples against other samples: the scatter in the data is just too large at this time. Only by taking the rotation activity correlation over many orders of magnitude does the general trend become apparent.

The rapid rotators in the Pleiades may be an example of this latter sampling effect since, within the cluster, an increase in the v sini value of almost an order of magnitude seems to yield little, if any, increase in x-ray emission. The same can be said to hold true for the Orion PMS stars, though we are on more shaky ground here: the Orion PMS are apparently in their post T Tauri stage (Smith, Pravdo, and Ku 1983), but are presumably still contracting toward the main sequence. Smith, Pravdo, and Ku (1983) suggest that a magnetic saturation effect similar to that proposed by Hartmann et al. (1983) for the Pleiades Ca II data and by Cruddace and Dupree (1983) for the W UMa's is also seen in the Orion X-ray data. Yet a similar lack of correlation exists in, for example, the Hyades data alone.

One may quibble with plotting the Orion PMS and W UMa stars on the same graph as main sequence cluster stars, but I believe such a comparison serves an important purpose. It is telling us that a further complication exists for young or very rapidly rotating stars. Because of differences in subphotospheric structure, either related to stellar age or possibly tidal effects in contact binaries, branches may exist on this x-ray "H-R" diagram of activity against rotation. For each branch, the generation of magnetic flux or mechanical energy may be limited to some maximum value, which either prevents the generation of more loop structures or limits their pressure, or perhaps simply limits the total heating rate of the corona.

Such speculations on the efficency of the dynamo and coronal heating mechanisms are supported by only very immature theories. It is to be hoped that as the x-ray, CaII, and UV line observations become more numerous, much more detailed constraints can be placed upon these theories.

Main Sequence Evolution

Since for the Hyades (6-8x10^8 yr), the Pleiades (7 x 10^7 yr), and, if we are allowed some license, the Orion PMS stars (1-4 x 10^6

yr; Smith, Pravdo, and Ku 1983), the ages are reasonably well known
(but see Stauffer 1983b), we may plot the observed ranges of x-ray
luminosity for each cluster against cluster age (Figure 3). The solar
values are also plotted. Note that the only cluster effectively
sampled below the L_x mean for the solar-type (F8-G5) stars is the
Hyades.

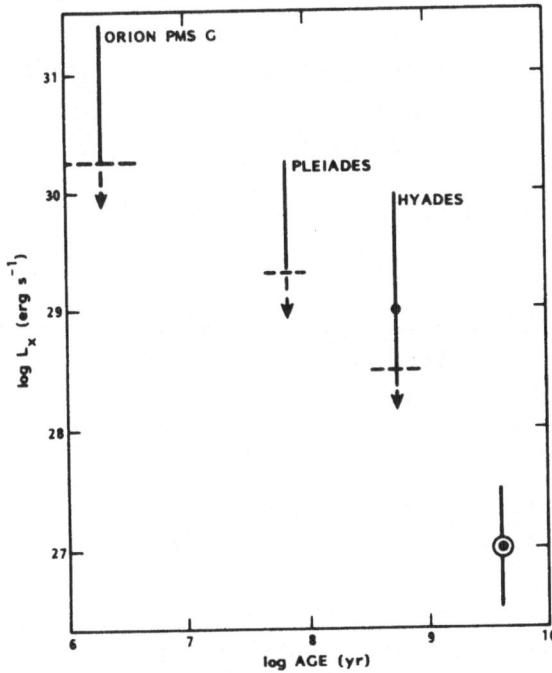

Figure 3. Correlation
between X-ray Luminosity and
Cluster Age. Dashed lines
indicate detection
thresholds.

The trend of the data is clear: we can certainly say that the
maximum L_x observed decays with increasing age. Unfortunately, the
younger Pleiades and Orion clusters are so sparsely sampled that a
comparison of mean L_x values is not particularly fruitful. If we
accept the results of Micela et al. (1983) that roughly 50% of the
Pleiades G stars were detected, then the median L_x is close to the
x-ray detection threshold; hence the Pleiades-Hyades Sun comparison
possibly indicates the same saturation effect proposed by Hartmann et
al. (1983) for Ca II.

Remember, however, that the solar point is only a single sample
of solar-age star. Returning to Figure 2, we see that, if a similar
scatter in L_x vs age exists as for L_x vs rotation in solar type stars
of low activity, the "saturation" effect may not be as pronounced as
at first glance. Until the L_x's for a number of younger clusters and
the ages of low activity stars are better determined, the precise form
of the age-activity relation remains uncertain.

Evolution of Variability and Flare Activity

Thus far we have discussed possible evidence for the evolution of the quiescent x-ray luminosity of cool stars. By examining variability differences among cool stars of different ages, we may hope to constrain properties of the coronal loop structures on such stars, and possibly obtain more clues as to the nature of the dynamo mechanism compared to that of the solar dynamo. Hence, if the stellar dynamo evolves in a qualitative as well as quantitative sense as stars age on the main sequence, such variability studies may be the key to understanding this evolution.

We may break down coronal variability into three rough categories:

1) Rotational modulation of the integrated x-ray luminosity as active regions on the star pass into and out of our field of view (similar to the Ca II rotational modulation studies of Baliunas et al. 1983),

2) Observation of long-term stellar activity cycles in the manner pioneered by Wilson (1978), again in Ca II emission, and

3) Observation of flaring activity. Although visible light flares have long been observed in dMe and similar stars, the photospheric radiation of relatively 'hot' solar type stars has effectively prevented ground-based observations of events similar to solar flares. Thus x-ray and UV emission line observations are the primary data base for flaring activity in solar-type stars, meager as this database is (see Haisch 1983 and Stern 1983 for reviews).

Rotational Modulation

The rotational modulation of the full-disk solar soft x-ray flux is a well known phenomenon . Because at most a few to 10% of the solar surface is covered with active regions, even at solar maximum, the rotational modulation is quite pronounced over the ~27 day solar rotation period. During solar minimum, this variation can approach an order of magnitude or more (see, e.g. Kreplin 1970). Evidence from Ca II rotational modulation data suggests that for the more active cool

stars the apparent rotational modulation may decrease. The variability
in Ca II in such stars may be largely stochastic in character (Vaughan
and Preston 1980).

There is only limited evidence from Einstein to see if the same
holds true in soft x-rays, since few solar type stars could be monitored for a significant fraction of their rotation periods.   One
binary system, HD165590, consisting of a G dwarf and likely a K dwarf
companion, was monitored by Einstein for about 1/2 an orbital period.
The projected equatorial velocity is ~75 km s-1, in good agreement
with a ~ 1 R ⊙ dwarf in a synchronously rotating binary system with
period 0.88 day. As reported by Stern and Skumanich (1983), HD 165590
shows little, if any, rotational modulation of its x-ray flux; indeed,
what appear to be small flares or other very rapid variability
characterize the soft x-ray light curve of this system (see Figure 4).

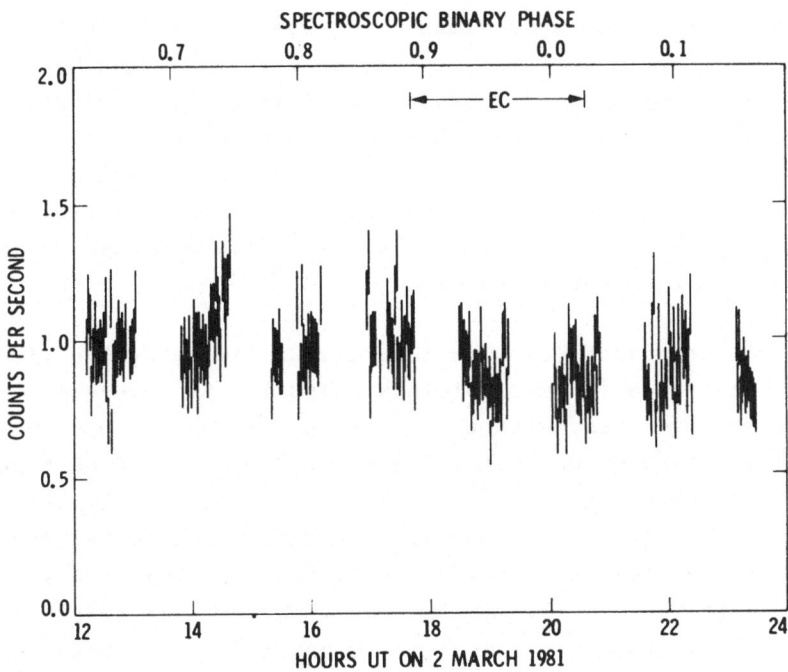

Figure 4. X-ray Light Curve of HD 165590.   "EC" indicates region of
primary eclipse (after Stern and Skumanich 1983).

At least in this case, then, we are reasonably certain that the distribution of x-ray active regions on the stellar surface is fairly uniform. One possibility is, of course, that the stellar surface is fully covered by active regions: we cannot, however, rule out a large number(say, > 100 or so) small, high pressure active regions with a significantly smaller surface coverage factor, since our point-to-point statistical variation is estimated to be no better than 10%.

In either case, it is clear that the distribution of magnetic loop structures on the stellar surface is much more uniform than the sun. Although one example does not prove a point, the corroborating evidence from the Ca II data suggests that the x-ray luminosity evolution of solar type stars may well involve a substantially different level and distribution of magnetic flux generation for the youngest and most active stars.

Stellar Coronal Activity Cycles

The existence of the approximately 11 year solar activity cycle is a well-known phenomenon. Similar length activity cycles have been observed in Ca II by Wilson (1978), and more recently by Vaughan and his collaborators. If the fundamental properties of the stellar dynamo change for very young or rapidly rotating stars,as suggested by some authors (e.g. Durney, Mihalas, and Robinson 1981), then a search for coronal activity cycles in stars of the Hyades age and younger may be used to constrain such dynamo models.

An attempt to search for such long term activity cycles with Einstein was made by Stern and Zolcinski (1983), who observed a number of Hyades solar type stars with the Einstein Imaging Proportional Counter over a period of up to 1-1/2 years. Roughly half of t`he stars reobserved over this time period show no evidence of variability within the statistical accuracy of measurement (which ranged from about 20-50%); the remainder show variability on a long time scale by no more than a factor of 2-3. It is thus apparent that strong cyclic activity in the Hyades stars cannot be occurring on much shorter timescales than a solar cycle. Since the original Hyades survey of Stern et al.(1981) exhibited a roughly 1-1/2 order-of-magnitude spread in the x-ray luminosity of solar-type stars, the implication is that either these stars have longer, possibly solar-like, activity cycles, or that the luminosity spread is intrinsic.

Flare Activity

Solar flares invariably occur in solar active regions. As suggested above, younger or more rapidly rotating stars may have more active regions than the sun. One can suppose that , all other parameters being equal, stellar flares on the most x-ray luminous stars should be higher in frequency or perhaps more luminous than solar flares. In fact, the largest stellar flares observed on Einstein were more than 1000 times the peak x-ray luminosity of solar flares (Stern 1983, Haisch 1983).

It is important to note that stars with the largest flaring x-ray luminosities are not necessarily classical "flare stars". Classical flare stars brighten dramatically in the U or B bands on time scales of minutes or less. Virtually all are of spectral type dKe or dMe, indicating Ca II emission lines in their visible spectra. Yet any rapid photospheric heating which produces a "bluer" effective photospheric temperature is easily concealed in the photospheric continuum of a relatively hot (e.g., G type) star.

In contrast, x-ray observations of flares are not hampered by this effect, which selectively limits observations of optical flares to the coolest stars. The x-ray flaring luminosity of the sun, for example, is often an order of magnitude or more above its quiesecent x-ray luminosity. Hence x-ray or UV observations are much more sensitive flare detectors than visible light monitoring.

Large x-ray flares have been observed in several Hyades stars (Stern and Zolcinski 1982), pre-main sequence stars in the Rho Oph dark cloud (Montmerle et al 1983), and at least one RS CVn binary system, Sigma Cor B (e.g.,Agrawal et al. 1983). Also, a number of flares in classical dMe flare stars have been detected in x-rays (Haisch 1983). A summary of stellar flare observations and derived properties compared to the sun is given in Table 1. The flare in the Hyades binary HD 27130 (Stern, Antiochos and Underwood 1983) stands out as one of the largest ($> 10^{31}$ erg s^{-1}). The flare emission measure and derived volume of this flare is > 1000 times the largest solar flares. In their paper, Stern et al. suggested the reason for this was the involvement of many active regions in the flare.

Although the flare data are, as might be expected, few in number, the implication that young stars (or those in rapidly rotating binary systems) produce substantially larger flares than those seen in the sun is borne out by the available data. However, a large number of these flares are from binary systems, such as two of the Hyades flares, those in HD27130 and HD27691. These two systems have orbital periods of 5.6 and 4.0 d respectively; their rotation periods, if they are in synchronous orbits, are roughly comparable with the average rotational velocity of a solar type Hyades star, about 5 days. Thus separating out the effects of binarity and youth will be difficult until we have a larger sample of flares in single rapidly rotating stars.

Solar and Stellar Flares
Thermal X-Ray Plasma Properties

Log of	Sun	Flare Stars	HD 27130	Pre-M.S.	RS CVn
τ_{rise} (s)	2-3	2-3	< 3.3	2.4	< 4
τ_{decay} (s)	3-4	2-3	3.4	~ 3	~ 4
L (peak, x erg s^{-1})	< 27-28	27-31	> 31	< 30-34	31-33
E_x (total, erg)	< 30-31	31-33	> 34.5	< 34-36	35-38
T_{max} (K)	7.0-7.5	7.1-7.6	7.5-8.0	7.0-7.3	> 7.5 ?
EM (cm^{-3})	< 50	51-54	>53.4-54	53-55	55-56
N_e (cm^{-3})	10.5-12	11-11.6	11.6	10-11.5	
V (cm^{-3})	25-28	27-28	30.6	30-33	

Table 1.   Solar and Stellar Flare Properties (after Stern, Underwood, and Antiochos 1983).

Some authors have suggested that the interacting magnetic fields in RS CVn binary systems could trigger flare activity ((Simon, Linsky, and Shiffer 1980; Linsky, 1983). Since many,if not all dMe stars are binaries, activity in these systems could also be related to binarity. Whether large scale reconnection takes place or binarity induces a greater stress on already twisted field lines, such a hypothesis is at least plausible. Although it is tempting to believe that the peak x-ray luminosities of flares in very active single stars are comparable to those in the binary systems, no unambigous example exists for solar type stars. What is clear is that such young rapid rotators have the potential to produce large flares, perhaps because of a stellar surface in which active regions are in much closer proximity to each other than on the sun.

Summary and Outstanding Questions

The Einstein results have clearly demonstrated a strong correlation between quiescent x-ray luminosity, stellar rotation, and age for solar type stars. The picture is still complicated, however, for stars of the Pleiades age and younger. The youngest stars also appear to have either a larger number of higher pressure active regions than the sun, or a larger stellar surface coverage with solar-like pressure. Young or rapidly rotating stars show a marked increase in peak flare luminosity, though here nearby binary companions may play an important role in other ways than just enforcing tidal synchronism.

X-rays are thus becoming an established indicator of stellar age for single stars, in the same way Li abundance is used. However, the lack of well developed models for the dynamo or for x-ray heating is holding back interpretation of the x-ray results. Outstanding questions include the influence of initial conditions in stellar atmospheric structure when a star arrives on the main sequence (see, e.g. Stauffer, 1983b) possible differences in the nature of the dynamo for fast and slow rotators (e.g. Durney, Mihalas, and Robinson 1981); and especially, the question of whether the x-ray heating or generation of magnetic flux can saturate at some level as a result of limitations in the dynamo efficiency related to convective zone properties.

EXOSAT, ROSAT, EUVE, and AXAF will provide some of the answers, as will the UV spectrographs on Space Telescope and COLUMBUS. One may hope that in twenty or thirty years, perhaps, we will understand the physics and evolution of stellar activity as well as we now understand stellar nuclear evolution.

Acknowledgments

I thank Dave Soderblom, Sallie Baliunas, John Stauffer, Lee Hartmann, Leon Golub, and Jean-Pierre Caillault for helpful conversations and their permission to use unpublished data. This work was supported in part by NASA contract NAS5-23758 and the Lockheed Independent Research Program.

References

Agrawal,P.C., Rao, A.R., Riegler, G.R., and Stern, R.A., 1983, Proceedings 18th International Cosmic Ray Conference, Bangalore, India, Vol.1

Ayres,T.R., and Linsky, J.L., 1980, Ap.J., 245, 671.

Baliunas, S.L., et al. , 1983, Ap.J., in press.

Belvedere,G., Chiuderi, C., and Paterno, L., 1981, Astr. Ap., 96, 369.

Bonnet, R.M., and Dupree, A.K., eds., 1981, Solar Phenomena in Stars and Stellar Systems, D.Reidel, Dordrecht:Holland.

Caillault ,J.P. and Helfand, D.J., 1981, B.A.A.S., 13, 811.

Cruddace, R.G, and Dupree, A.K., 1983, preprint.

Dorren, J.D., and Guinan, E.F, 1983, these proceedings.

Durney,B.R., Mihalas,D., and Robinson, R.D., 1981, Pub. Ast. Soc. Pacific, 93, 537.

Fossat, E., 1981, in Solar Phenomena in Stars and Stellar Systems, Bonnet and Dupree, eds., Dordrecht:Holland, p.75.

Golub et al ., 1983, Ap.J., 271, 264.

Hartmann, L., 1983, Adv. in Space, Research, Vol 2., No. 9.

Hartmann L., Soderblom, D., Noyes, R.W., Burnham, N.,and Vaughan, A.H., 1983, Ap.J., in press.

Kraft, R.P, 1967, Ap.J., 150, 551.

Kreplin , R.W., 1970, Ann. Geophys., 26, 567.

Leibacher, J.W., and Stein, R.F., 1981, SAO Special Report 392, Vol I (Proceedings of Second Cambridge Cool Star Workshop), 23.

Micela , G. et al. 1983, these proceedings

Montmerle, T., Koch-Miramond, L., Falgarone, E., and Grindlay, J., 1983, Ap.J., in press.

Noyes , R., 1981, in Solar Phenomena in Stars and Stellar Systems, Bonnet and Dupree, eds., D. Reidel, Dordrecht:Holland, p.1.

Noyes, R., Baliunas, S.L., Belserene, E., Duncan, D.K., H orne, J., and Widrow, L., 19983, these proceedings.

Pallavicini , R., Golub, L., Rosner, R., Vaiana, G.S., Ayres, T.R., and Linsky, J.L, 1981, Ap.J., 248, 279.

Rosner, R., Golub, L., and Vaiana, G., 1983, preprint.

Simon, T., Linsky, J.L., and Schiffer, F.H., 1980, Ap. J. 239, 911

Skumanich , A., 1972, Ap.J., Ap.J. 171, 565.

Smith, M.A., Pravdo, S.H., and Ku Wm. H.-M., 1983, Ap.J., 272, 163.

Stauffer, J.P., 1983a

Stauffer, H.P., 1983b, these proceedings.

Stern, R.A., 1983 , Adv. in Space Res., 2, 39.

Stern, R.A., and Skumanich , A., 1983, Ap.J., 267, 232.

Stern, R.A., Underwood, J.H., and Antiochos ,S.K., 1983, Ap.J.
(Letters), 264, L55.

Stern, R.A., and Zolcinski,M.-C., 1983, in Activity in Red Dwarf
Stars, Byrne, P.B.,and Rodono, M., eds., D.Reidel,
Dordrecht:Holland, p.131.

Stern, R.A., Zolcinski, M.-C., Antiochos, S.K., and Underwood, J.H.,
1981, Ap. J., 249, 647.

Sturrock P.A., and Uchida, Y., 1981, Ap.J. 246, 331.

Vaughan , A.H.,and Preston , G.W., 1980, Pub. Astron. Soc. Pacific,
90, 267.

Vaiana, G.S., et al., 1981, Ap.J., 245, 163.

Wilson, O.C., 1966, Ap.J. 144, 695.

Walter, F.M., and Bowyer, S., 1981, Ap.J. 245, 671.

Walter ,F.M., 1982, Ap. J. 253, 745.

Zolcinski,M.C., Antiochos, S.K., Walker, A.B.C., and Stern, R.A.,
1982, Ap. J. 258, 177.

CORONAL STRUCTURE AND ROTATION IN LATE-TYPE STARS

C.J. Schrijver, R. Mewe
The Astronomical Institute, Space Research
Laboratory, Utrecht, The Netherlands

F.M. Walter
Joint Institute for Laboratory Astrophysics, University of
Colorado, and National Bureau of Standards, Boulder, CO

Soft X-ray spectra, obtained with the HEAO-2 IPC (Giacconi et al. 1979), are analysed for 34 late-type stars. The sample includes single stars, binaries (many of which are named RS CVn systems), two FK Com stars, and a BY Dra star. We fitted observed spectra (0.15-4.keV) to single-temperature model spectra (Mewe and Gronenschild 1981) to obtain the effective coronal X-ray temperature T and the specific emission measure: $\zeta = \int_V n_e^2 dV/4\pi R^2$ (n_e electron density, V emitting volume, R stellar radius). The fits appear to be adequate approximations for most of the observed spectra. Errors in ζ are mainly due to counting statistics. These errors are much smaller than errors in the derived temperatures, so that only the latter are drawn in the figures.

There is no conspicuous relation between the coronal temperature and (B-V), although for dwarfs there may be a marginally significant trend of decreasing T with increasing (B-V) (Mewe et al. 1983, Schrijver et al. 1983). The stellar data indicate two branches in a diagram in which ζ is plotted against T (Fig. 1). We argue that at least two distinct coronal loop structures exist: one with a relatively high temperature ($\sim$20 MK), most pronounced on giants, and one with a relatively low temperature ($\sim$2 MK), seen mostly on dwarfs. For very active stars the temperature distinction disappears, either because the cool component vanishes, or because the temperature of the cool component rises until it equals that of the hot component.

The stellar rotation rate plays an important role in stellar activity. At first activity and rotation were thought to be related only through stellar age. Evidence that rotation rate, rather than age, determines stellar activity is provided by a study of synchronised binaries (Middelkoop and Zwaan 1981, Zwaan 1981), which display the same behaviour as single stars (Middelkoop 1982, Walter 1982, but see Vilhu 1983).

Figure 2 is an emission measure-period diagram. Periods were collected from CaII modulation data (Baliunas et al. 1983), if available. Orbital periods (Hall 1981, Hoffleit 1982) are used for synchronized binaries. Photometric periods (Chugainov 1976, Rucinsky 1981) and v sin i values

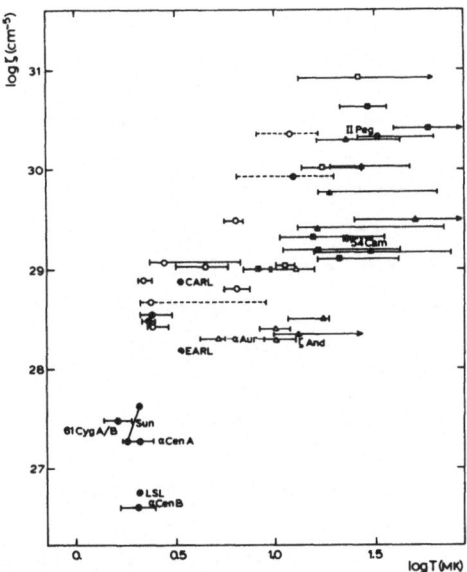

Fig.1 Specific emission measure
ζ vs coronal temperature T. The
data for the average Sun are
from Vaiana et al. (1976).

Luminosity class	II III	III-IV IV	IV-V V
B-V<0.6	△	□	○
0.6<B-V<0.8	△	□	◑
0.8<B-V<1.0	▲	◩	●
1.0<B-V	▲	■	●

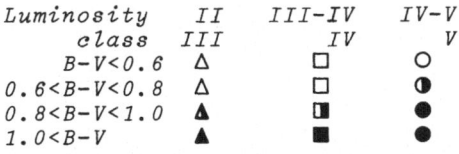

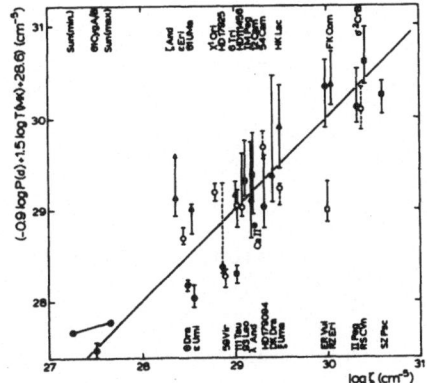

Fig.3 Specific emission measure resulting from Eq.(1), using period
and temperature, vs the specific
emission measure obtained through
spectral analysis. For λ And the
orbital period is used; the Ca II
H+K modulation period would yield
the point marked Ca II. Symbols
as in Figure 1.

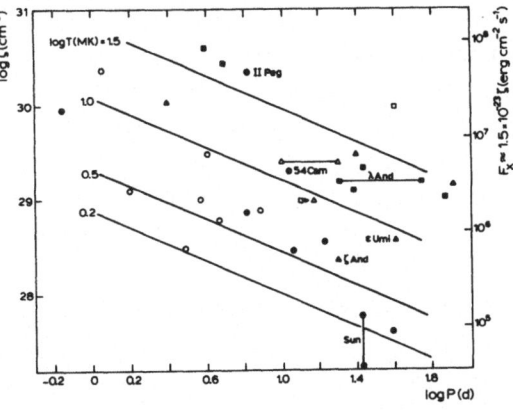

Fig.2 Specific emission measure ζ vs
period P. The scale on the right for
the X-ray surface-flux density is
given using $F_x = C_T$ ζ with an average
conversion factor given in the figure.
Symbols as in Figure 1. The lines
represent the result of a correlation
analysis using ζ, P and the effective
coronal temperature T as parameters.
Three points represent individual
structures on the Sun: CARL compact
active-region loop ($L\sim3$ 10^9 cm, $p\sim5$
dyn cm^{-2}); EARL extended active-region
loop, corresponding to an "equivalent"
active-region loop that matches the
average of an active region on the
Sun ($L\sim9$ 10^9 cm, $p\sim1.5$ dyn cm^{-2});
LSL large-scale loops connecting
active regions ($L\sim2$ 10^{10} cm, $p\sim0.2$
dyn cm^{-2}). These points are drawn for
surface filling factors equal to unity. The data are derived from observations made with the Skylab S054
instrument (Pallavicini et al. 1981).
The data for α Aur are from Mewe et
al. (1982).

(Hoffleit 1982, Kraft 1967, Smith 1978,
Soderblom 1982) (6 objects) complete the
data. Figure 2 suggests that a third parameter is involved in the relation
between ζ (or F_x, see Fig. 2) and the
rotation period P. A correlation analysis with ζ (cm^{-5}), T (MK) and P (days)
as parameters yields:

$$\log \zeta = (28.6\pm0.2) + (1.51\pm0.16) \log T + (-0.88\pm0.14) \log P \qquad (1)$$

(plotted in Fig. 3). The high correlation (0.88) is surprising considering

the diversity of the stars: the existence of coronae with very different average properties was inferred from an analysis of Ca II H+K and soft X-ray data (Schrijver 1983), as well as from a comparison of the results of spectral analyses with a model for magnetic loops (Schrijver et al. 1983).

From the tightness of the relation between the X-ray flux density and the Ca II H+K excess flux density Schrijver (1983) concludes that the flux levels of the outer atmospheres of stars with convective envelopes are uniquely determined by one activity parameter - probably a combination of several stellar parameters, the rotation rate being one of these. Relation (1) shows that at least two parameters are needed to describe the structure of stellar coronae. The above suggests a functional dependence of two of the three parameters in relation (1). Support for this inferred dependence may be found on the Sun: the effective coronal temperature is seen to increase with increasing ζ. Observations of other stars, throughout their activity cycles, are required to determine whether the variations of ζ and T with time obey relation (1), or whether the variations produce scatter about the relation.

References.

Baliunas, S.L. et al.: 1983, Astrophys. J., in press.
Chugainov, P.F.: 1976, Inf. Bull. Var. Stars, No. 172.
Giacconi, R. et al.: 1979, Astrophys. J. 230, 540.
Hall, D.S.: 1981, Solar Phenomena in Stars and Stellar Systems, eds.
    R.M. Bonnet and A.K. Dupree (Reidel, Dordrecht), 431.
Hoffleit, D.: 1982, The Bright Star Catalogue, Yale Univ. Obs., New Haven.
Kraft, R.P.: 1967, Astrophys. J. 150, 551.
Mewe, R., Gronenschild, E.H.B.M.: 1981, Astron. Astrophys. Suppl. 45, 11.
Mewe, R., Gronenschild, E.H.B.M., Westergaard, N.J., Heise, J., Seward,
    F.D., Chlebowski, T., Kuin, N.P.M., Brinkman, A.C., Dijkstra, J.H.,
    Schnopper, J.H.: 1982, Astrophys. J. 260, 233.
Mewe, R., Schrijver, C.J., Gronenschild, E.H.B.M., Zwaan, C.: 1983,
    Solar and Stellar Magnetic Fields: Origins and Coronal Effects, ed.
    J.O. Stenflo, (Reidel, Dordrecht), 205.
Middelkoop, F., Zwaan, C.: 1981, Astron. Astrophys. 101, 26.
Middelkoop, F.: 1982, Astron. Astrophys. 107, 31.
Pallavicini, R., Peres, G., Serio, S., Vaiana, G.S., Golub, L., Rosner,
    R.: 1981, Astrophys. J. 247, 692.
Rucinsky, S.M.: 1981, Astron. Astrophys. 104, 260.
Schrijver, C.J.: 1983, Astron. Astrophys., in press.
Schrijver, C.J., Mewe, R., Walter, F.M.: 1983, in preparation.
Smith, M.A.: 1978, Astrophys. J. 224, 584.
Soderblom, D.: 1982, Astrophys. J. 263, 239.
Vaiana, G.S., Krieger, A.S., Timothy, A.F., Zombeck, F.: 1976, Astrophys.
    Sp. Sc. 39, 75.
Vilhu, O.: 1983, Physica Scripta, in press.
Walter, F.M.: 1982, Astrophys. J. 253, 745.
Zwaan, C.: 1981, Solar Phenomena in Stars and Stellar Systems, eds. R.M.
    Bonnet and A.K. Dupree (D. Reidel, Dordrecht), 463.

A SEARCH FOR SECULAR LUMINOSITY VARIATIONS
IN THE PLEIADES CLUSTER

John A. Eddy and Peter B. Bandurian
High Altitude Observatory*
Boulder, Colorado 80307

and

Lee W. Hartmann
Center for Astrophysics
Cambridge, Massachusetts 02138

The Pleiades are a young open cluster of more than 3000 catalogued stars with apparent magnitudes between 3 and 17[1]. Their age, estimated to be about 50×10^6 years, is consistent with rapid rotation and enhanced surface activity[2,3]. They are probably the most extensively observed stellar association in the sky, and as such are a perennial test object for varied astronomical studies.

A number of early photometric studies have detected variability in members of the cluster[4,5], and there are many known flare stars[6]. In a recent photometric sampling of 19 randomly selected G and early K Pleiads, van Leeuwen and Alphenaar[7] found every one of them variable, in that each demonstrated a rotational modulation due to surface activity or non-axial symmetry of 5% to about 20%, with periods between 0.2 and 1.2 days. A similar photometric investigation with the now defunct Cloudcroft telescope by Radick et al.[8] and Radick et al.[9] identified at least 6 of 40 late, solar-type stars in the cluster as possible photometric variables, based on a statistical analysis of the scatter in sampled b and y magnitudes during two observing seasons. In the Cloudcroft studies the amplitude of variability was found to be an apparent function of spectral class, increasing to as much as 5% for variables later than class F, consistent with the findings of van Leeuwen and Alphenaar.

These studies suggest a search for related luminosity variations on longer time scales, both as an independent test of rotation and activity in the cluster and for what it might tell us of _secular_ luminosity variations in young stars of solar type.

We have initiated such a search using the historical plate collection of the Harvard College Observatory. Since we rely on uncalibrated photographic records we are limited to much coarser relative variations than are detectable in real-time photoelectric photometry. This is partially compensated by the possibility, in using a modern scanning microdensitometer, of sampling a far larger number of Pleiads, over at least 70 years of sky patrol.

Routine photographic coverage of the Pleiades region began at Harvard in 1885 and continued from stations in both northern and southern hemispheres at a rate of about 15 plates per year through 1952, with occasional sampling thereafter. HCO plates

of the region were exposed in several concurrent patrol programs, of varied optical
sophistication. Three patrol series are available to us: (1) a low-resolution, wide-
field series ("AC", "AM", and "CA" plates) made with a 1.5 inch, f/9 refractor, a
plate scale of 600"/mm and a limiting usable magnitude of about 11.5; (2) an intermediate resolution series ("B" and "I" plates) made with an 8 inch, f/6 system, a plate
scale of about 170"/mm, and a limiting usable magnitude of about 13.5, and (3) the
less frequently exposed, high-resolution plates of the "MC" series, made at f/5.2
with a 16 inch telescope, a plate scale of 98"/mm, and a limiting usable magnitude of
perhaps 15. For the Pleiades, the first series will reach spectral classes F and early
G, the second all of G and K, and the third, early M stars.

In this paper we report initial results from a pilot analysis of the lowest resolution series (1) for which we have concluded an initial survey of about one plate for
each available year between 1899 and 1950. In this we monitor the photographic magnitudes of about 150 Pleiads in the magnitude range from 4 to 11 (early G).

We utilize the Yale PDS microdensitometer at New Haven and our own photometric
reduction programs to convert the measured, integrated density of each identified two-
dimensional star image to a relative photographic magnitude. The latter is derived for
each identified star, as shown in Figure 1, by fitting the empirical plot of measured,
integrated density vs. the tabulated photographic magnitudes from Hertzsprung's catalog[1]. In this way we perform photometry relative to the ensemble average for the entire
set of star images, and surmount differences in plate sensitivity, exposure, development, and sky conditions.

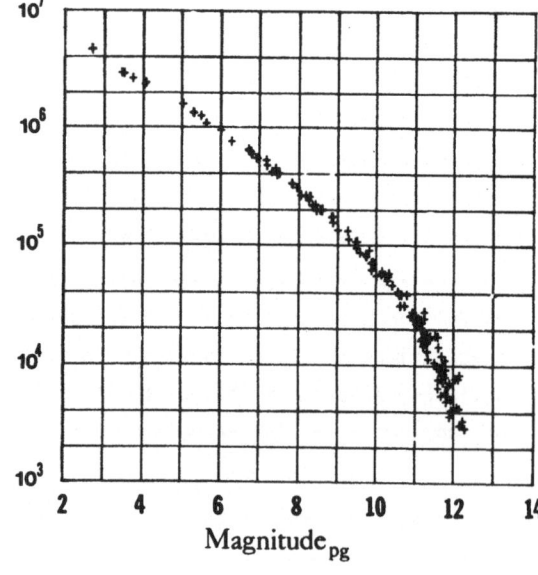

Volume
Density

Figure 1

Integrated density of two-
dimensional star images on an
AC patrol plate as a function
of the photographic magnitude
from the Hertzsprung catalog[1].
At about m=11 a practical
limit is reached.
A calibration curve is derived by fitting these empirical points with an
analytical curve.

One measure of the precision of the method is the consistency in recovering tabulated magnitudes. The average error is less than 0.05m, corresponding to a photometric accuracy of about ±5%. This estimate of accuracy is generally confirmed in subsequent plots of the temporal behavior of reference stars in the cluster.

Among the stars in the initial set is 28 Tau, Pleione, a known shell-star variable. Our survey readily recovers its known pattern of variability[10], of roughly constant brightness between 1890 and 1935, a subsequent, smooth 5-year decrease of about 0.4 magnitudes, and a partial recovery thereafter.

When variability is large and systematic as with Pleione we can recognize variability in the pattern of derived magnitudes. A more quantitative index for short-term variables, considering the random sampling of the historical plates, is the scatter about the mean. Figure 2 illustrates the secular behavior of two example stars from the initial survey: H2087 (numbered according to the 1947 Herzsprung catalog[1]), showing no variability within our limits of detection, and H2690, that gives evidence of considerable variation, with a deviation over 50 years of 0.116m.

The only star in our initial survey of these brighter, earlier Pleiads that was monitored and found variable (marginally) by Alphenaar and van Leeuwen[11] and the Cloudcroft group[9] is H727, of blue magnitude 9.7 and spectral type F9V. We would also have labelled it as possibly variable within 1-sigma limits of certainty, with a 50-year standard deviation of 0.084m.

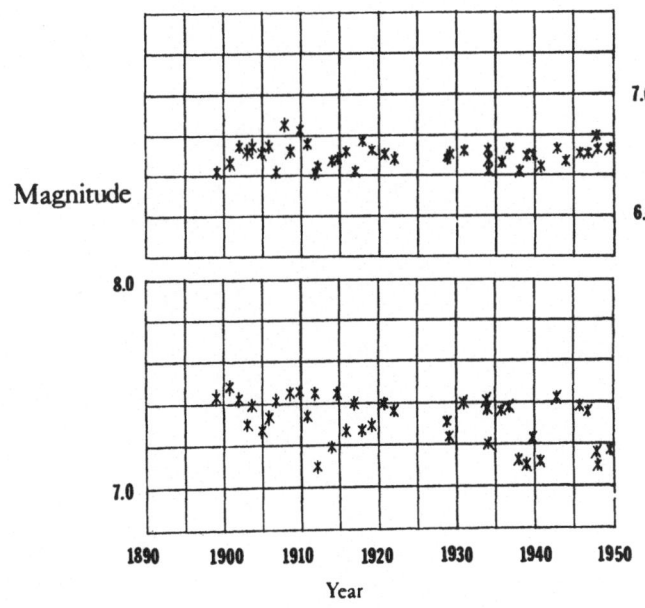

Figure 2

Derived magnitudes of two example stars for the period 1899-1949. H2087 (upper trace), a reference star, varies about the mean with a standard deviation of 0.055m, the approximate photometric accuracy of our reduction method. H2690 (lower trace, standard deviation 0.116m) exhibits a pattern expected of short-term, rotational variability.

Magnitude

7.0

6.4

8.0

7.0

1890   1900   1910   1920   1930   1940   1950

Year

Although we have not completed the analysis of all 150 stars in the initial test, we feel that the feasibility of the method has been demonstrated. Our planned analysis of the higher resolution, deeper plates of HCO class (2) promises a greater yield, since we should reach most of the Pleiads of spectral classes G and K, where variability is more commonly expected.

REFERENCES

1. Hertzsprung, E. 1947, Leiden Annalen 19, part A.
2. Skumanich, A. 1972, Ap.J., 171, 565.
3. Stauffer, J.R., Hartmann, L., Soderblom, D.R., and N. Burnham 1983, Ap.J., in press.
4. Johnson, H. and R.I. Mitchell 1958, Ap.J., 128, 31.
5. Robinson, E.L. and R.P. Kraft 1974, A.J., 79, 698.
6. Stauffer, J.R. 1983, Ap.J., submitted.
7. Van Leeuwen, F. and P. Alphenaar 1982, ESO Messenger, No. 28, 15.
8. Radick, R.R., Hartmann, L., Mihalas, D., Worden, S.P., Africano, J.L., Klimke, A., and E.T. Tyson 1982, P.A.S.P., 94, 934.
9. Radick, R.R., Lockwood, G.W., Thompson, D.T., Warnock, A. III, Hartmann, L.W., Mihalas, D., Worden, S.P., Henry, G.W., and J.M. Sherlin 1983, P.A.S.P., in press.
10. Burnham, R.Jr. 1978. Celestial Handbook, III, 1879. Dover, N.Y.
11. Alphenaar, R., and van Leeuwen, F. 1981, Inf. Bull. Variable Stars, No. 1957.

*The National Center for Atmospheric Research is sponsored by the National Science Foundation.

SOLAR FLUX VARIATIONS AND CONCEPTS FOR A STELLAR PHOTOMETRIC SATELLITE

Hugh S. Hudson
Center for Astrophysics and Space Sciences
UCSD, La Jolla, California 92093

Introduction

The flux of radiation from an astronomical object contains most of the information we are likely to get from it. The wavelength dependence of the radiation (its spectrum) gives us a remarkably rich array of clues about the nature of an object. The time dependence also contains many clues, but the study of the time variations has had severe limitations in the past both because of the lack of precision of measurement and because of the limited completeness of the time series available for study. Recent solar observations have made breakthroughs in both of these directions. This paper simply points out the significance of these observations for stellar astronomy.

The Solar Constant Observed from Space

The Solar Maximum Mission carried the first detector capable of making frequent and precise measurements of the solar constant, or in astronomical terms the bolometric magnitude of the Sun (Willson et al., 1981). The ACRIM (Activity Cavity Radiometer Irradiance Monitor) measures the flux essentially every two minutes, with a 60% duty cycle in an orbit of 96 minutes' period. These observations continued through most of 1980, during sunspot maximum, with an observed standard error per orbit of about 20 parts per million of the total flux. Cross-calibration of the three independent sensors of ACRIM has shown no long-term drift between their outputs, so the observed variations represent true solar variations on all time scales.

The measurement of total solar irradiance with this unprecedented precision and frequency has produced remarkable results. As Nature would have it, the largest variations of a few tenths of a percent lie just below the threshold of reliable ground-based photometry. Below this level, we have identified with confidence three separate classes of variation: the effects of solar active regions (sunspots predominantly), the granulation, and the global oscillations in the five-minute band. Other sources of variation are certainly present but have not been studied yet.

These different kinds of solar-constant variation produce different kinds of information about the solar interior: its structure, the nature of convection and magnetism, and the transport and storage of energy. For example, the sunspots appear to block the convective flow without the formation of an immediate "bright ring" (Hudson and Willson, 1981). This should result in the slow diffusion of the excess energy away from the site of the spot (Spruit, 1977). Instead, we observe a substantial amount of energy coming from the faculae on time scales that are intermediate between the spot lifetimes and the long thermal diffusion time predicted from the Spruit theory. Refinement of the analysis of the variations will tell us empirically about how and where the energy is stored.

The solar five-minute oscillations appear in the power spectrum of the total irradiance time series. They have amplitudes of up to a few parts per million of the total solar flux (Woodard and Hudson, 1983). The discovery of these oscillations (Claverie et al., 1979) in the Sun might be the beginning of "stellar seismology" as a probe of stellar interiors, since the Sun is observed as a star would be (without spatial resolution).

Motives for Precise Stellar Photometry and Concepts for the Observations

From the above it is clear that observations of the variations of the total solar flux are valuable if carried out frequently and precisely. Analogous observations would be equally valuable for other stars not known as variables. As for the vast variety of variable stars, there is no question about the advantages of obtaining complete time series with higher precision than that presently available. Systematic photometry with these characteristics would have innumerable applications. One typical application different from the solar variations discussed above would be the identification and measurements of low-mass companions of distant stars (Borucki et al., 1984). A Jupiter orbiting a distant Sun would produce an occultation deeper than one percent, a level very easily detectable against a background of stellar variations at the solar level. Even an Earth would be marginally detectable, so that a large range of low-mass eclipsing systems would be observed with a photometric system precise enough and pointed at the right stars.

The most direct and comprehensive means for obtaining this kind of information would be with a small satellite dedicated to photometry. Since many astronomers with a variety of objectives would use data from such a satellite, the International Ultraviolet Explorer would probably serve as a good model for an efficient program. In the following we describe two alternative concepts for a dedicated photometric satellite. Both of them would use essentially existing technology in a small spacecraft, so that costs would be relatively small.

All-sky Photometric Explorer. An array of CCD or CID cameras in a
slowly-spinning satellite could essentially survey the whole sky, with
different cameras having different choices of field of view, angular
resolution, color or polarization, etc. The Steward Observatory transit
telescope (McGraw et al., 1982), though aimed at ground-based observations of supernovae and extragalactic variables, provides some ideas
about how it might work. An All-sky Photometric Explorer would attempt
to observe the largest possible number of stars, for the longest possible
period, with the maximum achievable precision. We note that the solar-
type variations could be observed on other main-sequence stars brighter
than about tenth magnitude, based upon the photon statistics limit.

Planetarium Camera. A telescope continuously pointed in the same
direction would feed individual sensors, probably discrete silicon di-
odes, each mounted on the focal surface with its own field optics at the
right location for the stellar image. This idea was suggested by R.
Angel, J. McGraw, P. Stockman, and N. Woolf, and has an obvious advantage in that discrete detectors could much more easily be made to produce
the "micromagnitude" precision necessary for stellar seismology, for
example. On the other hand only a limited number of stars could be surveyed conveniently - perhaps of order 1,000 - and the camera would have
to be pointed stably in the proper direction. The satellite would therefore require expensive 3-axis stabilization.

These ideas are obviously very preliminary. However I think that
the objective of systematic precise photometry is of very broad significance for astronomy, and that this demonstrated fact will encourage the
development of better ground-based photometric systems and also further
thinking about the advantages to be obtained by going into space.

References

Borucki, W., Scargle, J., and Hudson, H., 1984, in preparation.
Claverie, A., Isaak, G.B., McLeod, C.P., van der Raay, H.B., and Roca
    Cortes, T., Nature, 282, 591.
Hudson, H.S., and Willson, R.C., 1981, The Physics of Sunspots (eds.
    L.E. Cram and J.H. Thomas), p. 434.
McGraw, J.T., Stockman, H.S., Angel, J.R., Epps, H., and Williams,
    J.T., SPIE, 331, 137, 1982.
Spruit, H., 1977, Solar Phys., 55, 3.
Willson, R.C., Gulkis, S., Janssen, M., Hudson, H.S., and Chapman, G.A.
    1981, Science, 211, 700.
Woodard, M., and Hudson, H., 1983, Nature (to be published).

SOLAR-CYCLE TEMPERATURE VARIATIONS
IN SUNSPOTS

P. Maltby, F. Albregtsen, and O. Kjeldseth Moe
Institute of Theoretical Astrophysics
University of Oslo, Norway
and
R. Kurucz and E. Avrett
Harvard-Smithsonian Center for Astrophysics

Abstract: The observed umbra/photosphere intensity ratio varies from the beginning to the end of each solar cycle by ~ 30% at 1.6 μm and by factors > 2 at visible wavelengths. We present the intensity ratios measured in 10 wavelength bands extending from 0.387 to 2.35 μm for 22 large sunspots observed during the period 1968-82, thus covering most of solar cycles 20 and 21. These results together with new observations of umbral limb darkening, and available data on photospheric absolute intensities, are used to estimate the dependence of the relative umbral intensity, and the absolute umbral intensity, on wavelength, heliocentric angle, and phase of the solar cycle. These umbral intensities are used to determine preliminary sunspot models which show the temperature as a function of depth in early, mid-, and late phases of the solar cycle. In the model calculations we use an extensive new compilation of atomic and molecular line data, allowing us to carry out the analysis by means of a detailed synthesis of the observed spectral bands.

1. Introduction

During the last two decades, improved observational methods and better correction methods for scattered light have led to increasingly reliable information regarding sunspot intensities. Whereas earlier measurements showed considerable disagreement between the results of different observers, the present accuracy (see Albregtsen and Maltby, 1981a for a review) allows us to investigate the causes for the relatively small changes in intensity from one sunspot to another. By limiting the discussion to large sunspots, i.e., having umbral radii > 5 seconds of arc, it appears that the umbra/photosphere intensity ratio varies with the phase of the solar cycle (Albregtsen and Maltby, 1978; 1981b) in the sense that the umbral intensity increases almost linearly with time throughout each solar half-cycle; see Figure 1. Schüssler (1980) and Yoshimura (1983) have suggested different explanations for this effect.

Our aim in this paper is to consider the time variations of umbral properties by constructing one-component umbral models for the early, mid-, and late phases of the solar cycle, taking into account the difference in limb darkening and spectral line behaviour between the umbra and the photosphere.

2. Observations and Data Reduction

The data are based on sunspot intensity measurements on approximately 600 days during the period June 1967 to December 1982. The off-axis mirror system of the tower telescope of the Oslo Solar Observatory was used in combination with two broad-band pinhole photometers giving a total of 10 wavelength regions in the spectral range 0.387-2.35 μm. By limiting the data to 22 large sunspots observed during good seeing conditions we are able to determine the umbra-photosphere intensity ratio with an accuracy of 0.015 in each wavelength region.

These results have been used by Albregtsen, Joras, and Maltby (1984) to derive the center-limb variation of the umbra/photosphere intensity ratio. In contrast to results reported by previous authors (e.g., Rödberg, 1966; Wittmann and Schröter, 1969) we find in most wavelength regions a small but significant decrease in the umbra/photosphere intensity ratio towards the limb. We suggest that improper corrections for scattered light prevented earlier detection of this decrease.

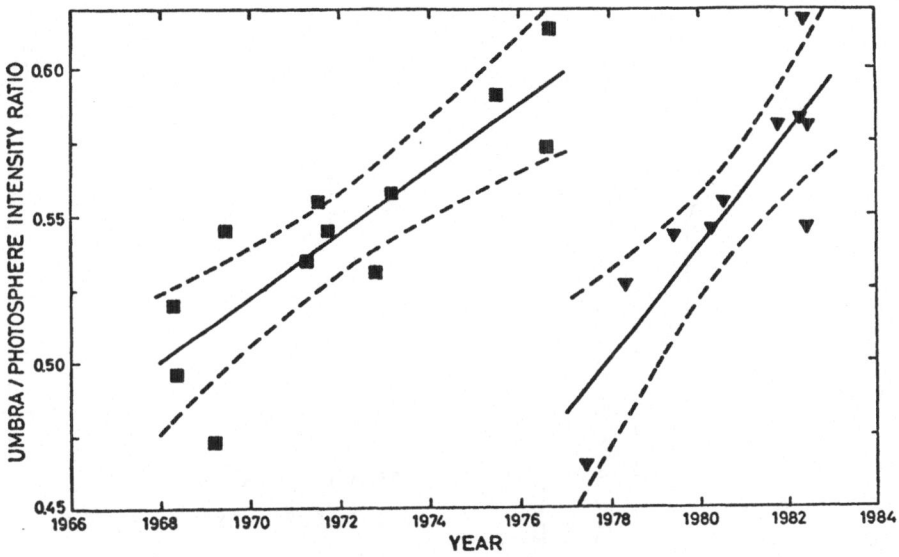

Figure 1 -- Umbra/photosphere intensity ratio at 1.67 μ m as corrected for center-limb variation, plotted as a function of time. Observations cover cycle 20 (squares) and the first part of cycle 21 (triangles). All large sunspots observed during good seeing conditions are included. Least squares regression lines are given, together with their 95% confidence limits.

Here the data are used to derive improved values for the disk-center intensity ratio as a function of phase in the solar half-cycle. The center-limb variation can be represented by the function $\phi_u(\mu) = a(\lambda)+b(\lambda)\mu$ (Albregtsen, Jorås, and Maltby, 1984). Using this observed center-limb variation we determine what the intensity of each observed spot would have been if the spot had been located at disk center. Then the time dependence can be represented by $\phi_u(\mu = 1) = c(\lambda) + d(\lambda)\, t/t_o$. Here $\phi_u(\mu = 1)$ is the umbra/photosphere intensity ratio at disk center, t is the time elapsed since the last minimum in the cycle, and t_o is the duration of the solar half-cycle. We have chosen $t/t_o = 0.1, 0.5,$ and 0.9 as representative values for the early, mid-, and late phases of the solar half-cycle. Figure 2 shows $\phi_u(\mu = 1)$ in each of the 10 wavelength regions for these values of t/t_o.

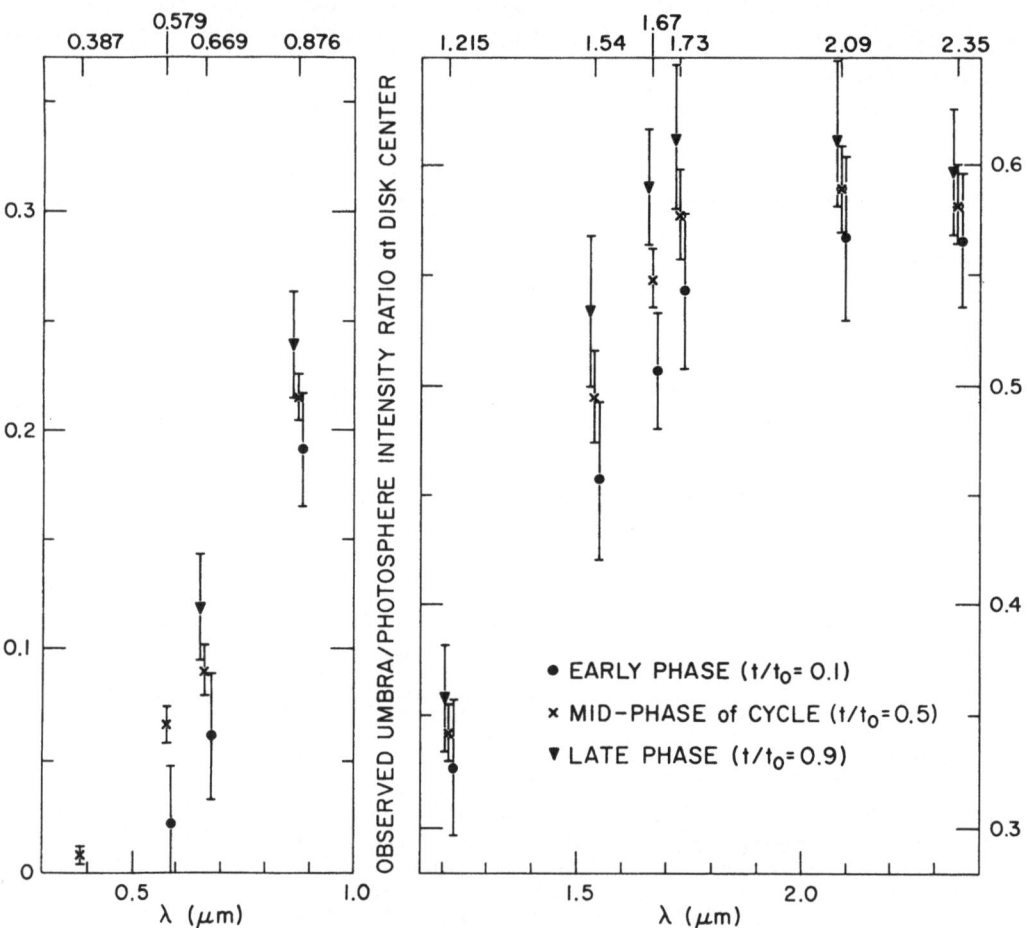

Figure 2 -- Observed umbra/photosphere intensity ratio, normalized to disk center, in 10 wavelength bands from 0.387 to 2.35 μm at early, mid-, and late phases of the solar half-cycle.

3. Model Calculations and Results

Model calculations for the umbra as well as the photosphere have been carried out both in Oslo and Cambridge. Previous discrepancies in sunspot modeling between Albregtsen and Maltby (1981a) and Avrett (1981) were caused by differences in partition functions, chemical composition, and molecular hydrogen equilibrium. These differences have now been resolved.

We have carried out model calculations 1) with continuum opacity sources alone, and 2) including the opacities from an extensive new compilation of atomic and molecular lines (preliminary results are described by Avrett and Kurucz, 1983). The calculated continuum intensities require multiplication by band-average/continuum ratios before the calculated results can be compared with the observed band averages. The spectrum-synthesis calculations require adjustments based on the extent to which the synthetic spectra properly include all observed lines. The details of these calculations will be described elsewhere.

In the model calculations we determine the umbral temperature distribution based on a fit to the wavelength dependence of the umbral intensity normalized to disk center, both on an absolute scale and relative to the photosphere.

Our preliminary conclusions are as follows:

1) The umbral temperature appears to increase by $\sim 300K$ from the beginning to the end of each solar half-cycle.

2) The umbral models have temperature gradients that lie between radiative equilibrium models and the minimum gradient model of Zwaan (1975).

3) The umbral models are consistent with the observed decrease in the umbra/photosphere intensity ratio towards the limb.

Acknowledgements: This work was sponsored in part by the Norwegian Research Council for Science and the Humanities, and by the National Aeronautics and Space Administration, Grant NSG-7054.

References

Albregtsen, F., Joras, P.B., and Maltby, P. 1984, Solar Phys., in press.
Albregtsen, F. and Maltby, P. 1981a, in The Physics of Sunspots, L. Cram and J.H. Thomas, eds., Sacramento Peak Obs., p. 127.
Albregtsen, F. and Maltby, P. 1981b, Solar Phys., 71, 269.
Avrett, E.H. 1981, in The Physics of Sunspots, L. Cram and J.H. Thomas, eds., Sacramento Peak Obs., p. 235.
Avrett, E.H. and Kurucz, R.L. 1983, Nordic Astronomy Meeting, Ø.Hauge, ed., Inst. of Theoretical Astrophys. Report, in press.
Rödberg, H. 1966, Nature, 211, 394.
Schüssler, M. 1980, Nature, 288, 150.
Wittmann, A. and Schröter, E.H. 1969, Solar Phys., 10, 357.
Yoshimura, H. 1983, Solar Phys, 87, 251.
Zwaan, C. 1975, Solar Phys., 45, 115.

HRTS ULTRAVIOLET IMAGES OF THE SOLAR CHROMOSPHERE

AND TRANSITION ZONE

K.P. Dere, J.-D.F. Bartoe and G.E. Brueckner
E. O. Hulburt Center for Space Research
Naval Research Laboratory
Washington, D.C. 20375

Observations of the Sun by the NRL High Resolution Telescope and Spectrograph (HRTS) provide a detailed picture of the fine structure and dynamics of the solar chromosphere and transition zone. The instrument obtains stigmatic photographic spectra with a resolution of 1" along a 1000" slit in the 1170-1710 Å wavelength range. During its third rocket flight, the slit was rastered across the Sun in 2" increments to produce two-dimensional arrays of line profiles in a 10" x 800" region extending from the south polar limb to near Sun center. Most of the observed features are typical of the quiet Sun. The wavelength coverage was limited to the region near 1550 Å which includes strong lines of C I, C IV, Si I, Si II and Fe II as well as the continuum. The Si I lines are used to reference a wavelength scale which is accurate to about 0.01 Å or 2 km s^{-1}. From the observed line profiles, the total line intensity, the line shift and line width are calculated and used to form images of these quantities.

The most notable structural element is the chromospheric network which delineates the boundaries of the supergranular cells. In general, the continuum and chromospheric and transition zone lines are all bright in the network. On a smaller scale, Hα observations show that the network consists of assemblies of dark mottles. The HRTS images of the Fe II intensities mimic the dark Hα structures as do the C I intensities to a lesser extent. The continuum, which is associated with the temperature minimum region, is formed at the base of the mottles. The C IV intensities outline elongated (3"-5") transition zones that are extensions of the mottles. Pressures typical of the quiet Sun indicate that the volume occupied by the transition zone is quite small. The most likely structure is that of a thin filament with a diameter of less than 100 km. Such extended structures with their low temperature gradients are clearly incompatible with current one-dimensional energy balance models which are summarized by the Rosner, Tucker and Vaiana scaling law. The low temperature gradients along the magnetic field lines indicate a lack of conductive flux to compensate the transition zone radiative losses so that some form of local heating is necessary.

Based on a wavelength scale derived from the narrow Si I lines, the C IV lines have a net average redshift of 0.027 Å (5 km s^{-1}). Downflows of up to 20 km s^{-1} are often observed. Since the C IV flow velocities show no correlation with the C IV intensities, the normal supergranular flow patterns seen in the chromosphere are not present at transition zone levels. Small transition zone explosions with velocities on the order of ±100 km s^{-1} are commonly found and occur isotropically over the solar surface. The most prominent chromospheric Doppler shifts are seen in the cell centers in the C I lines. Upflows and downflows of 10 to 20 km s^{-1} are seen in small arc-second areas. Doppler shifts in the network are much smaller (<3 km s^{-1}).

THE NSO INITIATIVE TO PURSUE THE SOLAR-STELLAR CONNECTION
ON THE McMATH SOLAR TELESCOPE

Myron A. Smith and David B. Jaksha
National Solar Observatory, P.O. Box 26732, Tucson, AZ 85726

I. Introduction

The main purpose of this talk is to acquaint participants of this conference and their students of the initiative the new National Solar Observatory is taking to make available the McMath Solar Telescope at Kitt Peak for fully scheduled nighttime observations which pertain to "solar-stellar-connection" astronomy. This facility is intended to be mainly, though not exclusively, devoted to high resolution spectroscopy. By committing a substantial part of the scheduled time to synoptic programs carried out largely by resident observers, we hope to guarantee that programs requiring a number of observations over an extended period of time can finally be carried out at a national facility. This option does not yet exist, even to astronomers having excellent (read: competitive) observing facilities elsewhere. We fully expect that the availability of the McMath for routine nighttime synoptic spectroscopy will allow solar-stellar inquiry to expand into entirely new fields over the next several years.

II. Instrumental Layout

Stellar spectroscopy will be carried out in the near future at the old spectroheliograph ("SHG") port of the McMath telescope, which is some 15 feet south of the main solar spectrograph. Further down the line, a decision will be made to remain at this location or to move our operation to the main spectrograph, or even to a new "stable stellar spectrograph", designed and under review by Richard Dunn. Our spectrograph is vertically situated and looks up at an f/54 beam reflected from the third mirror of the McMath. This beam encounters, in turn, a wide-field TV guiding system, image stablizer, a local field TV-guiding aperture plate, image slicer, and order-sorting filter, before encountering the spectrograph proper. While this may sound like a bewildering arrangement, each piece of equipment serves a necessary function for nighttime work with this telescope, and as long as properly aligned, the system permits a relatively high throughput. We are now building a flatfield/comparison assembly, which will illuminate the front of the image stabilizer, so that the beam will closely mimic the stellar beam seen at the image slicer.

The image slicer is composed of five 400 μm by 2mm slices, which are stacked end to end; the slicer projects onto 5" by 5" on the sky. A system of transfer optics in the slicer assembly changes the optical beam to the f-ratio of the spectrograph, f/24. The spectrograph is a 1:1 system comprised of 20 cm collimator and camera mirror with focal lengths of 455 cm. A grating turret houses two gratings, which can be pivoted conveniently around the turret center. We now have available a Bausch and Lomb cervit-backed grating blazed at 1.2 μm, with θ = 45.2° and a groove spacing of 1200 gr./mm. Our second grating is a 79 gr./mm., tan θ = 2 echelle. Order sorting is accomplished with colored or interference filters. The camera mirror focuses the beam on a large field lens, which is located in the plane of the image slicer. The light is then diverted horizontally by a flat mirror. Light is finally refocused onto the detector by one of a number of commercial Nikon lenses we have available. The spectroscopic resolution of this system is defined by the slicer widths, and is 3 to 5 x 10^4 for the Bausch and Lomb grating, and 7 to 9 x 10^4 for the echelle. Our detector is a copy of the highly successful system designed for the coudé spectrograph of Lick Observatory (Vogt 1981). It utilizes a "bare" Reticon RL 1872F/30 chip with a single row of 1872 pixels, of dimensions 15 μm by 750 μm. Our system has a measured noise of 560 electron/hole pairs per readout. We are restructuring our data transmission system to allow for multiple dark reading following the stellar integration; this procedure will effectively cancel residual noise resulting from a "dark" subtraction. The array operates between -110°C and -155°C, and we believe it will be viable throughout the spectral region between the K-line and λ10830.

Any bare Reticon is of course optimized for high illuminance, i.e. high signal-to-noise observations of bright stars. For moderate S/N values, e.g., ~100, other contemporary detectors are faster, but do not necessarily have the 1-inch long field and/or the small pixels that this detector has. For the future, we contemplate getting a second detector more suited to lower S/N and lower resolution observations of fainter stars.

III. Performance

As an astrophysical illustration, Figure 1 is a section of the spectrum around the Li I λ6707 wavelength, of the G0 Ib star β Aqr. The upper limit for our nondetection of this feature in this star agrees with Luck's (1977) results using the Reticon system of the 107-

inch telescope of McDonald Observatory, and places this star among the
"low-lithium fringe" of G-type supergiants. To make this observation
we utilized the B & L grating in order 2 with a 200mm Nikon lens.
With this configuration, the measured spectroscopic resolution is 4.4
x 10^4 and the wavelength sampling is 35.6m A per pixel.

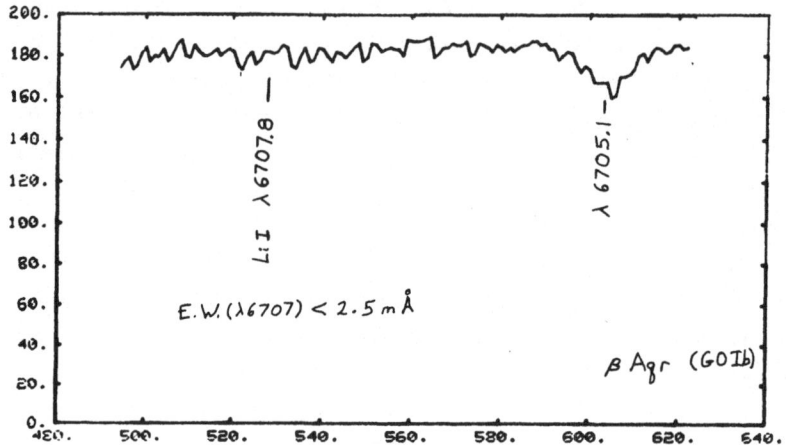

Figure 1.   Li λ6707 line region in β Aqr.

From a few stellar spectra that are suitable for such purposes,
we have attempted to make a speed comparison with the Reticon systems
at the McDonald 82-inch and 107-inch systems. Our checks show that
our system is already nearly comparable in speed to these two Texas
systems, with our image slicer and superior detector noise
compensating for our smaller effective telescope aperture (60
inches). We expect an additional speed gain of a factor of 2 or so to
result from improvements in our optical alignments and from the
multiple-dark readout procedure. We expect to have much more
information available about the system's speed by the end of this
fall.

IV.   Future Plans

We are now soliciting observing proposals for nighttime programs
that can be carried out by guest investigators during the first two
quarters of 1984. A full time observer will be hired on April 1st, in
addition to an already present half-time observer to assist

investigators during their runs or to supplement their programs after they leave. The primary purpose of this shakedown phase of our operation will be to exercise the system and explore its capabilities in magnitude limit, S/N, resolution, wavelength range, and wavelength stability. Observing proposals for the first quarter are due by the end of this month and should be sent to Myron Smith; the second quarter deadline will be January 15th.

In the second half of the year we will evaluate the full capabilities of our spectroscopic system and we will accept the first proposals for truly synoptic programs. These will generally run a year or longer. The ground rules for this mode of operation, which can be expected eventually to take up the majority of the scheduled observing time, have not yet been firmed up. However, it is likely that we will encourage such programs to originate from more than one investigator in order to ensure their long term continuity. We may also ask investigators to observe occasionally themselves for these programs, say, once a year to alleviate our own manpower constraints. Otherwise, we intend to utilize full time observers to carry out these programs remotely, in a mode similar to Cloudcroft Observatory's conducting of photometry programs over recent years.

References

Luck, R. E. 1977, Ap. J., 218, 752.
Vogt, S. S. 1981, Soc. Photo Optical Instr. Engineers, 290, 70.

LABORATORY IDENTIFICATION OF THE $^3P_{1,2} - {}^5S^o_2$ INTERSYSTEM LINES OF S III.

Peter L. Smith,[1] C.E. Magnusson,[2] and P.O. Zetterberg[2]

(1) Harvard-Smithsonian Center for Astrophysics
(2) Physics Department, Lund University

The spin-changing, $3s^2 3p^2 \ ^3P_{2,1} - 3s3p^3 \ ^5S^o_2$, "intersystem" lines of S III (S^{2+}) were identified recently for the first time from low-resolution, IUE, ultraviolet spectra of emission from the Io torus (Moos et al. 1983). The two wavelengths were determined to within 0.6 Å and 1.0 Å. No triplet-quintet intersystem lines of S III had been found prior to this astronomical discovery. The transition prob-abilities, collision strengths, and therefore, intensities of these lines could be expected to be comparable to those of other intersystem lines currently used in determinations of electron density and tempera-ture in astrophysical plasmas. Therefore, laboratory spectra were studied in order to provide accurate wavelengths for these lines.

The spectra were obtained at the Atomic Spectroscopy Department of Lund University as part of a continuing program of study of the atomic spectra of the elements, in particular of those with $Z \lesssim 30$. The sliding-spark light source, spectrograph, and experimental method were similar to those of Magnusson and Zetterberg (1974) and Pettersson (1983).

The laboratory spectra showed two lines, at 1728.939 and 1713.117 Å. The wavelengths were measured on ten spectra; the uncer-tainty is estimated to be ± 5 mÅ at the 95% level of confidence. The S III ground term $^3P_1 - {}^3P_2$ splitting obtained from the laboratory wavelengths of the intersystem lines is 534.2 ± 0.3 cm^{-1}. Lines in the resonance multiplets connecting to the $3s3p^3 \ ^3D^o$, $3s^2 3p3d \ ^3P^o$, and $3s^2 3p4s \ ^3P^o$ terms were also used to measure this value; a least squares calculation for the triplet system lines gave 534.5 ± 0.3 cm^{-1}. Both values agree with those obtained from observations of the 18.71 μm $^3P_1 - {}^3P_2$ fine structure line in nebulae by Baluteau et al. (1976) and Greenberg, Dyal, and Geballe (1977) and with an unpublished value of 534.45 ± 0.10 cm^{-1} obtained from spectroscopic study of the resonance line multiplets by V. Kaufman at the U.S. National Bureau of Standards.

The line-list for the NRL solar spectrum from Skylab shows an unidentified line at 1728.94 ± 0.010 Å, in exact coincidence with the stronger member of the S III] doublet (Doschek et al. 1976). The second S III] line at 1713.117 Å is blended with a strong Fe II line at 1712.997 $\pm$ 0.010 Å (Cohen 1981). The stronger S III] line also coincides with an unidentified line at 1828.91 $\pm$ 0.05 Å in the line list for the slow nova, RR Tel (Penston et al. 1983). Nussbaumer (1983) reports that the S III] doublet is seen with the expected separation in another spectrum of RR Tel and in one of V 1016 Cyg; the 1713.1 Å component appears to be a barely resolvable blend.

A more detailed description of this work will be published in the Astrophysical Journal, Letters to the Editor.

The authors gratefully thank many colleagues for advice and for communicating results prior to publication. This work was supported in part by NASA Grants NSG-7304 and NGL 22-007-006 to Harvard College and by the Swedish Natural Science Research Council.

References

Baluteau, J.-P., Bussoletti, E., Anderegg, M. and Moorwood, A.F.M. 1976, Astrophys. J. 210, L45.

Cohen, L. 1981, An Atlas of Solar Spectra Between 1175 and 1950 Å Recorded on Skylab with the NRL's Apollo Telescope Mount Experiment, NASA Reference Publication 1069.

Doschek, G.A., Feldman, U., Van Hoosier, M.E., and Bartoe, J.-D.F. 1976, Astrophys. J. Suppl. 31, 417.

Greenberg, L.T., Dyal, P. and Geballe, T.R. 1977, Astrophys. J. 213, L71.

Herter, T., Briotta, D.A., Jr., Gull, G.E., Shure, M.A., and Houck, J.R. 1982, Astrophys. J. (Letters) 259, L109.

Magnusson, C.E. and Zetterberg, P.O. 1974, Physica Scripta 10, 177.

Moos, H.W., Durrance, S.T., Skinner, T.E., Feldman, P.D., Bertaux, J.L. and Festou, M.C. 1983, Astrophys. J. (Letters) 275, in press.

Nussbaumer, H. 1983, personal communication.

Penston, M.V., Benvenuti, P., Cassatella, A., Heck, A., Selvelli, P., Macchetto, F., Ponz, D., Jordan, C., Cramer, N., Rufener, F. and Manfroid, J. 1983, Mon. Not. R. Astr. Soc. 202, 833.

Pettersson, J.E. 1983, Physica Scripta, in press.

TRANSITION REGION - CORONAL MODELS: A COMPARISON WITH XUV OBSERVATIONS

B.C. Monsignori Fossi, M. Landini
Arcetri Astrophysical Observatory
F. Paresce
Space Telescope Science Institute
R.A. Stern
Lockheed Palo Alto Research Laboratory

In the last ten years a lot of accurate observations of X-ray and Far-UV emission from stars have been possible owing to the instruments on board of IUE and Einstein satellites. In order to exploit in the best way, the detailed informations contained in many IUE and Einstein observations and to get profit in the future on the observations of the EXOSAT, ROSAT and EUVE instruments, we have evaluated the spectral distribution from thin plasmas from 1 to 2000 Å, which takes into account the more recent progress on atomic data and uses an homogeneous computation all over the spectrum (Landini and Monsignori Fossi, 1983). We believe useful to use this computation:

1) to investigate a detailed modelling of coronal structures starting from models already known for the Sun;

2) to evaluate, from the observing data, the main parameters which define the structure of the transition region-corona in stars;

3) to produce the spectral energy distribution emitted from some stars in order to show the agreement within the spectral range already observed and to give a prediction in those spectral bands where no observation is now available.

With this aims in mind we have computed the total power emitted in a spectral band $\Delta\lambda$:

$$I = \int_{\Delta\lambda} \int_{V} \Phi(\lambda,T) \; (n_e^2 \; dv) \; d\lambda \tag{1}$$

where $\Phi(\lambda,T)$ is the spectral emissivity computed following Landini and Monsignori Fossi (1983) and $n_e^2 \; dv$ is the differential emission measure which strictly depends by the transition region-coronal model.

The transition region corona of a magnetic loop structure is computed evaluating stationary solutions of the mass, momentum and energy balance equations. Convenction, radiative losses and conduction from a fully ionized plasma are included together with an heating function which may be defined as input data. Radiative losses are calculated numerically according to the computed spectrum. Loops are assumed to be shaped in semicircular form with constant section (s) and that the structure is in hydrostatic equilibrium (gravity component along the loop is considered); the heating function is assumed to be uniform (Landini and Monsignori Fossi, 1982). The main features of the model may thus be described via the temperature (T_{cor}) at top of the loop, the pressure (P_O) at the base, where $T_O = 2 \; 10^4$ K is assumed, and the full length L of the loop. The elementary volume in eq. (1) is given by $dv = s \cdot dz$ where z is the position coordinate along the loop, it is noted that the section s will be proportional to the star surface via filling factor (ff). The integration of the model gives density (pressure) and temperature profiles.

In order to produce the total emitted power I observed from a star we must select three parameters: the coronal temperature (T_{cor}), the base pressure (P_O) of the magnetic loop and the filling factor (ff). But these three parameters cannot be uniquely determined: the observed emitted power I is roughly proportional to $P_O \cdot ff$ and we can select, for each value of coronal temperature, only the value of $P_O \cdot ff$ which reproduces the observations.

Generally (i.e. Golub et al., 1982) one other constraint needs to separate P_O

and ff. In the following we discuss the solutions which correspond to a given variation from the base pressure (P_0) to the top pressure (P_{cor}) of the loop [i.e. any value of $\ln = (P_0/P_{cor})$ defines the fraction of pressure scale height corresponding to the length of the loop (cfr. Landini et al., 1983)]. For each model, defined by a coronal temperature and the corresponding base pressure, the predicted line intensity I is evaluated and the comparison with the observed emission gives the proper value of the filling factor.

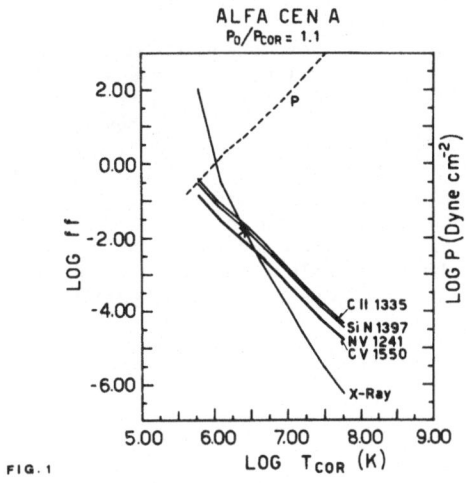

In Fig. 1 for the star α Cen A we show the allowed selection of ff and P_0 values that reproduces observations in the case $P_0/P_{cor} = 1.1$ (model 1): the dashed line indicates for each coronal temperature (T_{cor}) the base pressure (P_0) which produces model with 10% of pressure variation. The observational data are taken from Ayres et al. (1982).

We have also investigated models with other pressure ratios; for instance $P_0/P_{cor} = 2.7$ (i.e. the half loop length L/2 has the same magnitude order of the pressure scale height) (model 2). In Table 1 the results for the two different models are given for stars of different level of activity.

FIG. 1

Table 1

star	T_{cor}	P_0	ff	L	E_i	T_{cor}	P_0	ff	L	E_i
		Model 1						Model 2		
Alfa Cen A	2.5E6	4.9	.015	2.8E9	1.3E-2	3.8E6	1.1	.056	4.4E10	1.6E-4
BD 15:640	5.0E6	1.3E1	.13	7.6E9	1.6E-2	7.0E6	2.2	.8	1.4E11	1.3E-4
BD 14:693	2.0E7	2.7E2	.006	2.5E10	1.8E-1	2.5E7	14.9	.11	1.2E11	1.6E-4
HD 165590	1.9E7	2.5E2	.05	2.4E10	1.6E-1	2.4E7	14.5	.9	9.8E11	1.4E-4

Moreover for α Cen A we have looked for the largest differences that the two above models generate in the emitted surface fluxes (erg $cm^{-3}s^{-1}A^{-1}$) in selected bands of 1 A ; the ratio between the emissivity produced with model 2 (I_2), and the other one (I_1) is shown in Table 2.

Table 2

Band(A)	Ions	$\log I_2/I_1$	Band(A)	Ions	$\log I_2/I_1$
3.5–4.5	S 15–16	2.38	16.5–17.5	Fe 17	0.91
4.5–5.5	S 15;Si 13	2.24	17.5–18.5	O 7	1.12
5.5–6.5	Si 13–14	1.76	169.5–170.5	Ni 14	0.62
6.5–7.5	Si 13	1.61	283.5–284.5	Fe 15	1.06
7.5–8.5	MG 11–12	1.19	334.5–335.5	Fe 16	1.11
8.5–9.5	Mg 11	1.03	360.5–361.5	Fe 16	0.98
14.5–15.5	Fe 17–18–19	1.14			

α Cen A

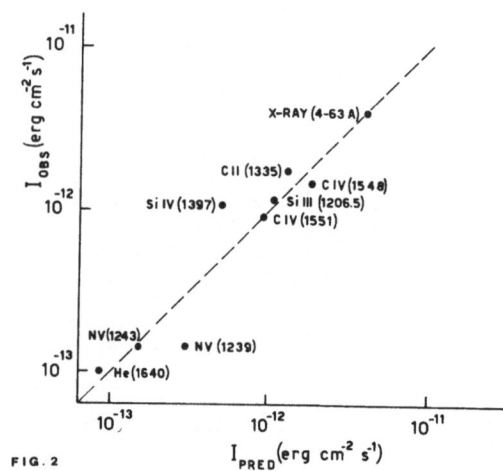

FIG. 2

In Fig. 2 for α Cen A the surface fluxes observed in selected UV lines with IUE and X-ray region with the Imaging Proportional Counter (IPC) of Einstein Observatory are plotted versus the predicted one by the models. Fig. 2 refers to model 1, but model 2 reproduces these data with the same accuracy.

In Fig. 3 for α Cen A the surface fluxes distribution in the spectral region 1-2000 A from the transition region-corona is shown; they are predicted with the model $P_0/P_{cor} = 1.1$, $P_0 = 4.9$ (dyne/cm^{-2}), $T_{cor} = 2.5 \ 10^6$ K, $L = 2.8 \ 10^9$ cm, and ff = 0.015. Also absorption from interstellar medium is taken into account. The atomic hydrogen column density has been evaluated assuming a mean particle density of $7 \ 10^{-2}$ (cm^{-3} pc^{-1}).

FIG. 3

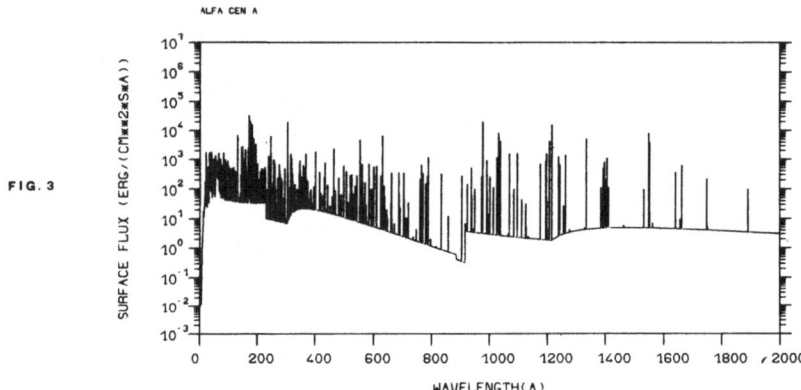

This model analysis produces also the power which is necessary to supply (column E_i of Table 1) in the assumption of uniform energy distribution along the loop. This information may help in studying the heating mechanism when the procedure is applied to stars of different spectral type and evolutionary stage.

REFERENCES

Ayres, T.R., Simon, T. and Linsky, J.L. 1982, Ap.J. 263, 791.
Golub, L., Harnden, F.R. Jr., Pallavicini, R., Rosner, R. and Vaiana, G.S. 1982, Ap.J. 253, 242.
Landini, M. and Monsignori Fossi, B.C. 1982, Astron. Astrophys. 102, 391.
Landini, M. and Monsignori Fossi, B.C. 1983, Physica Scripta, In press.
Landini, M., Monsignori Fossi, B.C., Paresce, F. and Stern, R.A. 1983, Ap.J., submitted.

X-RAY TRANSIENTS AND FLARES FROM ACTIVE STELLAR SYSTEMS

Carol Ambruster and Kent S. Wood
E.O Hulburt Center for Space Research
Naval Research Laboratory
Washington, D.C. 20375

Many X-ray flares detected to date from active cool stars were initially discovered as fast ($\lesssim$ 3 hour) transients[1,2,3]. A survey of the HEAO A-1 Sky Survey Experiment 0.5-20 keV data for fast transients confirms the tendency for a significant fraction of such events to be associated with active stellar systems. Eight months of the mission (which include one full scan of the sky) were searched, yielding 10 transients of which three are identified with flares on cool stars. Of the remaining sources[4], all but one[5] are at least consistent in their X-ray properties with a stellar flare origin.

We discuss below the three fast transients which we have identified with active stellar systems. They are EV Lac, a dM4.5e flare star (d = 5 pc)[6], HD 8357, an RS CVn type binary (d $\sim$ 70 pc)[7], and EQ Vir (BD-7°3646), a dK5 flare star (d = 16 pc)[8].

DISCUSSION

The identification with EQ Vir is the result of a computer search of approximately 30 catalogs for counterparts to a relatively faint (3.3 Uhuru Flux Units (UFU)) transient event on 13 January 1978. The flare star is the only likely X-ray source in the $0°.2 \times 8°$ line of position error box. The absence of detectable quiescent emission in either one or four day summations of X-ray counts is consistent with the observed quiescent luminosities from dMe stars[7,9] and the distance of EQ Vir. The flux from the transient event converts to an X-ray luminosity, L_x, of 7.4×10^{30} ergs s^{-1}, in good agreement with the range of 10^{27}-10^{31} ergs s^{-1} observed for flares from dMe stars[3,7,10].

A complete search of the A-1 data for the winter 1977 observations of EV Lac and HD 8357 revealed a second, smaller flare in both cases, as well as variable quiescent emission. The A-1 light curves for these two sources are shown in Figures 1 and 2; they are the first to show unambiguously more than one flare plus intervening quiescent emission from active stellar systems. Important supplementary observations were contributed by the HEAO A-2 experiment, which detected a $\sim$ 50 UFU source (comparable in flux to Flare 2 in Figure 1) at the EV Lac position 56 minutes prior to our Flare 1[11], and the HEAO A-3 experiment, which saw the second, brighter, flare from HD 8357 at roughly the same intensity as the A-1 detection, but one hour later[12]. The smaller flare from EV Lac, therefore, may signify either flare decay or some form of low level, post flare activity. Post-peak emission is also found after the second EV Lac flare, where low level (∿ 4 UFU) detections are seen three and four hours after flare peak.

The luminosities, L_x, of the two EV Lac flares (which occurred on 25 and 30 December 1977) are $2.9 \pm 0.6 \times 10^{29}$ and $2.2 \pm 0.1 \times 10^{30}$ ergs s^{-1}, with a mean quiescent luminosity, L_q, of ∿ 4×10^{28} ergs s^{-1}. The luminosities for the two HD 8357 flares (which occurred on 11 and 13 January 1978) are $7.2 \pm 1.2 \times 10^{31}$ and $4.0 \pm 0.2 \times 10^{32}$ ergs s^{-1}. For both stars, these luminosities lead to L_{peak}/L_q ∿ 50 for at least one of the flares, the largest seen from either class of star. Further data on these two stars, as well as an instrument description, are given elsewhere[7].

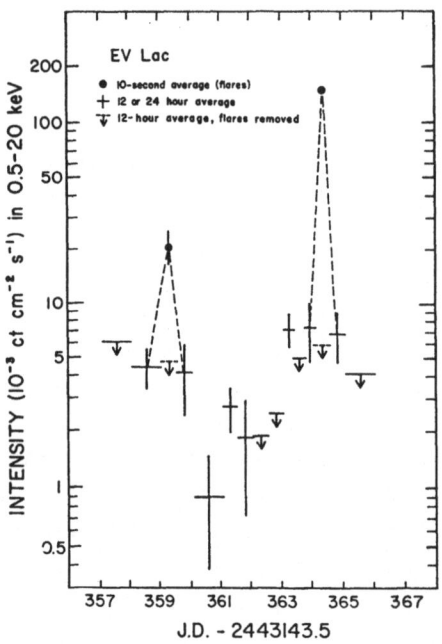

Figure 1. 0.5-20 keV light curve curve for EV Lac.

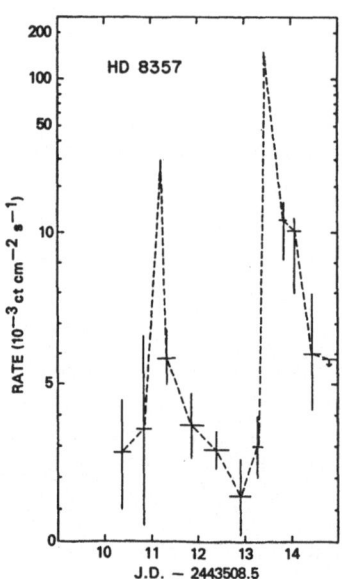

Figure 2. 0.5-20 keV light curve for HD 8357. The quiescent flux data points represent 12 hour data summations.

The light curves shown in Figures 1 and 2 permit us to calculate the time averaged flaring luminosity, $\langle L_f \rangle$, for the ∽ one week period of observation of the two stars. The scanning properties of the A-1 experiment are such that we would expect all major flare events lasting longer than about one hour to have been detected. Since emission from (or related to) three of the four flares in Figures 1 and 2 was seen lasting two and four hours, we consider it unlikely that any significant flaring activity was missed for either object; that is, the A-1 flare coverage was, to first order, continuous. We calculate the average flaring luminosity by: $\langle L_f \rangle = (\Sigma L_f \tau_f)/T$, where the τ_f are decay timescales and T the total observing time. For EV Lac we find $\langle L_f \rangle = 2 \times 10^{28}$ ergs s^{-1}, or a factor of 10 higher than the optical U-band value of 1.55×10^{27} ergs s^{-1} [13]. For HD 8357, the X-ray $\langle L_f \rangle = 5 \times 10^{30}$ ergs s^{-1}; there is no optical value for comparison.

Finally, we find the ratio $\langle L_f \rangle/L_q$ ∽ 0.5 for both EV Lac and HD 8357. (We estimate this ratio, as well as the results for $\langle L_f \rangle$, to be good to a factor of a few.) The statistical distribution of flares in time, whether Poisson distributed or grouped, has not yet been conclusively determined[13,14], nor has the effect on flaring rates of possible long term stellar cycles. Nevertheless, a similar value for $\langle L_f \rangle/L_q$ suggests at least some overlap in the relative amount of magnetic energy release going into large scale flaring as opposed to quiescent emission in these two stars. Yet a dMe star and RS CVn binary differ significantly in spectral type, temperature, mass and evolutionary phase. If the X-ray $\langle L_f \rangle/L_q$ ratio is found to be ∽ 1 for other active stellar systems, this places important constraints on flare models for convective, late type stars.

REFERENCES

1. Schwartz, D.A., et al. 1981, MNRAS 196, 95.
2. White, N.E., Sanford, P.W., and Weiler, E.J. 1978, Nature 274, 569.
3. Kahn, S.M., et al. 1979, Ap.J. 234, L107.
4. Ambruster, C., and Wood, K.S. 1983, Proceedings of the Santa Cruz Summer Workshop on High Energy Transients, in press.
5. Ambruster, C., et al. 1983, Ap.J. 269, 779.
6. van de Kamp, P., and Worth, M.D. 1972, A.J. 77, 762.
7. Ambruster, C., Snyder, W.A., and Wood, K.S. 1983, submitted to Ap.J.
8. Pettersen, B.J. 1976, Catalog of Flare Star Data, Report No.46, Inst. of Theor. Astrophysics, Blindern, Oslo.
9. Johnson, H.M. 1981, Ap.J. 243, 234.
10. Haisch, B.M., et al. 1980, Ap.J. 240, L107.
11. J.H. Swank, private communication.
12. Garcia, M., et al. 1980, Ap.J. 240, L107.
13. Lacy, C.H., Moffett, T.J., and Evans, D.S. 1976, Ap.J. Supp. 30, 85.
14. Pazzani, V. and Rodono, M. 1981, Ap & Space Sci. 77, 347.

LONG-TERM BEHAVIOR OF STELLAR FLARE ACTIVITY IN AD LEO

Bjørn R. Pettersen
McDonald Observatory and Department of Astronomy
University of Texas at Austin, Austin, TX 78712.

INTRODUCTION

AD Leo is a dM4e solar neighborhood red dwarf on the main sequence. It has an effective temperature of 3500 K, radius of 0.44 $R_\odot$ and a bolometric luminosity of 0.024 $L_\odot$ (Pettersen 1980). The mass-luminosity relation (Pettersen 1983) implies a mass of 0.44 $M_\odot$. Vogt et al.(1983) determined v(rot)sin i=5.0 km s^{-1} from high-resolution spectroscopy. Photometry of starspots indicates a rotation period of 2.7 days (Sandmann, private comm.) implying that the rotation axis of AD Leo is inclined 38° to the line of sight.

FLARE ACTIVITY

Four per cent of the U-filter flux received from the star is due to flare activity (Pettersen et al.1983). Observations since 1971 (Pettersen et al.1983 and references therein, Panov et al.1982) have been collected to investigate the long term behavior of flare activity in AD Leo from 1971 to 1982. Cumulative flare energy distributions for each observing season were used to avoid detection effects, by determining the smallest flare energy for which the flare detection is no longer complete. Flares larger than 10^{30} erg fullfill this requirement, and all flares larger than this limit has been detected by the observers.

RESULTS

Flares on AD Leo are Poisson distributed with time (Pettersen et al.1983). The observed number of flares per unit of time(N/T) above the detection limit thus has a formal uncertainty of $\sqrt{N}$/T. Table 1 gives pertinent data for each observing season in the first four columns. The flare frequency and its error estimate is plotted versus time in Fig. 1. The average flare frequency for the entire data set is 0.53 flares per hour, based on 255.3 hours of U-filter monitoring with 135 flares larger than

Table 1: Flares with U-filter energy above 10^{30} erg.

Observing season	N	T(hours)	N/T ± √N/T	E_i	C_i^2
Jan-Feb 1971	7	16.828	0.42 ± 0.16	8.9	0.41
Jan-Apr 1973	18	89.91	0.20 ± 0.05	47.7	18.49
Jan-Feb 1974	21	21.116	0.99 ± 0.22	11.2	8.58
Nov 1974	14	19.227	0.73 ± 0.19	10.2	1.42
Feb-Apr 1979	20	23.008	0.87 ± 0.19	12.2	4.99
Apr-May 1979	19	32.967	0.58 ± 0.13	17.5	0.13
Oct 1979-Jan 1980	20	34.299	0.58 ± 0.13	18.2	0.18
Feb-Mar 1982	16	17.945	0.89 ± 0.22	9.5	0.03

10^{30} erg, of which 101 were observed at McDonald Observatory. To determine if variability is present in Fig. 1 we compare the observed flare
frequency to the null hypothesis that the flare frequency remained
constant during the 11 year interval. The expected number of flares
within each observing season, using 0.53 flares/hour, is given in column
5 of Table 1. The statistical test parameter in column 6 is defined by

$$\Sigma C^2 = \sum_{i=1}^{m} \frac{(N_i - E_i)^2}{E_i}$$

where N_i is the number of flares actually observed during observing
season i, and E_i is the corresponding expected number of flares from the
null hypothesis. A large value of the test parameter is found, and the
χ^2-test concludes that variability is indeed present. However, most of
the contribution to the large value comes from the unexpectedly low
number of flares observed in 1973 by Indian, Italian and Japanese observers.
If this data point is excluded the test parameter reduces its value by
54%, and the case for variability is reduced correspondingly.

Left with this situation we can only recommend that more observations
be made over the next few years, with the intention of both increasing
the time base line and improving the statistics. The seasonal number
of observed flares should be increased in order to lower the formal
errors of the flare frequency.

ACKNOWLEDGEMENTS

It is a pleasure to acknowledge interesting discussions with Drs. David
S. Evans and William H. Sandmann. Financial support was made available
by the Norwegian Research Council for Science and the Humanities, and
by a senior Fulbright award.

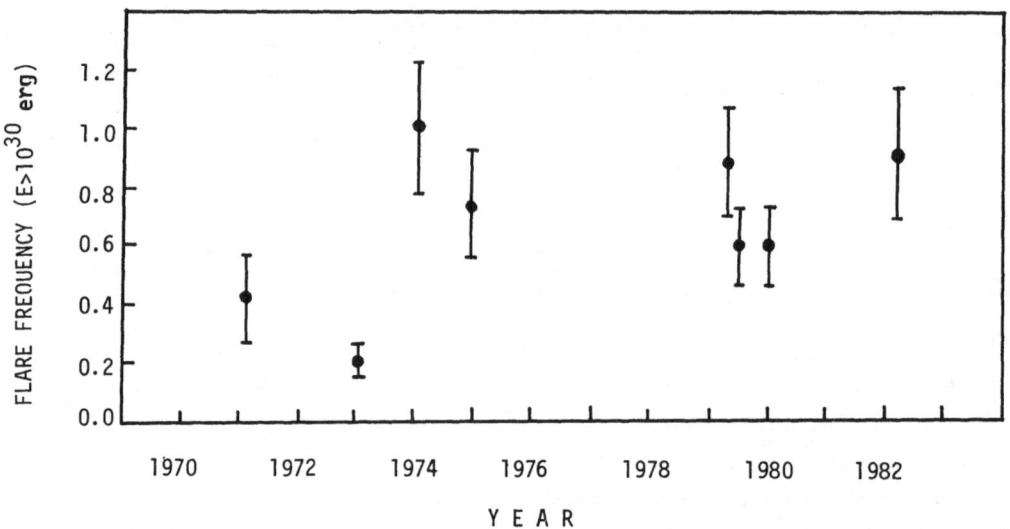

Fig. 1. The flare frequency of AD Leo as a function of time.

REFERENCES

Panov, K. P., Grigorova, M., Tsintsarova, A., 1982, Inf. Bull. Var. Stars No. 2220.

Pettersen, B. R., 1980, Astr. Ap. 82, 53.

Pettersen, B. R., 1983, IAU Colloq. No. 71, Activity in Red Dwarf Stars (Dordrecht:Reidel) eds. P. B. Byrne and M. Rodono, p. 17.

Pettersen, B. R., Coleman, L. A., Evans, D. S., 1983, Ap. J. suppl., in press.

Vogt, S. S., Soderblom, D. R., Penrod, G. D., 1983, Ap. J. 269, 250.

RADIO OBSERVATIONS OF ACTIVE STARS: DIRECT EVIDENCE FOR POLARITY REVERSALS?

D. M. Gibson[1]

Joint Institute for Laboratory Astrophysics, University of Colorado and
National Bureau of Standards, Boulder, Colorado 80309

INTRODUCTION

Radio astronomical measurements of source spectra and polarization can provide significant insight into the physical conditions, e.g. the electron density, n_e, the magnetic field, B, and the particle energy, γ ($=E/m_oc^2$), in the emitting regions. Such observations of active stars, e.g. the RS CVn binaries and dMe flare stars, can be particularly useful because they provide information on coronal conditions which is different from, but complementary to, that obtained by X-ray observations. Observations at other bands (UV, optical, IR) allow us to "connect" the corona to the stellar surface and, in effect, provide additional boundary conditions for our interpretations.

In evaluating models for stellar radio emission one has to 1) take into account the implications the model has for radio emission that may arise from other mechanisms, 2) check to see whether the source physics and structure are consistent with observations at other bands. In this paper, I evaluate the homogeneous gyrosynchrotron model for stellar radio emission proposed by Owen et al. (1976; hereafter called the OJG model) and find it difficult to reconcile with models for the coherent radio emission from the same stars. Furthermore, I find the statistics of stellar radio polarization to be inconsistent with this model as well. Finally I suggest that a phenomenological model based on the "leading-following spot" (LFS) geometry seen in active regions (AR) on the Sun serves to explain most of the peculiarities observed in radio star spectra and polarizations and, in fact, provides an explanation for similar peculiarities seen in the rapidly varying coherent emission.

PROBLEMS WITH THE OJG MODEL

Currently there are two classes of radio emission models that are used to explain what appear to be two very different types of observed radiation.

In the first, the emission is slow-varying (timescales ~tens of minutes to hours) mildly circularly polarized ($\pi_c \leq 20\%$) and broadband. The spectrum may "turn over" in the microwave region. (In fact, there exists only one set of observations by Hjellming and Brown (see Hjellming and Gibson 1980) that shows the latter; the spectra of UX Ari and HR 1099 turn over at about 5 GHz.) These observations have generally been interpreted (cf. Owen et al. 1976) as self-absorbed gyro-synchrotron emission wherein the source is becoming optically thick at the longer wavelengths.

---

[1] JILA Visiting Fellow (1983-1984); on leave from Dept. of Physics, New Mexico Tech., Socorro, NM.

The fractional polarization at the shorter wavelengths would indicate the energy of the emitting electrons ($\gamma \approx 1/\pi_c \approx 5\text{-}10$). The OJG model assumes the B-field is homogeneous and isotropic; the inferred field strengths are $B \sim 30$ G. Similar analyses by Spangler (1977) argue for 10^8 cm$^{-3} < n_e < 10^9$ cm^{-3} in the emitting region.

The second type of emission is highly variable (timescales $\lesssim$ few seconds), highly circularly polarized ($\pi_c \sim 100\%$), and quite possibly monochromatic. Melrose and Dulk (1982) argue that this coherent emission can be explained by an electron-cyclotron maser operating at the second harmonic of the gyrofrequency. Here the inferred $B \sim 600$ G.

A. Inconsistency with Coherent Emission Processes

From the point of view of the models alone the two emission processes may be inconsistent. The optical depth of the coronae as inferred from the incoherent source model is $\tau \gtrsim 1$ and the effective source size is large. This would markedly diminish the flux arising from the presumed lower levels where the higher B's are characteristic. In other words, the large, low-density, low-field corona could totally absorb the coherent emission.

Whether or not one subscribes to the electron-cyclotron maser model, it seems certain that the highly-variable, highly-circularly-polarized emission is coherent. Virtually all coherent emission models produce emission at the fundamental or first few harmonics of the gyrofrequency $\nu_B = 2.8 \times 10^6$ B Hz or the plasma frequency $\nu_p = 9 \times 10^3$ $n_e^{1/2}$ Hz or some combination of the two. These conditions imply for coherent emission at ~ 5 GHz that $600 < B < 1800$ and/or $3 \times 10^{10} < n_e < 3 \times 10^{11}$, respectively. With what is known about active stars from X-ray, UV, and optical observations (cf. papers in Activity in Red-Dwarf Stars, IAU Colloq. #71), it is difficult to believe that the above conditions occur at large distances from the stellar surfaces. Therefore, the observation that coherent emission arises from low levels and is not smothered by the incoherent emission region suggests the latter probably arises in a similar region. The temporal association of the two types of emission, i.e. we usually see highly-circularly-polarized emission at the beginning of flaring episodes or at times when it appears that re-energization is taking place, suggests the emitting particles and therefore the emitting regions are substantially the same.

B. Curiosities in the Circular Polarization (CP)

If one examines the handedness of CP of active stars, one finds the following trends:

1) Those stars that exhibit substantial CP in their incoherent emission tend to be seen more pole-on, whereas those that are substantially unpolarized are seen equator-on. Equator-on stars include the eclipsing binaries Algol, AR Lac, RT Lac, and YY Gem. Pole-on stars include the spectroscopic binaries UX Ari, HR 1099, and 39 Cet and the single stars UV Ceti (L726-8B), L726-8A, and YZ CMi.

2) The polarized stars show no particular preference for the hand of CP in which they exhibit their activity. For example, HR 1099, UV Ceti, and L726-8A have always been observed to be right-hand circularly polarized (RCP) while UX Ari and 39 Ceti have only exhibited substantial LCP. YZ CMi was originally observed to be LCP (cf. Fisher and Gibson 1982), but more recently most of its activity has been in RCP (Gibson 1983).

3) When both coherent and incoherent emission are seen from a given star the hand of polarization has been the same. The only exception I know of is YZ CMi (Gibson 1983).

Trend (2) is the difficult one to understand in terms of the OJG model. If active stars are like the Sun, and from all indications they are, then the strong magnetic fields arise in active regions where the net magnetic flux is _zero_. There is as much upward- as downward-directed flux. Thus, on the whole, if AR's look like simple dipoles, a large number of them added together will have the same effect as the "homogeneous, tangled field" assumed by OJG. In such a situation, the RCP would dominate the left because of propagation effects when the source _appears_ optically thin, i.e. at high frequencies. As shown in trend (2), there is no such effect. In fact, the source cannot be treated as homogeneous since critical parameters such as B and n_e are functions of height. In addition, the idea that the source becomes optically thin (throughout) is not applicable since it must become optically thick at some height.

A "LEADING/FOLLOWING SPOT" MODEL

The discussion above hints at one possible solution to the dilemma posed by having a large, homogeneous, low-field source which neither reproduces the observed polarization properties nor allows radiation that is almost certainly produced at lower levels to escape. The solution is to adopt a new source geometry, in particular one with enough of an asymmetry to reproduce the observed spectral and polarization properties. One such geometry is offered to us by the Sun and is based on the "leading/following spot" configuration seen in AR's (see Fig. 1). It has been known for some time (Stenflo 1968) that the fractional photospheric area covered by the polarity characteristic of the leading spots is substantially smaller than that covered by the following polarity. However, since flux is conserved this means the strength of the B-field in and above leading spots is larger. This asymmetry has the following effects:

1) The source becomes optically thick for a given frequency at higher altitudes above the leading spot compared to the following spot, since the principal absorption mechanism is gyroresonance absorption. The local brightness temperature will increase with height regardless of the emission mechanism as long as the source is optically thick.

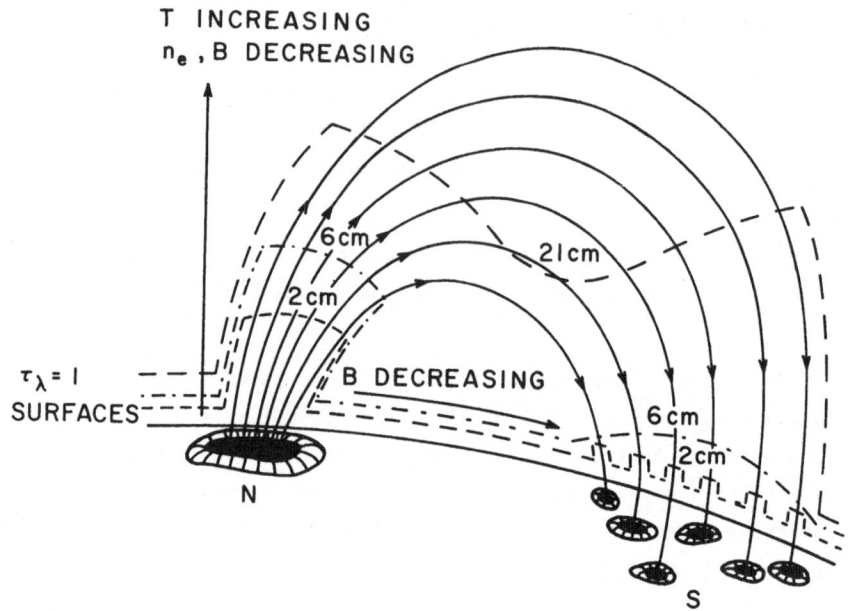

Fig. 1. A typical "loop" structure found in the Sun and similar stars. The radio emission must arise above the $\tau_\lambda = 1$ surface appropriate to that wavelength. The most intense emission at any wavelength will arise in the region of strong magnetic field.

2) The effective size (cross-sectional area) of the emission region will increase with increasing wavelength. This is primarily due to the "ballooning" effect in the magnetic field once it has risen above the chromosphere. The "effective" source spectrum for the entire AR will depend on the ratio of effective areas and emissivities for the source at those wavelengths.

3) The percent polarization at a particular wavelength will depend on the ratio of the effective areas appropriate to each polarity of the B-field and their emissivities. There will be a tendency for the source to be brighter above the outward directed field (because of transfer effects), but this could be a minor effect compared to the asymmetry as shown in the figure. In particular, this asymmetry could even be responsible for a change in the hand of circular polarization mimicking the effect seen in the OJG model.

4) We would no longer "know" the energy of the emitting particles, i.e., the relation $\gamma \approx 1/\pi_c$ would not be valid, since π_c would be predominantly a function of the source geometry. The presumed "incoherent" emission could, in fact, arise from the sum of lots of little coherent sources.

The LFS model would also have an interesting effect on the hand of CP expected from the electron-cyclotron maser model (Melrose and Dulk 1982). As fast particles stream back and forth along the loop(s) they will be "mirrored" more easily above the leading spot and will "precipitate" more easily over the following spots. The higher concentration of upward moving particles above the leading spot (and probably a longer coherence length) will favor masing above leading spots. This emission would share the same hand of CP as the more intense hand of CP of the incoherent emission, in agreement with what is observed.

SECULAR CHANGES IN POLARITY: IMPLICATIONS AND EVIDENCE

The LFS model would have a most intriguing effect if it is correct. The observed CP of a star would be an integrated effect of all of its AR's. Since one would expect the number of AR's in the northern magnetic hemisphere to equal the number of AR's in the southern magnetic hemisphere, one would expect to see significant CP only from those stars that are tilted with respect to our line of sight, e.g., tidally-locked spectroscopic binaries. This seems to be observed. Equator-on systems would not be expected to exhibit significant CP, nor do they.

If other stars undergo polarity reversals as part of their stellar cycle, as does the Sun, then it may be possible to see these reversals in the hand of CP of the microwave emission. Has such a reversal been observed? Possibly. Only the stars mentioned before have had observations reported on their CP. Algol was observed heavily from 1974-1976. UX Ari and HR 1099 have been observed a number of times since 1976. AR Lac has been observed since 1978, and UV Cet and YZ CMi since 1979. The others only have recent observations. Only YZ CMi seems to have undergone a CP reversal. If believable, and if all types of active stars have similar cycle times, the present statistics favor rather long cycle times ($\gtrsim$10-15 yr). However, much more systematic polarization and spectral monitoring of active stars must be done before evidence for polarity cycles will become convincing.

REFERENCES

Fisher, P. L. and Gibson, D. M. 1982, in Second Cambridge Workshop on Cool Stars, Stellar Systems, and the Sun (SAO Spec. Rept. 392), eds. M. S. Giampapa and L. Golub, Vol. 2, p. 109.
Gibson, D. M. 1983, Proc. Southwest Reg. Conf. Astron. Astrophys., IX, in press.
Hjellming, R. M. and Gibson, D. M. 1980, in Radio Physics of the Sun, IAU Symp. 86, eds. M. R. Kundu and T. E. Gergely (Dordrecht: Reidel), p. 209.
Melrose, D. B. and Dulk, G. A. 1982, Ap. J., 259, 844.
Owen, F. N., Jones, T. W., and Gibson, D. M. 1976, Ap. J. (Lett.), 210, L27.
Spangler, S. R. 1977, A. J., 82, 169.
Stenflo, J. O. 1968, in Structure and Development of Solar Active Regions, IAU Symp. 35, ed. H. Kiepenheuer (Dordrecht: Reidel), p. 47.

CAPELLA HL

Thomas R. Ayres

Laboratory for Atmospheric and Space Physics,
University of Colorado, Boulder, Colorado 80309

I. Introduction

Capella HL (=Gliese 195AB: dM2+dM4) is a little-known visual pair (a ≅ 50 AU; P ≅ 400 years) that is a physical companion to the well-known spectroscopic binary Capella (α Aurigae A: G6 III + F9 III) (Heintz 1975). The separation of the two systems is about 10^4 AU and the orbital period is tens of thousands of years.

Both Capella and Capella HL have space velocities which are parallel to that of the nucleus of the nearby Hyades cluster, and the evolutionary age of the yellow giants of Capella, 3.5×10^8 million years, is comparable to the cluster turnoff time. Therefore, Capella HL represents the nearest of the red dwarfs that with certainty can be considered a bona fide member of the Hyades moving group. Accordingly, Capella HL is an example of a star that can be used to explore the relationships among age, rotation, and chromospheric activity for stars of the lower main sequence, without the usual bias of magnitude-limited surveys of young galactic clusters. It is hoped that such studies will clarify the mechanisms responsible for the generation, spatial concentration and dissipation of the surface magnetic fields believed to be at the heart of stellar "activity".

II. Observations

Johnson (1983) has compared *Einstein* x-ray fluxes of nearby red dwarf stars having well-determined space velocities. A subset of Johnson's list, including Capella HL and the well-known flare star UV Ceti, share the space motion of the Hyades nucleus. Capella HL is comparable in x-ray luminosity to the other M dwarfs in the 'Hyades' group, while the mean L_x of the group is intermediate between the brightest of the red dwarf coronal sources--the young disk stars AU Mic and AT Mic-- and the faintest of the red dwarfs detected in the 0.1-4 KeV band.

In addition to the *Einstein* measurement of Capella HL, low dispersion (5Å), ultraviolet (1150-3000 Å) spectra of the binary have been obtained with the International Ultraviolet Explorer. A catalog of the observations is provided in Table 1 and the ultraviolet spectrum is illustrated in Figure 1.

In Figure 2, the C IV and Mg II fluxes of Capella HL are compared with UV measurements of other M dwarfs from the survey of Linsky et al. (1982).

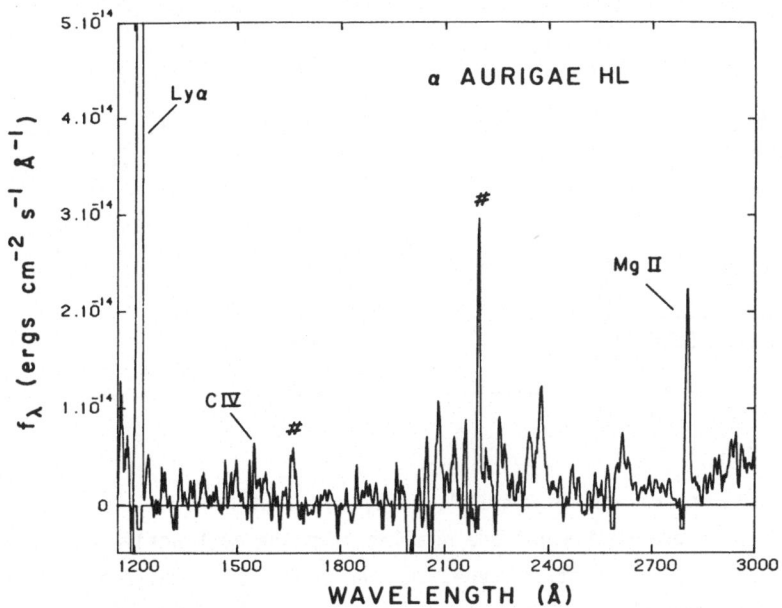

FIGURE 1. Ultraviolet spectrum of Capella HL based on 75-minute LWR and 320-minute SWP exposures. Prominent cosmic ray "hits" are marked "#", and the HI Lyα feature is mostly geocoronal emission. Aside from the strong Mg II λ2800 emission, few lines can be identified reliably, although a 2.5σ feature appears at the wavelength of the C IV doublet (λ1550).

FIGURE 2. A C IV - Mg II correlation diagram for red dwarfs. The emission line intensities have been normalized to the bolometric fluxes of each star. The shaded curve depicts the power law dependence (with slope 1.5) exhibited by the analogous fluxes of G and K dwarfs from the survey by Ayres, Marstad, and Linsky (1981). The red dwarfs might obey a similar relationship, but displaced upwards. The error bars are ±1σ, and the two pairs of measurements of Capella HL are indicated.

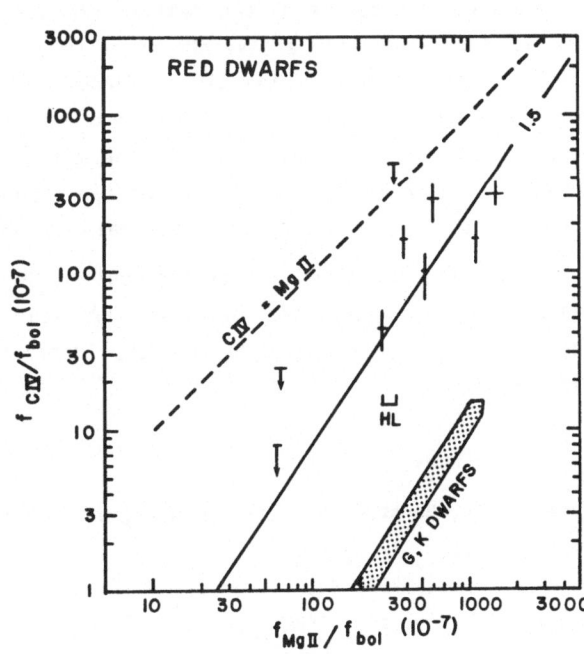

Table 1. Catalog of IUE Observations

Image No.[1]	Date (1983)	Exposure time (minutes)	Line flux $\pm 1\sigma$ (10^{-13} ergs cm^{-2} s^{-1})
LWR 15688	9 April	20	4.1±0.6 (Mg II)
SWP 19686	9 April	30	<6 (3σ) (C IV)
LWR 16734	4 September	75	3.3±0.2 (Mg II)
SWP 20926	4 September	320	0.5±0.2 (C IV)

[1] LWR= 2000-3000Å; SWP= 1150-2000Å

III. Discussion

Capella HL, at an age of 3.5×10^8 years, has luminosities in soft x-rays, C IV and Mg II up to an order of magnitude smaller than the most active of the young disk field stars of similar spectral type, but an order of magnitude larger than the least active of the old disk red dwarfs that have been observed by Einstein and IUE.

Therefore, Capella HL exhibits an intermediate level of UV and x-ray emission, qualitatively consistent with the several studies that have demonstrated a systematic decline of chromospheric and coronal activity with advancing age among main sequence stars. In fact, the x-ray and UV emission levels of Capella HL probably are more representative of the typical red dwarfs of the Hyades main sequence than are the few exceedingly active dMe stars detected in the existing magnitude-limited survey of the nuclear region of the Hyades (Stern et al. 1981). Accordingly, systems like Capella-Capella HL, which contain proxy tracers of stellar age, can provide an alternative way to explore the age-rotation-activity relations until the advent of x-ray and UV instruments of far greater sensitivity than Einstein or IUE permits the direct study of the lower main sequences of nearby open clusters.

This work was undertaken at the Boulder IUE RDAF, operated under contract NAS5-26409, and was supported by grant NAG5-199 from NASA to the University of Colorado. I thank E.W. Brugel and C.L. Imhoff for their assistance with the observations.

IV. References

Ayres, T.R., Marstad, N.C., and Linsky, J.L., 1981, Ap. J., 247, 585.
Heintz, W.D., 1975, Ap. J., 175, 411.
Johnson, H.M., 1983, Ap. J., in press (15 October).
Linsky, J.L., et al., 1982, Ap. J., 260, 670.
Stern, R.A., et al., 1981, Ap. J., 249, 647.

LITHIUM DEPLETION IN SOLAR-TYPE STARS

David R. Soderblom
Harvard-Smithsonian Center for Astrophysics
60 Garden Street, Cambridge MA 02138

The theme of this conference is, essentially, stars with surface convective zones. Most of the phenomena discussed here, such as chromospheric and coronal activity, or angular momentum loss, occur because these stars - representing a broad range of temperature and mass - have surface convection. But despite the fundamental importance of convection for what we study, we know little of it, except in general terms.

Lithium depletion in late-type dwarfs (and lithium dilution in giants) offers one of the few direct manifestations of convection; the Li abundances of stars can tell us about the physical conditions at the base of the convective zone. For this reason, a detailed investigation of the Li abundances of late-type stars is appropriate. The existing observations reveal the broad trends of Li depletion, but most of the data are not good enough to address detailed questions. Recent advances in detector technology, especially the Reticon and CCD, enable us to obtain spectra that are ideal for Li studies. And better analysis techniques, such as spectrum synthesis (e.g., Lambert, Dominy, and Sivertsen 1980), mean that better results are obtained from the observations.

Thus I feel that an examination of the theory of Li depletion is premature until observers have made their best efforts to get the best data possible. We have at hand the means to observationally restrict the processes that can produce the depletion that we see. When combined with the stellar structure information that we expect to gain from stellar seismology, this can lead to a genuine understanding of convection in late-type stars. for discussions of the theory of Li depletion, see Schatzman (1969, 1977), Vauclair (1973), Straus, Blake, and Schramm (1976), and Vauclair *et al*. (1978).

I shall describe the observations that exist, emphasizing the most recent since previous reviews have also treated this subject (Wallerstein and Conti 1969; Boesgaard 1977). I shall dwell on a few subjects in particular to point out some of the problems encountered in studying Li. Before doing so, let me begin by defining a convenient abbreviation. I shall discuss logarithmic Li abundances by number, on a scale where $\log N(H) = 12.00$, and I will use "Li" to mean $\log N(Li)$.

1. Basic Problems in Studying Lithium Abundances

For most elements, one can partially compensate for the inaccuracies of measuring individual spectral lines by constructing a curve-of-growth (COG) from many lines. For Li, we can observe only one feature, a resonance doublet of the neutral atom at 6708 Å (although a secondary feature at 6103 Å is occasionally seen in very cool supergiants). At least this feature is in a relatively uncrowded spectral region, well suited to observations of cool stars with red sensitive detectors. But this line is frequently weak and contaminated with CN features in older, Li-depleted stars, or saturated in younger stars (if stronger than 50-100 mÅ), so that modest errors in the equivalent width (W_λ) lead to larger errors in the abundance. In stars later than about G5V, Li is rarely seen, but if it's present, it's often confused with an Fe I line about 0.3 Å blueward. Thus to obtain even a reliable W_λ, high resolution spectra with good signal-to-noise are needed. For an illustration of some of these problems, see Figure 1 of Cayrel *et al.* (1983).

Given W_λ, we wish to analyze it to determine the Li abundance. The Li atom is simple, so that the gf values are known, and the thermal velocity of Li is about 3.7 km s^{-1} for the Sun, so microturbulence is unimportant. But the ionization potential of Li is only 5.39 eV, so most of the Li is in the form of Li II, which is essentially invisible. Thus the derived abundance is sensitive to the star's temperature and thermal structure. Although non-LTE effects don't seem to be important (Müller, Peytremann, and de la Reza 1975; Duncan 1981), the choice of model atmosphere does affect the results. Many of these problems are minimized by reducing and analyzing observations in a consistent way, but these problems must be remembered when comparing the results of different observers. (Unfortunately, estimates of Li or Li/Ca are sometimes published instead of W_λ's, so that these observations cannot be reanalyzed.) Even if systematic effects exist in the absolute abundances, we can at least compare stars to each other if the data are analyzed consistently, which is mostly what we want to do anyway. The most recent

COG's are those of Duncan and Jones (1983).

Giampapa (1984) has recently pointed out that a star with an active chromosphere, like a young solar-type star, may not show its true Li abundance because of changes in the star's atmosphere. His solar observations are not conclusive, but do show that significant effects may occur. It is not clear in what sense the apparent Li abundance should be changed, since plages show less of a Li feature, due to greater photoionization (Giampapa 1984), but spots show enhanced Li (Traub and Roesler 1971), due to their being cooler. An observing program is currently underway to test the importance of a chromosphere on the apparent Li abundance by monitoring an active star over a rotation period to see if changes in the Li line strength occur, and, if they do, whether those changes are in- or out-of-phase with the chromospheric variations.

If we've managed to get a good abundance, we wish to interpret it: what does it say about the star? In particular, Li abundances have been used to estimate ages for solar-type stars by calculating the depletion time scales from cluster observations (Duncan 1981; Soderblom 1983a). This time scale (τ) is *very* sensitive to stellar mass ($\tau \propto M^3$, Soderblom and Dravins 1984), and appears to depend on metallicity as well (Spite and Spite 1982a). Thus Li is mostly useful as a qualitative age indicator, not a quantitative one.

2. Basic Lithium Observations - What We Know With Certainty

a) The Sun has very little Li; Li = 1.0 ± 0.1 (Müller, Peytremann, and de la Reza 1975). Only a few rare earths are less abundant by number, none by mass. Although different observers have obtained different W_λ's and abundances over the years (see Grevesse 1968), the observations of Brault and Müller (1975), and the above analysis, are definitive.

b) The material from which the Sun formed has lots of Li. Carbonaceous chondrites, thought to represent primordial solar system material, have Li = 3.34 ± 0.07 (Nichiporuk and Moore 1974).

c) The interstellar medium (ISM) also has lots of Li. As for stars, determining the ionization balance is crucial, but there is the additional complication of depletion onto grains. Previous determinations (summarized in Duncan (1981) and Soderblom 1983a)) were at or below the meteoric abundance, hence consistent with it. White (1983) has recently obtained CCD observations in several lines-of-sight, and gets Li = 3.4 everywhere, to within a factor of two.

d) Very young stars also have lots of Li. T Tauris are difficult
to observe and analyze because they are faint, show spectral veiling,
and have uncertain T_{eff}'s and inhomogeneous atmospheres. But clearly
T Tauri's have $Li \simeq 3.0$ (Bonsack and Greenstein 1960; Zappala 1972),
essentially the primordial abundance.

e) Stars in young clusters have significant Li abundances as well.
The abundance decreases with decreasing mass or advancing age (Herbig
1965; Zappala 1972).

f) Field stars show a range of Li abundances (Herbig 1965), but
none are greater than the meteoric abundance, and most are less. Li is
rarely seen in stars later than G5V. Li abundances appear to decline
with age, but there are some old, metal-deficient stars that have preserved their Li (Spite and Spite 1982a), as well as some F dwarfs that
have prematurely depleted theirs (Duncan 1981; Soderblom 1983a). The
latest dwarf in which Li is seen is Gliese 182, dM0.5e (Bopp 1974;
de la Reza, Torres, and Busko 1981).

g) These remarks refer to the ^{7}Li isotope. In the Sun and the ISM,
$R \equiv {}^6Li/^7Li \lesssim 0.04$ (Traub and Roesler 1971; Ferlet 1982), while $R = 0.08$
for the earth and meteorites (Krankowsky and Müller 1967), and $R \lesssim 0.1$
in stars (Cohen 1972). This isotope ratio has no direct bearing on Li
depletion, but has important consequences for the cosmic source of Li.
There is not the space here to discuss this fascinating subject, so the
reader is referred to, e.g., Audouze and Tinsley (1974) and Spite and
Spite (1982b). It is worth noting, however, that this low abundance of
^{6}Li rules out the possibility that Li is created by cosmic ray spallation
in the ISM, despite the fact that this process accounts for the observed
Be and B isotopic abundances (Meneguzzi, Audouze, and Reeves 1971).
The apparent uniformity of Li abundances everywhere argues against highly
localized sources (Duncan 1981). Novae are a promising source (Starr-
field et al. 1978).

We can confidently draw these conclusions from this data:

a) The Sun has definitely depleted Li during its main sequence lifetime, because the material from which the Sun formed is rich in Li. Thus
we can rule out the possibility that young stars appear to have lots of
Li because the Galactic Li content has increased by two orders of magnitude over the Sun's lifetime. We can also rule out the autogenic hypothesis, in which young stars *form* Li at a high rate, but all stars deplete it quickly (in a convective turnover time, perhaps). Although
some stars may form Li in their atmospheres (Canal 1974), it would be

hard to reconcile that process with the observed uniform abundance. Also, more ^{6}Li than ^{7}Li is produced in solar flares (Hultqvist 1977).

b) The Li abundance of proto-stellar material hasn't changed much in 4 Gyr - if anything it may have gone down slightly. The overall metallicity of the Galaxy hasn't changed much during this time either; one must go further back to find stars with [Fe/H] < -0.5 (Twarog 1980). This could mean that Li production and the formation of heavy elements are related, but probably not. One reason is that stars with [Fe/H] ≃ -2 don't have Li 2 dex less than the meteoric abundance. Also, we are confident that we know how these elements are produced, at least in general terms; Li and Fe arise in very different ways.

c) Li depletion is probably related to convection in late-type dwarfs. It has been suggested that the observed decline in Li abundance with age could occur if stellar winds carry away enough surface material to cause dilution of the remainder with material that was previously below the convective zone. However, one would need to remove $\gtrsim 10^{-2}$ $M_\odot$ in ~ 1 Gyr, or ~10^{-10} $M_\odot$ yr^{-1} to reproduce the observed Li depletion of solar-type stars, four orders of magnitude greater than the present solar wind mass flux. This mass flux would have to be even larger for less massive stars to bring about their accelerated Li depletion. If we assume that Li depletion is related to convection, we can at least understand qualitatively the observed dependence of the depletion time scale on mass and metallicity, even if we do not understand the details.

3. Observations of Stars in Clusters

A cluster is presumably a population of uniform composition and age (although we might like to test these assumptions). Even without converting W_λ's to Li, or colors to temperatures, the scatter about the mean relation, and comparisons between clusters, tell us much about Li depletion.

Based on the existing data, we postulate the following model. Stars are born with Li ≃ 3.4. Some depletion occurs during pre-main-sequence (PMS) contraction, but only by about a factor of two at 1.0 $M_\odot$ (Duncan 1981). Main sequence depletion is very sensitive to mass, and also depends somewhat on the metallicity; the lower the stellar mass, the faster the depletion. The convective zone (CZ) of a late-type star changes little during its main sequence lifetime, so that Li depletion occurs at a constant relative rate, i.e., $Li \propto \exp(-t/\tau)$.

Given this model, we expect the following for high quality data for a group of uniform composition and age. There should be a one-to-one correspondence between color (or spectral type) and mass, so we expect a smooth, tight relationship between Li and color. If there is more spread than can be attributed to observational error, then one of our assumptions must be invalid. In particular, such a spread would probably be interpreted as being due to non-coeval star formation, since variations in metallicity within a cluster are unlikely.

Since CZ's grow thinner as we progress up the main sequence, and disappear somewhere near T_{eff} = 7000 K (Baker 1963), we expect F dwarfs in clusters and the field to show little or no Li depletion, except that which can be attributed to the PMS phase.

If we examine clusters of the same age, we expect the same run of Li with color. Differences in Li in the more massive stars would suggest unequal initial Li abundances, while differences in Li among the less massive stars would presumably be due to different depletion rates (perhaps due to different compositions) or different cluster ages.

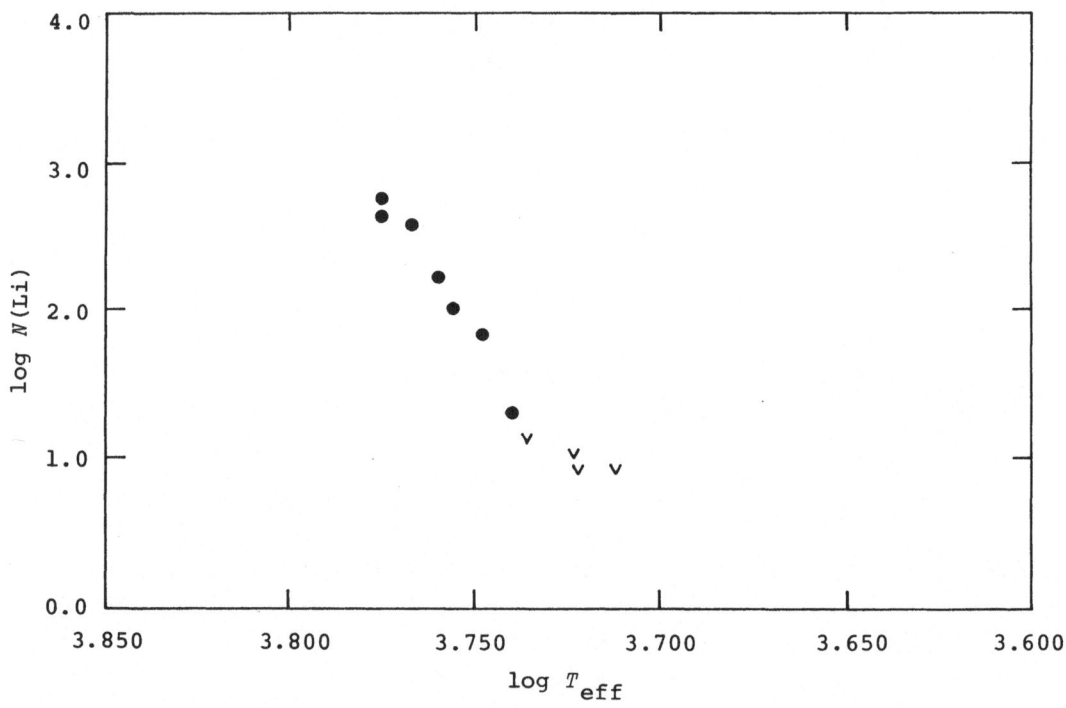

Figure 1. Lithium abundance versus temperature for Hyades dwarfs. The equivalent widths are from Cayrel *et al*. (1983), analyzed with the curves-of-growth of Duncan and Jones (1983).

a. The Hyades, Coma, and Praesepe

The Hyades' membership is well determined, and the stars are fairly bright, therefore it is the best studied cluster (Wallerstein, Herbig, and Conti [WHC] 1965; Zappala 1972; Duncan and Jones 1983; Cayrel *et al.* 1983; also Duncan 1981; Soderblom 1983*a*). Cayrel *et al.* (1983) have recently used a Reticon on a high dispersion coudé camera to observe several late-type Hyads. For consistency, I have reanalyzed their W_λ's using the COG's of Duncan and Jones (1983) and the same T_{eff}'s I used in reanalyzing Zappala's (1972) data (Soderblom 1983*a*). These reanalyzed Li's don't differ significantly from those given by Cayrel *et al.* Features as weak as 2 mÅ were reported by Cayrel *et al.*, but I have conservatively taken 10 mÅ as their detection threshold.

These data of Cayrel *et al.* are shown in Figure 1. The slight scatter that exists about the mean relation can be attributed to the observations (in both coordinates). Both the run of Li with T_{eff} and the scatter are very different from WHC or Zappala. These latter curves

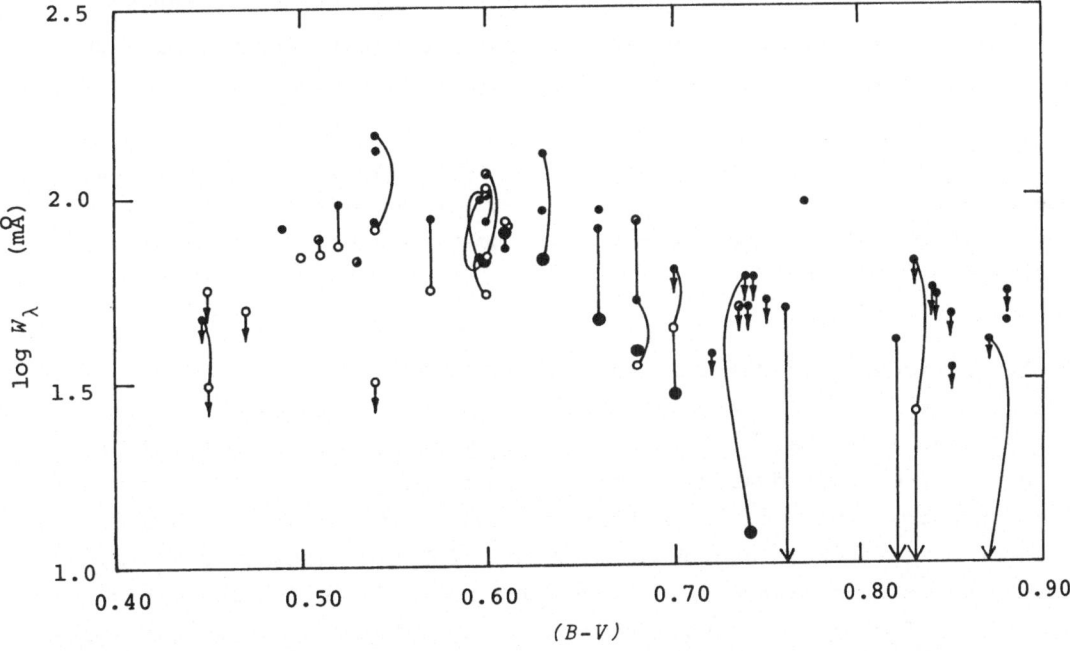

Figure 2. Published equivalent widths versus color for Hyades dwarfs. The large solid dots and the four arrows in the lower right are the data of Cayrel *et al.* (1983) that were used to produce Figure 1. The open circles represent data of Wallerstein, Herbig, and Conti (1965), while the small dots are data from Zappala (1972) and the half-filled circles are from Soderblom (1983*a*). Lines have been drawn to connect different observations of the same stars.

show more Li in the cooler stars, probably because of blending with the Fe I feature, as Cayrel *et al.* note. Although the Hyades deserves more detailed study, the minimal scatter in the data of Cayrel *et al.* shows what can be achieved with a high quality instrument.

The Reticon W_λ's of Cayrel *et al.* are compared to older photographic results in Figure 2. The large solid dots are the Cayrel *et al.* data that was used to construct Figure 1. The open circles are the data of WHC, the small dots Zappala (1972), and the half-filled circles are Herbig's data (listed in Soderblom 1983*a*). Vertical lines have been drawn to connect different observations of the same stars to show the range of reported W_λ's. Low dispersion "detections" in the cooler stars turn out to be misidentifications, and, in general, the Reticon gives smaller W_λ's than photographic plates did. The photographic data shown in Figure 2 are all from Lick Observatory, but were obtained from plates of different dispersion and emulsion, traced on different microphoto-meters. The *rms* scatter for different observations of the same stars (ignoring stars cooler than $B-V = 0.70$ and stars with upper limits to W_λ) is about 0.18 dex. This corresponds to about a 50% error in W_λ, and, for these saturated lines, the error in Li is about a factor of two.

Note that there are several hot stars with only upper limits to Li. These are examples of the Li-deficient stars mentioned below and discussed in Duncan (1981).

The existing data for Coma and Praesepe (Zappala 1972) are poor (Soderblom 1983*a*). These clusters also deserve careful study because they are nominally of the same age as the Hyades and so provide useful tests of our hypotheses.

b. The Pleiades - Is There an Age Spread?

Duncan and Jones (1983) have recently presented Li observations of a number of G and K dwarfs in the Pleiades. Based on the dispersion in Li abundances, they concluded that the oldest and youngest stars in that cluster may differ in age by as much as 0.4 Gyr. Although other data also indicate that late-type Pleiads are not as young as the early-type stars (Stauffer 1984), most estimates of the Pleiades age spread are a factor of 5 or so less than this. Figure 3 shows the Li data of Duncan and Jones.

There are several reasons why this apparent spread in Pleiades Li abundances may be at least partly due to other effects; Duncan and Jones discuss some of these. First, Duncan and Jones obtained their W_λ's using several different instruments and detectors; their data are inhomogeneous. Although their *rms* (σ/W_λ) is 0.21, the Hyades data presented above suggest that this could be larger; i.e., much of the scatter may be due to errors in the equivalent widths for these faint stars. However, as they illustrate in their Figure 2, some real differences seem to remain between stars of the same temperature.

Second, there may be errors in T_{eff} for these stars, leading to errors in *Li*. As Wilson (1963) noted, the relationship between color and spectral type in the Pleiades is not smooth - there may be non-uniform reddening. This is difficult to evaluate on a star-by-star basis, but I have lowered T_{eff} for two stars in Figure 3 by about 300 K each, based on Stauffer's (1984) *V-I* colors, which are less sensitive to reddening than *B-V*. Duncan and Jones used *R-I* colors to get T_{eff} when possible, but not for these two stars.

Third, late-type Pleiads have very active chromospheres (Wilson 1963). If chromospheric activity affects the apparent Li abundance

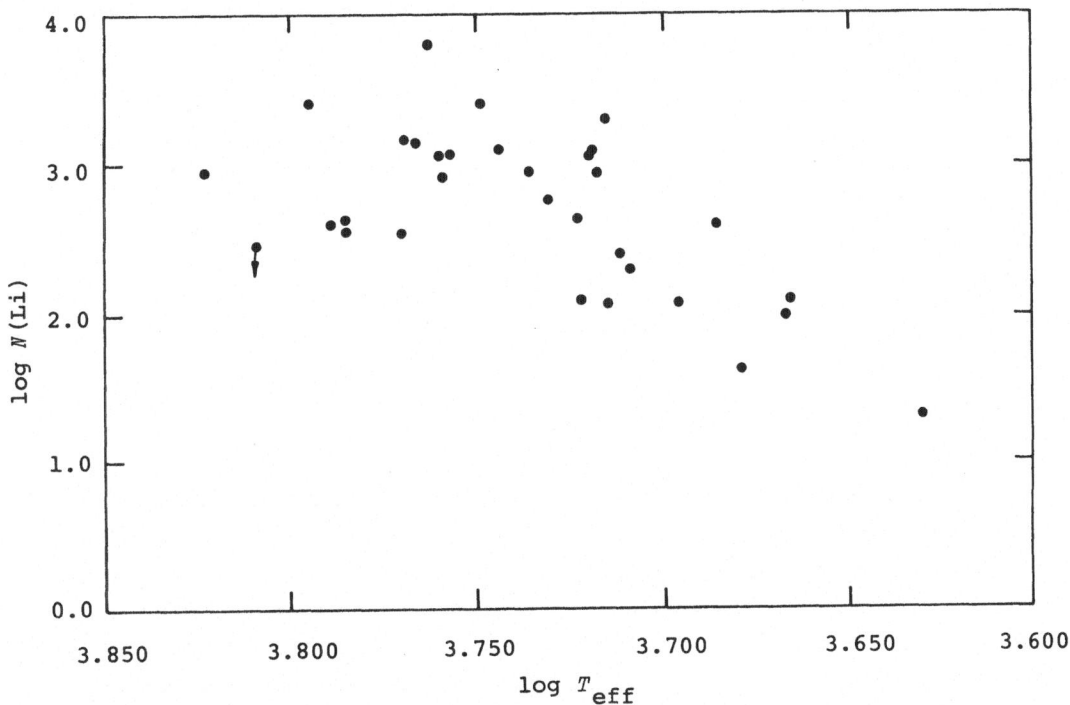

Figure 3. Lithium abundance versus temperature for Pleiades dwarfs. This is the same data as in Figure 5 of Duncan and Jones (1983), except that the temperatures of two stars have been lowered by about 300 K because of their *(V-I)* colors (Stauffer 1984), and the Li abundances have been changed accordingly.

(Giampapa 1984), these Pleiads should show it. This effect cannot yet be evaluated, but work is in progress to do so.

Other indirect evidence bears on this question. The process of Li depletion does not appear to be a convergent process, it is instead a proportional one. By this I mean that if we consider two stars of the same mass, but differing in Li by, say, a factor of two, then they will always differ by this factor. Angular momentum loss, by comparison, appears to be convergent because there is a feedback mechanism; rapid rotation leads to stronger magnetic fields, hence faster angular momentum loss. Li depletion would also be convergent if rapid rotation led to faster Li burning, but such is not the case (Soderblom 1983a).

My point is that if the Pleiades exhibit a spread in Li due to age differences among the stars, that spread will not diminish with time, and should therefore be evident in the Hyades data. The best Hyades data (Fig. 1) show no such spread.

We also know that lower mass stars deplete Li very quickly. Thus if there is a range of ages in the Pleiades, the spread in Li should increase for cooler stars. Although Duncan and Jones claim such an increase, none is apparent in Figure 3.

Finally, the age spread that Duncan and Jones derive depends on our understanding of Li depletion in low mass stars. Their data may say as much about that understanding as it does about star formation in this cluster.

The α Persei cluster is about the same age as the Pleiades, but does not appear to suffer from the same reddening problems. Study of Li in that cluster may help us understand the Pleiades.

c. Other Clusters and Binary Systems

One other useful cluster is the Ursa Major Group, a number of nearby stars with common kinematics. Soderblom (1983b) has recently studied this group whose age is midway between the Pleiades and Hyades. A number of kinematic members turn out not to be real members at all when spectroscopic age criteria are applied. The $bona$ $fide$ members have the Li abundances we would expect: intermediate to the Pleiades and Hyades (Soderblom 1983a; Soderblom and Jones 1984).

Binary systems are essentially "mini-clusters" since the two stars are presumably coeval and have the same composition. Wallerstein (1966) used binaries where the primary was slightly evolved to test the age

dependence of Li depletion in solar-type stars. He found that most
systems have Li abundances in accord with our expectations, but a few
could not be understood. Duncan (1984) has recently published a list
of such systems, and came to a similar conclusion in regard to the age
dependence of chromospheric emission. Lindroos (1981) has recently
looked at some K dwarfs that are companions of B stars and finds that
many have large Li abundances; this agrees with these systems being very
young.

If both components of a binary are late-type dwarfs, their Li abundances would define an isochrone in a Li vs. T_{eff} (or spectral type)
diagram, and so provide useful information even if the absolute age of
the system isn't known. There are, however, few binaries in which Li
can be detected in both stars. One is ξ Boo (Soderblom and Jones 1984).
Although this system has the spectroscopic attributes of a member of the
Ursa Major Group, it has been erroneously assigned to this group because
its space velocity is far from the group mean. The Li feature is strong
in star A, but weakly present in B in recent Reticon observations.

Soderblom and Dravins (1984) recently observed α Cen A and found a
Li abundance commensurate with the star's age and mass. No Li was found
in α Cen B.

A few binaries seem to have Li abundances that violate our model
for Li depletion. For example, Bonsack (1960) reported Li in 61 Cyg B
(K7V), but none in A (K5V). Li in either would be extraordinary for
such low mass stars in a kinematically old system, but we certainly don't
expect the less massive star to have Li if the more massive star doesn't.
Recent observations (Soderblom and Jones 1984) show that Bonsack's detection was spurious: neither star has a detectable Li feature.

Herbig (1965) reported that 16 Cyg A had Li ($Li \simeq 1.5$; Soderblom
1983a), while B had $Li < 1.3$. Since these two stars differ little in
spectral type (G1.5V and G2.5V; Soderblom 1983a), such a difference in
Li seems surprising; again, recent observations (Soderblom and Jones
1984) show that neither star has Li.

The ξ UMa system presents problems, though. The two components
have different spectral types (F8.5V and G0V [Keenan 1982] or F9.5V
and G2V [Harlan 1982]) and different colors (Soderblom 1983a). Star A
has a fairly strong Li feature ($W_\lambda \simeq 45$ mÅ; Herbig 1966), but B has no
more Li than the Sun (Soderblom and Jones 1984). These two stars do not
seem to differ enough in mass to account for their relative Li abundances. Both A and B are spectroscopic binaries: A is a long period system,
but B is a close binary ($P \simeq 4^d$), and so perhaps its chromospheric

activity (Wilson 1963) affects its Li abundance.

Other binaries that have been observed (Herbig 1965; Soderblom 1983a) do not conflict with our depletion model.

4. Clinkers - What's Wrong With the Picture?

Solar-type stars (about F8V and later) appear to deplete Li in accord with the model set forth at the beginning of § 3. Some stars do not.

First, there are metal-deficient stars that have preserved their Li. Spite and Spite (1982a) have found significant Li abundances ($Li \simeq 2$) in several Population II subdwarfs. There are also some stars with [Fe/H] $\lesssim -0.3$ that have preserved Li (Duncan 1981; Soderblom 1983a). This phenomenon can be understood qualitatively if metal-poor stars have thinner convective zones (CZ's) because of lower opacity in the outer layers. Given the sensitivity of the Li depletion time scale to mass (i.e., depth of the CZ), it is perhaps not surprising that low-metal stars have preserved Li. However, Duncan (1983) has examined more Pop II subdwarfs and has found several with metallicities as low as those of Spite and Spite, yet without detectable Li. More than metallicity appears to be at work here, and the cosmological conclusions of Spite and Spite (1982b) may be premature.

Mid-F dwarfs also deviate from our expectations, for a number of them have low Li abundances, even though their CZ's should be thin, and even though their main sequence lifetimes should be too short for significant Li depletion to occur. Many of these Li-deficient F stars are also Be-deficient (Boesgaard 1976), a phenomenon not seen at other spectral types.

These F dwarfs must deplete their Li quickly, for there are examples in the Pleiades and Hyades (Figs. 2 and 3; Duncan and Jones 1983). No satisfactory explanation exists. One concern is whether this phenomenon of premature depletion affects our conclusions about Li in solar-type stars. The distribution of Li abundances among G1-G3 dwarfs shows no excess of Li-poor stars (Soderblom 1983a), in fact this distribution agrees well with our model: uniform initial Li abundances, a uniform stellar birthrate, an exponential decay law, and gradual elimination of old stars from the solar neighborhood as their space velocities increase. For these reasons, Duncan (1981) and Soderblom (1983a) concluded that these F dwarfs deplete Li by a mechanism that's different from the one that operates in G dwarfs.

5. Lithium As an Age Indicator

After noting the correlation between Li abundance and age among solar-type stars, observers have tried to quantify the relation (Herbig 1965; Danziger 1969; van den Heuvel and Conti 1971; Skumanich 1972; Zappala 1972; Vauclair 1973; Duncan 1981; Soderblom 1983a). Let us first examine a simpler question: is Li at least a qualitative indicator of age?

For solar-type stars, the cluster observations (Figs. 1 and 3) indicate that a large Li abundance is a necessary condition for youth (age $\lesssim \frac{1}{2}$ Gyr), since there are no Li-deficient stars in these clusters later than about F8. Except for a few metal-deficient stars (which can be distinguished by their large space velocities), a large Li abundance is also a sufficient condition. Thus, among F8 to G5 dwarfs, if $Li \gtrsim 3.0$, the star is very young.

Li is rarely seen in stars G5V and later, not even in Pop II stars (Spite and Spite 1982a), so that one may safely judge a star to be young if Li is unambiguously present at all. For example, Danziger (1969) detected Li in only 1 or 2 of the field K dwarfs that he observed, although Lindroos (1981) has recently found some K dwarfs with Li by looking at companions of B stars. A few others have also been found by observing stars thought to be young on the basis of their chromospheric activity (Soderblom and Jones 1984).

Deriving quantitative ages from Li is much more difficult, although Duncan (1981) and Soderblom (1983a) have recently tried to establish Li-age calibrations. To reiterate, W_λ and T_{eff} must be well determined to get Li satisfactorily, and there is the unknown effect of chromospheric activity (Giampapa 1984). But even if Li is confidently known, a star's age is difficult to establish.

The main reason is that the depletion time scale is extremely sensitive to stellar mass, a quantity poorly known for most stars. To illustrate this, consider α Cen A, a star recently observed by Soderblom and Dravins (1984). This star is a spectroscopic twin to the Sun; in the absence of other information we would assume that it also has the Sun's mass. It has about twice the solar Li abundance, and is older than the Sun. However, this is consistent with the star's mass (10% greater than solar), and leads to an estimate of the e-folding time for that mass of 1.4 Gyr. For comparison, the e-folding time for the Sun's mass is about 1.1 Gyr. We happen to know the mass of α Cen A because it's in a visual binary.

Thus the *Li*-age relation can only be calibrated well for 1.0 $M_\odot$ (and maybe 1.1 $M_\odot$) because the Sun is the only old star for which we confidently know the mass, composition, age, and Li abundance. Cluster data can calibrate the relation for lower masses since Li is seen in only the youngest stars anyway, but this has not yet been done.

Li is most easily observed in young stars, but in those cases ages can only be roughly assigned because a modest range of *Li* (corresponding to observational uncertainties) represents a broad range in log *age*.

Li must be used judiciously as an age indicator for individual stars for there are many complicating factors. Some of these problems are lessened if we consider groups of stars since the group's average properties can be determined more confidently. Thus it is appropriate to judge a group of stars young because they have uniformly large *Li*, or old because *Li* is very small; we just can't say exactly how old or young they are.

Nearly all studies of lithium in late-type dwarfs owe a debt to George Herbig - this is no exception. I also wish to acknowledge the support of the Langley-Abbott Program of the Smithsonian Institution.

References

Audouze, J., Tinsley, B.M.: 1974, *Astrophys. J.*, 192, 487.
Baker, N.: 1963, *The Depth of the Outer Convection Zone in Main-Sequence Stars* (New York: NASA Inst. for Space Studies).
Boesgaard, A.M.: 1976, *Astrophys. J.*, 210, 466.
_____: 1977, in *Highlights of Astron.*, ed. E.A. Müller, v. 4, pt. II, 209.
Bonsack, W.K.: 1960, *Astrophys. J.*, 130, 843.
Bonsack, W.K., Greenstein, J.L.: 1960, *Astrophys. J.*, 131, 83.
Bopp, B.W.: 1974, *Publ. Astron. Soc. Pac.*, 86, 281.
Brault, J.W., Müller, E.A.: 1975, *Solar Phys.*, 41, 43.
Canal, R.: 1974, *Astrophys. J.*, 189, 531.
Cayrel, R., Cayrel de Strobel, G., Campbell, B., Däppen, W.: 1983, preprint.
Cohen, J.G.: 1972, *Astrophys. J.*, 171, 71.
Danziger, I.J.: 1969, *Astrophys. Lett.*, 3, 115.
de la Reza, R., Torres, C.A.O., Busko, I.C.: 1981, *Mon. Not. R. Astron. Soc.*, 194, 829.
Duncan, D.K.: 1981, *Astrophys. J.*, 248, 651.
_____: 1983, private communication.
_____: 1984, *Astrophys. J.*, submitted.
Duncan, D.K., Jones, B.F.: 1983, *Astrophys. J.*, 271, 663.
Ferlet, R.: 1982, *ESO Messenger*, No. 30, p. 9.
Giampapa, M.S.: 1984, *Astrophys. J.*, submitted.
Grevesse, N.: 1968, *Solar Phys.*, 5, 159.
Harlan, E.A.: 1982, private communication.
Herbig, G.H.: 1965, *Astrophys. J.*, 141, 588.
_____: 1966, *Lick Obs. Bull.*, No. 595.
Hultqvist, L.: 1977, *Solar Phys.*, 52, 101.

Keenan, P.C.: 1982, private communication.
Krankowsky, D., Müller, O.: 1967, *Geochim. Cosmochim. Acta*, 31, 1833.
Lambert, D.L., Dominy, J.F., Sivertsen, S.: 1980, *Astrophys. J.*, 235, 114.
Lindroos, K.P.: 1981, Stockholm Obs. Rep. 18.
Meneguzzi, M., Audouze, J., Reeves, H.: 1971, *Astron. Astrophys.*, 15, 337.
Müller, E.A., Peytremann, E., de la Reza, R.: 1975, *Solar Phys.*, 41, 53.
Nichiporuk, W., Moore, C.B.: 1974, *Geochim. Cosmochim. Acta*, 38, 1691.
Schatzman, E.: 1969, *Astrophys. Lett.*, 3, 139.
_____: 1977, *Astron. Astrophys.*, 56, 211.
Skumanich, A.: 1972, *Astrophys. J.*, 171, 565.
Soderblom, D.R.: 1983a, *Astrophys. J., Suppl. Ser.*, 53, 1.
_____: 1983b, in *The Nearby Stars and the Stellar Luminosity Function*,
    IAU Coll. 76, ed. A. Upgren, in press.
Soderblom, D.R., Dravins, D.: 1984, *Astron. Astrophys.*, in preparation.
Soderblom, D.R., Jones, B.F.: 1984, in preparation.
Spite, F., Spite, M.: 1982a, *Astron. Astrophys.*, 115, 357.
Spite, M., Spite, F.: 1982b, *Nature (London)*, 297, 483.
Starrfield, S., Truran, J.W., Sparks, W.M., Arnould, M.: 1978,
    Astrophys. J., 222, 600.
Stauffer, J.R.: 1984, *Astrophys. J.*, 280, in press.
Straus, J.M., Blake, J.B., Schramm, D.N.: 1976, *Astrophys. J.*, 204, 481.
Traub, W., Roesler, F.L.: 1971, *Astrophys. J.*, 163, 629.
Twarog, B.A.: 1980, *Astrophys. J.*, 242, 242.
van den Heuvel, E.P.J., Conti, P.S.: 1971, *Science*, 171, 895.
Vauclair, S.: 1973, in *Stellar Ages*, IAU Coll. 17, eds. G. Cayrel de
    Strobel and A.M. Delplace (Paris: Paris-Meudon Obs.).
Vauclair, S., Vauclair, G., Schatzman, E., Michaud, G.: 1978,
    Astrophys. J., 223, 567.
Wallerstein, G.: 1966, *Astrophys. J.*, 145, 759.
Wallerstein, G., Conti, P.S.: 1969, *Annu. Rev. Astron. Astrophys.*, 7, 99.
Wallerstein, G., Herbig, G.H., Conti, P.S.: 1965, *Astrophys. J.*,
    141, 610 (WHC).
White, R.E.: 1983, private communication.
Wilson, O.C.: 1963, *Astrophys. J.*, 138, 832.
Zappala, R.R.: 1972, *Astrophys. J.*, 172, 57.

NITROGEN ABUNDANCES IN DISK AND HALO DWARFS

Jocelyn Tomkin and David L. Lambert
Department of Astronomy
University of Texas
Austin, Texas 78712

ABSTRACT

High-resolution Digicon spectra of the NH band at 3360 Å have been
analyzed by spectrum synthesis to determine nitrogen abundances in 14
disk and halo F and G dwarfs with metal deficiencies covering the range
-2.3 ≤ [Fe/H] ≤ -0.3. We have determined carbon abundances from parallel
observations and analysis of CH. In all stars nitrogen closely follows
carbon and iron; i.e. [N/C] ≈ 0.0. We conclude that nitrogen is a
primary element.

OBSERVATIONS, ANALYSIS, AND RESULTS

The McDonald Observatory 2.7m telescope, coudé spectrograph, and a 936
Digicon (Tull, Choisser, and Snow 1975) were used to observe NH and CH
in the 14 program stars and in the Moon. The NH observations were
centered at 3360 Å and covered 120 Å at a resolution of 0.26 Å. For the
bright stars - those with V ≤ 7.0 - the resolution was increased to
0.13 Å. The CH observations were centered at 4310 Å and covered 60 Å
at a resolution of 0.12 Å. The signal-to-noise ratio was typically 100
to 1.

The program stars and their effective temperatures, gravities, and iron
abundances are listed in Table 1. They have temperatures and gravities
similar to those of the Sun, so scaled solar model atmospheres were
chosen as being most suitable.

We determined the nitrogen and carbon abundances by comparison of synthesized NH and CH spectra with the observed NH and CH spectra. Wavelengths, line identifications, accurate laboratory determinations of
dissociation energies, and band oscillator strengths are available for
both NH and CH. Numerous atomic lines contribute to both the NH and CH
spectral intervals. Accurate laboratory gf values were used for the
atomic lines.

Table 1

Atmospheric Parameters and Abundances

Star	T_{eff}	log g	[Fe/H]	[N/Fe]	[C/Fe]	[N/C]
HD 4614 (η Cas)	5730	4.3	-0.37	-0.1	-0.19	+0.09
HD 6582 (μ Cas)	5280	4.5	-0.85	-0.3	-0.2	-0.1
HD 19445	5830	4.0	-1.82	<0.0	0.0	<0.0
HD 63077 (171 Pup)	5720	4.1	-0.77	-0.3	-0.2	-0.1
HD 64090	5400	4.5	-1.60	-0.2	-0.2	0.0
HD 94028	5900	4.5	-1.56	0.0	0.0	0.0
HD 103095 (Gmb 1830)	5000	4.6	-1.37	-0.5	-0.5	0.0
HD 106516	6050	4.5	-0.85	0.0	-0.2	+0.2
HD 134169	5800	3.8	-1.00	0.0	0.0	0.0
HD 134678	6000	4.5	-0.70	-0.3	-0.3	0.0
HD 140283	5600	4.5	-2.30	-0.5	-0.2	-0.3
HD 207978 (15 Peg)	6150	4.0	-0.63	-0.3	-0.25	-0.05
HD 221377	6000	3.5	-0.90	-0.1	-0.3	+0.2
HD 224930 (85 Peg)	5300	4.3	-0.83	-0.6	-0.4	-0.2

We note that NH and CH are both hydrides and that their dissociation energies are almost identical ($\delta D_0^0 \approx 0.05$ eV). Thus, the circumstances of NH and CH line formation are very similar and errors of analysis for nitrogen and carbon abundances derived from NH and CH must, also, be very similar. This means the abundances of nitrogen relative to carbon are much more reliable than the nitrogen abundances alone; it is the rationale for the parallel observation and analysis of CH.

The matches of the synthetic and observed solar NH and CH spectra are excellent. The solar nitrogen and carbon abundances derived from their comparison are within 0.05 dex of the actual abundances. This agreement of the synthetic and observed solar NH and CH spectra, which was achieved without any forced fitting of either atomic or molecular lines, increases our confidence in the stellar nitrogen and carbon abundances. Figure 1 shows that the synthetic and observed spectra are also in good agreement for the program stars. We estimate the abundance errors associated with the matching of the synthetic to the observed spectra of ±0.05 dex for CH and ±0.1 dex for NH. This includes the uncertainty of continuum location.

Nitrogen and carbon abundances are given in Table 1. Both elements tend to be slightly more deficient than iron; the average [N/Fe] and [C/Fe] for the 14 program stars are -0.25 and -0.21, respectively. Nitrogen

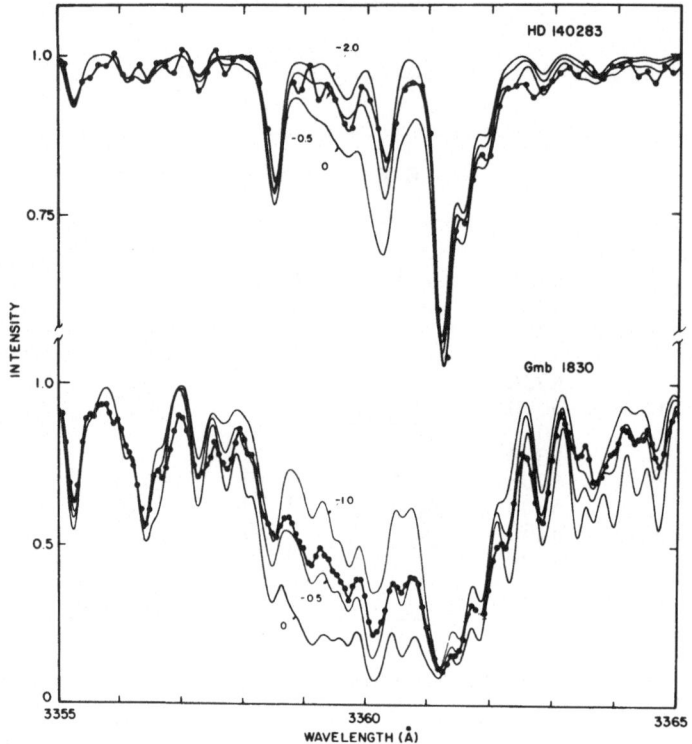

Figure 1 - The NH (0,0) bandhead in HD 140283 ([Fe/H] = -2.3) and Gmb 1830 ([Fe/H] = -1.37). Each observed spectrum is fitted with three synthetic spectra which are labelled with their [N/Fe]. The synthetic spectra with [N/Fe] = -0.5 provide good fits in both stars. (The 0.3 Å interval 3360.0-3360.3 Å, where the synthetic spectra fall below the observed spectra in all stars, including the Sun, has low weight.)

follows carbon closely; the average [N/C] is -0.02 and all [N/C] lie in the range ±0.3 dex. A full account of this work will be published (Tomkin and Lambert 1984).

This research has been supported in part by the National Science Foundation (AST 81-17485) and the Robert A. Welch Foundation of Houston, Texas.

REFERENCES

Tomkin, J., and Lambert, D. L. 1984, Ap. J. (April 1), in press.
Tull, R. G., Choisser, J. P., and Snow, E. H. 1975, Appl. Optics, 14, 1182.

FOUR W URSAE MAJORIS CONTACT BINARIES
IN THE OLD GALACTIC CLUSTER NGC 188

Sallie L. Baliunas
Harvard Smithsonian Center for Astrophysics
and
Edward F. Guinan
Villanova University

With an age of 5×10^9 years inferred from its color-magnitude diagram (Sandage and Eggen 1969), NGC 188 is the oldest known cluster. Its age alone provides a unique opportunity to explore stellar evolution and chemical evolution of the Galaxy. Moreover, the stellar population appears quite unusual. NGC 188 is the only old open cluster which seems to contain more than one contact binary system--in fact, there are four supposed W UMa binaries with main sequence G-K components presumed to be cluster members (Efremov et al. 1964). If the orbital axes are randomly oriented in space, a local space density of W UMa systems in the cluster is expected to be 20-30 times the mean space density. In addition to the unusually large number of contact systems predicted, the presence of these binaries in a cluster this age is puzzling since the lifetime of the contact phase of a W UMa system is only about 5×10^8 years (Vilhu 1982).

Spectra of these four faint short-period variable stars EP Cep, EQ Cep, ER Cep and ES Cep (See Table 1) were obtained in 1983 July with an intensified Reticon detector on the Multiple-Mirror Telescope (Baliunas and Guinan 1984). The spectra cover the wavelength range $\lambda\lambda 3850-4600$ with a resolution of about 1.2A. In each of the spectra of the stars, the photospheric lines are rotationally broadened in accord with the rapid rotation expected from these short-period W UMa systems. We resolved for the first time the photospheric lines of the individual stars in three of these systems, leaving no doubt that they are indeed W UMa systems.

We calculated the cross-correlation coefficients between the spectra of the contact binaries and slowly-rotating template stars of approximately the same color in NGC 188 (stars I-32 and I-33). From the cross-correlation coefficients we measured the radial velocities of the W UMa stars near the expected phase of velocity crossing with respect to the two other cluster stars. The velocities of the W UMa systems are consistent with cluster membership. In three of the systems, the spectra were obtained at phases where individual stellar components were resolved. Two Gaussians were fit simultaneously to the most significant portion of the blended cross-correlation peaks. The mass ratios (Table 1) of these three systems were determined by assuming that the velocity of the center of mass is equal to the mean radial velocity of the two cluster reference stars. The mass ratios in these three systems suggest the binaries are similar to other short-period, field W UMa systems designated W-type (Binnendijk 1970).

Our spectra confirm that these binaries are members of NGC 188 and contact binaries. Since the lifetime of the contact configuration is believed to be relatively short ($< 5 \times 10^8$ yr), the presence of W UMa systems in such an old cluster suggests that these binaries have evolved from detached or semi-detached systems which have undergone substantial angular momentum loss (Vilhu 1982). Mass loss and its associated magnetic torques may provide a mechanism for loss of orbital angular momentum (Huang 1966; DeCampli and Baliunas 1979).

TABLE 1. Characteristics of the W Ursae Majoris Systems in NGC 188

Sonneberg Number	Variable Star Name	m_V(max)[a]	$\langle B-V \rangle$[a]	$\langle B-V \rangle_o$[b]	Period[c] (days)	m_h/m_c[d]	Type
S8278	EQ Cep	+16.25	+0.90	+0.83	0.3069	0.37±0.11	W
S8279	ER Cep	+15.65	+0.83	+0.76	0.2857	0.68±0.08	W
S8280	ES Cep	+15.52	+0.88	+0.81	0.3424	0.13:±0.05	W
S8474	EP Cep	+16.6:	+0.9:	+0.83:	0.2897	--	--

[a]From Eggen and Sandage 1969.

[b]E_{B-V} = 0.07 adopted from Twarog 1978.

[c]From Kholopov and Sharov 1966.

[d]Ratio of masses of hotter and cooler components. Errors are standard deviations determined from velocities derived from cross-correlations with three different template stars (Baliunas and Guinan 1984).

References

Baliunas, S.L. and Guinan, E.F. 1984, Ap.J., submited.

Binnendijk, L. 1970, Vistas in Astronomy, Vol. 12, p. 217.

DeCampli, W.M. and Baliunas, S.L. 1979, Ap.J., 230, 815.

Eggen, O.J. and Sandage, A. 1969, Ap.J., 158, 669.

Efremov, Y.N., Kholopov, P.N., Kukarkin, B.V. and Sharov, A.S. 1964, IAU Info. Bull. Var. Stars, no. 75.

Huang, S.-S. 1966, Ap.J., 129, 331.

Kholopov, P.N. and Sharov, A.S. 1966, Astr. Circ. USSR, Nos. 427, 434, 452.

Sandage, A. and Eggen, O.J. 1969, Ap.J., 158, 685.

Twarog, B.A. 1978, Ap.J., 220, 890.

Vilhu, O. 1982, Astr.Ap., 109, 17.

Does the Physical State of a Mantle indicate the Stage of Evolution of a Star?

Anne B. Underhill
Laboratory for Astronomy and Solar Physics
NASA Goddard Space Flight Center

ABSTRACT

The physical state of the mantle of a star (its chromosphere, corona, and wind) is chiefly determined by the amount of nonradiative heat and momentum deposited in the mantle. Because the presence of magnetic lines of force buffeted by turbulent or convective elements in the photosphere seems to be a dominating factor for determining how much energy and momentum are deposited in the mantle, the question of the title can be rephrased to ask can the presence of magnetic fields in the mantle of a star and the presence of rotationally or convectively driven turbulence in the photosphere be related uniquely to any particular stage of evolution of a star. The answer to this question is important because many of the spectroscopic features which have been empirically selected as criteria for spectral type are products of the physical state of the mantle of the star rather than of its photosphere. The standard theory relates properties of the photosphere to the stage of evolution of a model star, not properties of the mantle. It is conditions in the envelope of the star that cause strands of magnetic flux to be created and expelled from the photosphere. The strands then expand, grow, and decay in the mantle of the star. Until the generation of the needed magnetic flux can be related uniquely and unambiguously to the age of the star, one cannot use the spectroscopic phenomena generated in the mantle as an unambiguous key to the stage of evolution of the star.

INTRODUCTION

What we observe from stars is the detailed energy distribution of the radiation field which emerges from the star. From this we try to deduce the main properties of the star and the stage of evolution of the star. The latter step is achieved by identifying a real star with a model star which has been evolved through significant periods of time. The identification process consists of comparing observed spectral details with parts of the predicted spectrum of the model star.

Classifying stars according to spectral type (empirically assigned according to the relative intensities of prominent spectral lines in absorption and emission) is a first step. Thereby the stars are sorted into bins such that, hopefully, all the stars in one bin will have the same mass, effective temperature, radius, and composition. If the spectral type corresponding to each bin can be related uniquely to a model star at a selected stage of evolution, the logical cycle has been completed.

However, standard, classical model stars generate purely absorption-line spectra. Thus stars of those spectral types which show emission lines in their spectra demand modified model stars which can generate emission lines both in the visible and in the ultraviolet regions of the spectrum. The question which I want to discuss is whether, at this time, we can establish a firm logical connection between stage of evolution and the kind of <u>ad hoc</u> modifications which it is necessary to make to the model star in order to predict emission-line spectra such as are observed. I speak only of single stars. The presence of gas streams in a binary system may add extraneous emission lines to the spectrum of the pair of stars.

Cool stars which have rather broad emission lines in the visible part of their spectra are usually classified as T Tauri stars. The emission spectrum may be variable. The ultraviolet spectra of all cool stars are dominated by emission lines, some lines corresponding to ions which are present in plasmas having electron temperatures of the order of 10^4 K, others corresponding to electron temperatures of at least 10^5 to 10^6 K. Clearly, modified model stars are required if one is to interpret accurately the emission and absorption line spectra of all cool stars. Can one relate the needed modifications uniquely to the stage of evolution of a standard, classical model star?

PROPERTIES OF STANDARD MODEL STARS

Model stars have an adopted composition and mass. They are divided into two parts: the interior and the boundary layer. The interior consists of a core where a radiation field is generated and an envelope which supports the boundary layer or atmosphere, and which transmits the radiation field. The boundary layer may be divided into a photosphere which may be modeled successfully by classical model-atmosphere procedures, and a mantle which comprises the chromosphere, corona and wind of the star. The major difference between the photosphere and the mantle is that the deposition of nonradiative energy and momentum in the mantle determines the physical state of the mantle, rather than the transmitted radiation stream from the interior doing so. The mantle appears to be inhomogeneous, and it is not spherically symmetrical whereas the photosphere usually can be modeled using plane-parallel-layer geometry.

Observations of the Sun indicate that the heating of the corona and the propulsion of a wind are largely determined by how efficiently strands of magnetic flux transfer kinetic energy from the turbulent or convective motion in the photosphere to the low-density plasma of the corona. The inhomogeneous magnetic field acts as a catalyst for transferring energy and momentum (arising from motion in the envelope and photosphere of the star) to the plasma of the mantle. The physical state of the relatively dense chromosphere of the Sun appears to be determined chiefly by the density of propagating acoustic waves which has been generated in the envelope of the Sun. Presumably the same is true for the chromospheres of cool stars of spectral types different from G2 V.

MODIFIED MODEL STARS

Electron temperatures and densities in the mantles of cool stars are inferred by various diagnostic methods from the relative strengths of the emission lines in the spectrum. Some information is inferred also about the wind from the star, its density and maximum rate of flow, chiefly from absorption features in the visible spectrum, but also from the infrared and radio spectrum. Thus it is possible to infer the chief characteristic properties of the mantles of different types of cool star. How are these properties related to the stage of evolution of the star?

This question may be answered by postulating explicit mechanisms for transferring energy and momentum from the envelope of the star to the mantle in the required amounts. Empirically it appears that the amount and rate of change of the amount of magnetic flux emerging locally from the photosphere of the model star is an important factor. Only if this factor can be correlated in a unique and monotone way with evolutionary age can one say that the observed physical state of the mantle of a cool star indicates the stage of evolution of a cool star.

At present it is still true that the most reliable indicators which we have for the evolutionary age of a cool star are its radius and effective temperature, given that the mass and composition of the star are known. Until an adequate theory of the generation of magnetic flux in the envelope of a star and of the rate of expulsion of magnetic flux from the photosphere of a star is in hand, we cannot use the character of the spectrum from the mantle of the star as a reliable key to the stage of evolution of the star.

In order to use the radius, effective temperature, and mass as indicators of the stage of evolution of a star, we must be able to correlate the assigned spectral type uniquely with radius, effective temperature, and mass. This is possible only if the criteria which are used for spectral classification are predominantly functions of the parameters T_{eff} and log g which are used to specify the character of a model atmosphere and the spectrum which emerges from it, and if luminosity class can be correlated in a unique and monotone manner with luminosity or visual absolute magnitude. If some of the empirically selected spectral classification criteria, whether for spectral class or luminosity class, are sensitive to the physical state of the mantle of the star, some ambiguity may occur in the hoped for correlation between spectral type and evolutionary age. The ambiguity occurs because the values of the classification criteria are being influenced by factors which cannot yet be correlated unambiguously with the stage of evolution of the star.

A REVIEW OF THE FIRST OBSERVATIONS OF STARS USING THE INFRARED ASTRONOMICAL SATELLITE (IRAS)[*]

C.A. Beichman
Jet Propulsion Laboratory
Pasadena, CA

H. H. Aumann
Jet Propulsion Laboratory
Pasadena, CA

B. Baud
Kapteyn Institute
Groningen, Netherlands

P. Clegg
Queen Mary College
London, United Kingdom

F. Gillett
Kitt Peak National Observatory
Tucson, AZ

F. Olnon
Stichting voor Ruimte
Leiden, Netherlands

M. Rowan-Robinson
Queen Mary College
London, United Kingdom

ABSTRACT

The primary aim of the IRAS mission is a survey of the entire sky at infrared wavelengths of 12, 25, 60 and 100 μm. Some of the preliminary results of the IRAS mission relating to stars include: 1) a characterization of the properties of the infrared sky based on a careful survey of a 900 sq. deg. of sky; 2) the presence of young stars of roughly solar mass within the dense cores of dark clouds; 3) the existence of a cloud of large dust particles around the A0 V star, Vega, implying, perhaps, the existence of a pre-planetary system around that star; and 4) the discovery of the infrared counterparts to a large number of stellar OH masers.

INTRODUCTION

The Infrared Astronomical Satellite (IRAS) revolves in a circular, 900 km polar orbit far above the attenuation and thermal emission of the terrestrial atmosphere. The scientific instrument (Neugebauer et al. 1984, hereafter Paper I) consists of a liquid helium cooled, 57 cm Ritchey-Chretien telescope with a focal plane array of 62 detectors operating in four broad wavelength bands centered at 12, 25, 60 and 100 μm. A second focal plane instrument, the Chopped Photometric Channel (CPC), operates at 60 and 100 μm with greater spatial resolution, 1.2', than the survey array.

---

[*] The Infrared Astronomical Satellite was developed and is operated by the Netherlands Agency for Aerospace Programs (NIVR), the U. S. National Aeronautics and Space Administration (NASA) and the U. K. Science and Engineering Research Council (SERC).

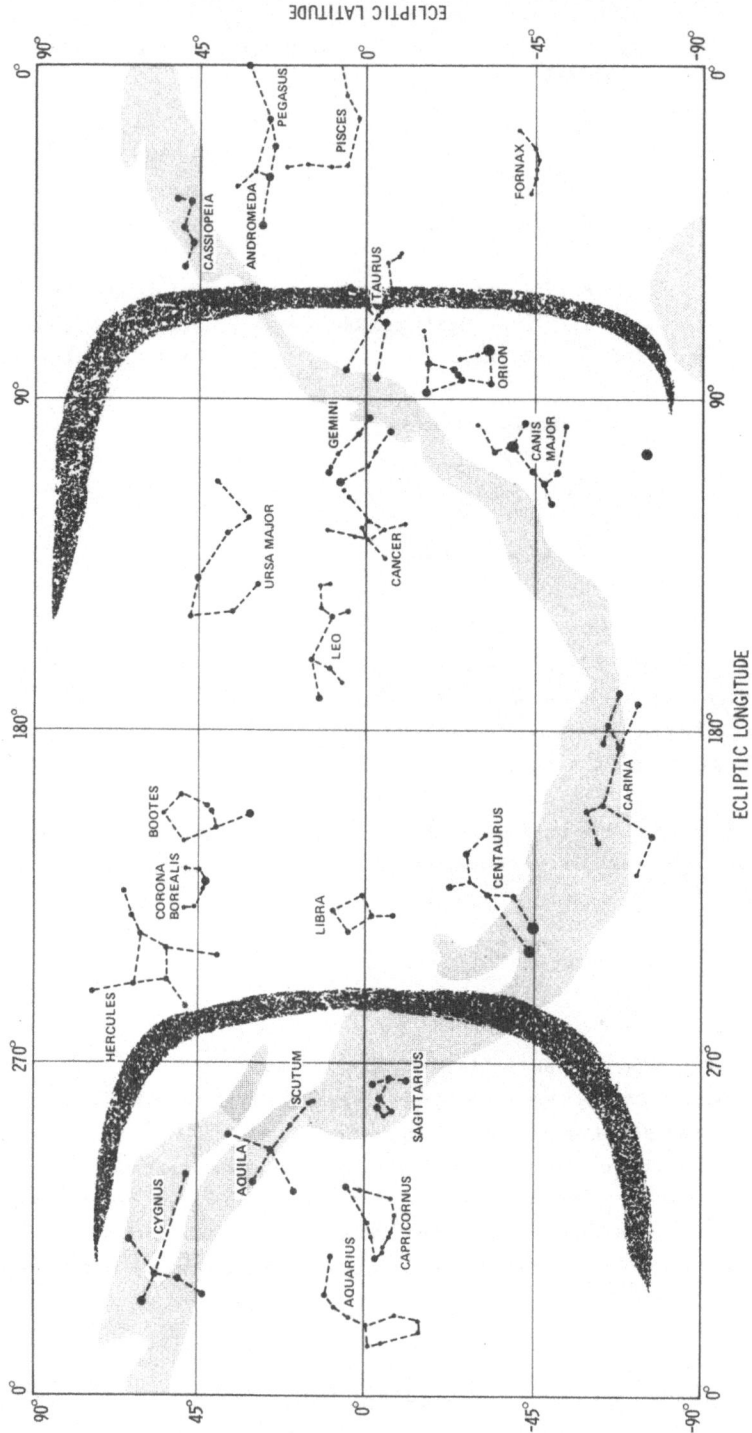

Figure 1. A schematic representation of the sky in ecliptic coordinates showing the portion of the sky observed as part of the IRAS mini-survey (dark shading). The galactic plane is shown lightly shaded.

The duration of the mission, the sensitivity and stability of the instrument have exceeded pre-launch expectations with the result that the goal of an infrared all-sky survey will be achieved with a high degree of completeness and reliablity. In addition, a large number of special observations have been made toward known objects of astronomical interest. The results described here come from an analysis of both survey and pointed observations.

THE MINI-SURVEY

As soon as the protective cover was ejected from over the telescope, a series of scans of a limited region of sky was conducted as a test of the overall survey strategy and of the data processing software. The 900 sq. deg of the "mini-survey" (Rowan-Robinson et al. 1984) consists of two strips of sky centered at ecliptic longitudes 60 and 252 deg and extending from ecliptic pole to pole (Figure 1). The selection of the region was dictated by the sky that was available immediately after launch and is not optimum astronomically, particularly since no sky was observed above a galactic latitude of 40 deg. However, the results of the mini-survey are interesting both for what they reveal about the infrared sky and for what they indicate about the sensitivity, completeness and reliability of the IRAS catalog.

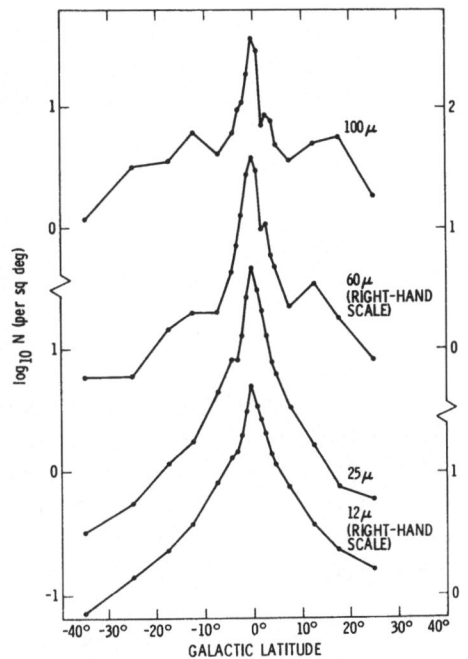

Fig. 2 Source counts as a function of galactic latitude.

The completeness of the survey drops off rapidly below flux density limits of 0.4, 0.4, 0.6 and 2.5 Jy at 12, 25, 60 and 100 μm, respectively. The corresponding source densities in the range of galactic latitude 40 deg > |b| > 20 deg are 1.1, 0.4, 0.65 and 1.25 sources per sq. deg in the four wavelength bands. Table 1 gives the 1 sigma noise equivalent flux densities (NEFD) for a point source in the four wavelength bands as well as estimates of the completeness and reliability of the catalog for sources seen on at least 4 out of 6 possible orbits. The completeness is defined as the probability that a source with more than a certain signal-to-noise ratio (SNR) will be included in the point source catalog, while the reliability is the probability that a source in the catalog above a certain SNR is a real astronomical object. The values in the table are for |b| > 20 deg.

Figure 2 shows the distribution of sources as a function of galactic latitude

in the four wavelength bands. Within 2-5 deg of most of the galactic plane the
survey will be confusion-limited at all wavelengths, at 100 μm the survey will
be confusion-limited for |b|≤ 40 deg.

Table 1. Survey Completeness and Reliability
|b|>20 Deg

Wavelength (μm)	NEFD (Jy)	Completeness		Reliability	
		SNR>5	SNR>9	SNR>5	SNR>9
12	0.10	0.99	1.00	0.98	1.00
25	0.07	0.97	0.99	0.78	0.99
60	0.07	0.90	0.95	0.98	1.00
100	0.30	0.82	0.86	0.96	0.98

By requiring positional agreement better than 1', it is possible to
identify IRAS sources with objects in astronomical catalogs. Different symbols are
used to plot various classes of astronomical objects in the color-color diagrams for
|b|>10 deg given in Figures 3a and 3b. Lines are shown for blackbody and power
law spectra. A large variety of astronomical sources are found within the mini-
survey region:

1) The limiting 12 μm magnitude of the survey is approximately 5
mag, so that hot stars (V-[12]=0) as faint as m_V=5, and cool stars (V-[12]=3-4) as
faint as m_V=8-9 should be detectable. At |b|>20 deg more than half of the 12 and
25 μm sources can be identified with SAO stars, typically K and M stars with
photospheric temperatures of 2000-3000 K. Many of the remainder of the
sources are likely to be stars fainter than catalog limits.

2) About a quarter of the mini-survey sources seen at 60 and 100 μm above
|b|>20 deg can be identified with cataloged galaxies. Soifer et al. (1984) found
that the surface density of galaxies emitting more than 0.5 Jy at 60 μm and
brighter than m_V=18 mag is approximately 0.25 per sq. deg. The remainder of the 60
and 100 μm sources include galaxies fainter than 18 mag (Houck et al. 1984), stars
with circumstellar dust shells, pre-main sequence objects associated with
Galactic dust clouds and some sources with, as yet, no counterparts at other
wavelengths (Houck et al. 1984).

3) Over much of the sky the concept of a smoothly varying background with
occasional point sources breaks down at 100 μm. A major finding of IRAS (Low et

232

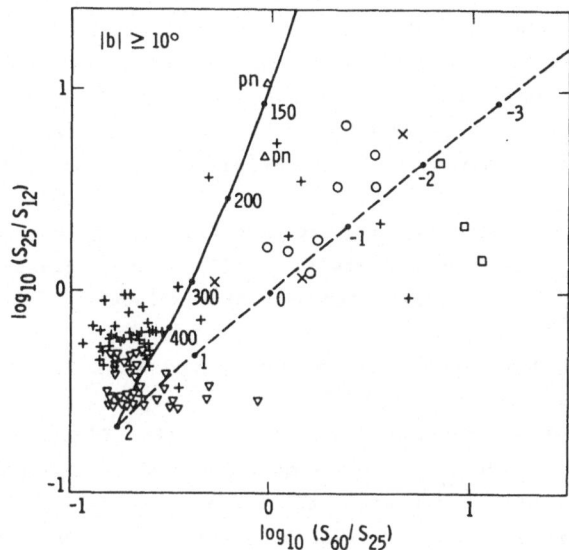

Figure 3a. 12,25, 60 μm color-color diagram for sources found in the mini-survey above galactic latitude |b|> 10 deg. Loci of power law spectra and blackbody emission at various temperatures are given. Different symbols are used depending on the tentative identifications of the objects with known astronomical sources: (squares), galaxies; (inverted triangles), late type stars; (crosses), early type stars; (upright triangles), radio sorces; (pn), planetary nebulae; (circles), pre-main sequence objects in molecular clouds; (plusses), not identified.

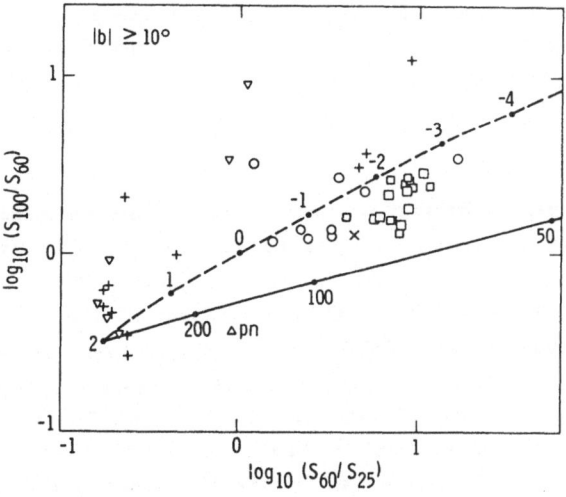

Figure 3b. As above except for 25, 60 and 100 μm.

al. 1984) is the presence of complex, extended emission seen at 100 μm and, in places, at 60 μm as well. In the mini-survey above |b|>10 deg, this "infrared cirrus" manifests itself as a large number of "point" sources, many of which are no more than clumps of emission on a 1–4' scale within regions of more extended emission. The color temperature of this emission ranges between 25– 35 K.

4) Closer to the galactic plane than |b|>10 deg most of the IRAS sources are unidentified, although normal stars, H II regions, planetary nebulae, stars with dust shells and young stars embedded within molecular clouds are found.

In summary, an analysis of the mini-survey has shown that it will be possible to acheive a primary aim of IRAS, namely a complete and reliable catalog of point sources at four infrared wavelengths. The final number of sources is expected to exceed 250,000 of which some 20,000 will be external spiral galaxies. The absolute positional uncertainty of an IRAS source will be less than 30" and the relative photometric accuracy will be better than 20 percent. The catalog is expected to be available around September, 1984.

PRE-MAIN SEQUENCE OBJECTS

A major theme of the IRAS mission, and of infrared astronomy generally, is star formation. To minimize the effects of confusion with background sources and to maximize the sensitivity to low luminosity objects we have concentrated our search for young, newly forming stars to dense, dark clouds lying well out of the galactic plane and located within a few hundred parsecs of the earth. The dark clouds Barnard 5 in the Taurus–Perseus region (Beichman et al. 1984) and Chamaeleon I (Baud et al. 1984) in the southern hemisphere contain a number of different classes of infrared sources.

Barnard 5

The central 0.5 deg of Barnard 5 contains four compact infrared sources, whose energy distributions are shown in figure 4. B5–IRS 1 is bright at all IRAS wavelengths and coincides with an object found by Benson, Myers and Wright (1984) at 2–5 μm. The source is located within the core of the cloud as defined by observations in CO (Young et al. 1982) and NH_3 (Myers, Linke and Benson 1983; Benson 1983). An estimate of the total luminosity of the object is approximately 10 $L_\odot$, for a distance to the cloud of 330 pc (Herbig and Jones 1983). The energy distribution from 2–100 μm resembles those of the T Tauri star HL Tau (Cohen 1983) and L1551–IRS 5 (Beichman and Harris 1981; Freidlund et al. 1980), an object tentatively identified as an embedded T Tauri star. Dust temperatures between 30–800 K are required. It is difficult to specify exactly the evolutionary stage of B5–IRS 1, but theoretical models suggest that a source of this luminosity and range of temperatures corresponds to a solar type star in its first 10^5 yr of existence (e.g. Stahler, Shu and Taam 1980).

Two other sources in B5, denoted IRS 3 and IRS 4, are detected at 12, 25 and 60 μm but not at 100 μm due to confusion from the molecular cloud. Both are associated with 100–200 K dust and emit about 0.5 L_Θ in the IRAS bands. Two visible stars are located within 1′ of IRS 4. Near infrared and visual observations will be required before more can be said about these objects. It is likely, however, that these are young stars, separate from the bulk of B5, but with a circumstellar shell of dust that is bright in the infrared. Similar shells of material have been seen toward a number of T Tauri and emission line stars (e.g. Harvey, Gatley and Thronson 1979; Harvey and Wilking 1982; Cohen 1983).

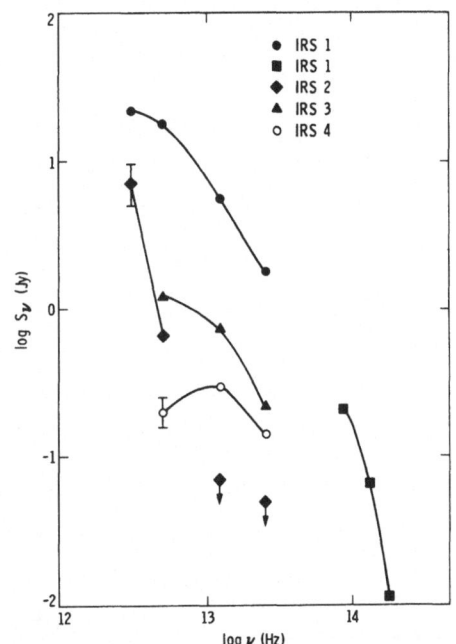

Fig. 4. Energy distributions for sources within B5. Near infrared data from Benson et al. 1984.

B5–IRS 2 is a compact 60 μm source, less than 1′ in diameter, located within, but near the edge of the dense part of the cloud seen in ^{13}CO (Young et al. 1982). At 100 μm it is hard to measure the brightness of the source, but an estimate of the 100 μm flux density yields a color temperature of 25 K and a physical temperature of 22 K for grains emitting with an emissivity proportional to frequency. Standard assumptions about grain properties (e.g. Hildebrand 1983) permit one to set a lower limit to the total mass of molecular gas associated with IRS 2 of 0.25 M_Θ. Dust colder than 15–20 K would escape detection by IRAS and could result in a considerably increased mass for the object. From the limits to the size (0.1 pc) and mass of the source, the H_2 number density exceeds 10^4 cm^{-3}. The temperature of the gas inferred from CO observations is 15–20 K (Young et al. 1982). The Jeans mass corresponding to this temperature and density is about 7 M_Θ, considerably greater than the limit set by the IRAS observations. There are two possible interpretations for IRS 2. First, it could be a dense condensation within B5 in the earliest stages of gravitational collapse into a star, similar to the object found by Keene et al. (1983) in the dark cloud B335. Second, B5–IRS 2 could be a compact, but relatively diffuse clump of material at the edge of the cloud heated by the interstellar radiation field and, perhaps, by the nearby nebula IC 348. Keys to understanding this object will be infrared observations at higher spatial resolution and observations in molecular lines excited at densities greater than 10^5–10^6 cm^{-3}.

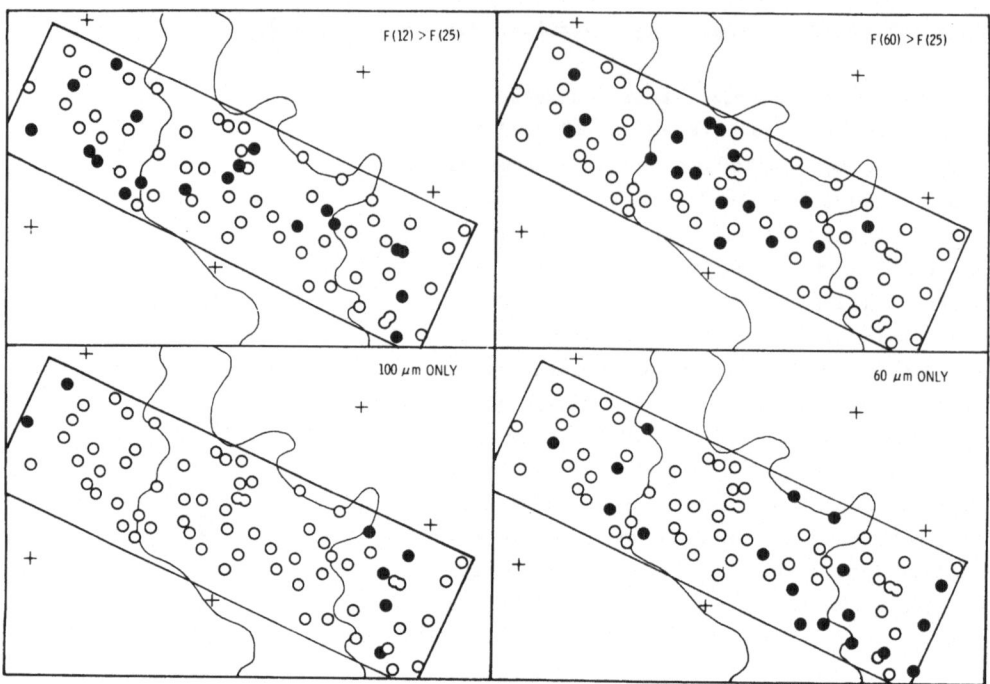

Figure 5. Schematic reprsentation of IRAS sources within the Chamaeleon I dark cloud. The curved line shows the extent of the visual extinction. The rectangle shows the area scanned by IRAS. Compact sources are shown as open or filled circles. Each panel highlights a different infrared spectral type as discussed in the text.

Chamaeleon I Cloud

A set of pointed observations designed to achieve the best sensitivity possible with IRAS was made of a 1.5 by 0.5 deg region within the Chamaeleon I dark cloud (Baud et al. 1984). This region is 215 pc away and includes a number of reflection nebulae and T Tauri stars and is similar to, but smaller than, the Taurus molecular cloud region. The cloud has been surveyed in the near infrared (Hyland, Jones and Mitchell 1982) but not yet in molecular lines. IRAS observations yielded 70 distinct, compact sources shown as open and filled circles in figure 5. The dark lines show the visual outline of the cloud, while the rectangle demarcates the area observed. Three classes of sources are found. Each panel of figure 5 highlights as filled circles sources of a specific infrared spectral type.

1) 19 "hot" objects emit more at 12 μm than 25 μm and can usually be identified with a visible star on the SERC print. These are field stars distributed uniformly across the region and appear to be unrelated to the cloud.

2) 17 "warm" sources are stronger at 60 μm than at 25 μm and mostly fall within the confines of the dark cloud. A careful examination of the SERC plate and the infrared data suggest that two of these objects are external galaxies, consistent with the known density of infrared bright galaxies (Soifer et al. 1984; Young et al. 1984). Color temperatures for the remaining sources are in the range 70-200 K, much hotter than typical galaxies and consistent with emission from dust around emission line stars and newly forming stars still embedded within the cloud. These sources resemble the objects IRS 1,3 and 4 found in Barnard 5.

3) 25 "cold" sources are seen only at 60 and/or 100 μm. These objects are typically found away from the center of the cloud, have color temperatures less than 40 K and, in some cases, can be associated with small regions of enhanced visual extinction. None are identified with known emission line stars or HH objects. Objects of this type may be small globules of material located at the edge of the cloud, heated by the interstellar radiation field, similar, perhaps, to B5-IRS 2.

Nine sources do not fit the above classification scheme due to the effects of confusion or to detection in only one band.

All of the sources found at 2 μm by Hyland, Jones and Mitchell (1982) in the region of sky overlapping the IRAS observations and identified by them as pre-main sequence (PMS) stars are detected in this deep survey; most are "warm" sources. Only 5 - 20 % of the total luminosity of sources identified with a near infrared PMS star is radiated longward of 12 μm, suggesting that these objects are not deeply embedded within the cloud, although they are still associated with dust shells.

A full inventory of the young stellar sources within the Chamaeleon dark cloud awaits the analysis of scans over the rest of the cloud. The observations to date have identified far-infrared excesses from as many as 15 objects with properties similar to known PMS stars.

FAR INFRARED OBSERVATINS OF
MAIN SEQUENCE STARS INCLUDING α LYRAE

Vega (HR 7001, α Lyrae) is one of the brightest stars in the northern hemisphere. IRAS observations have revealed that this star has a strong excess of far-infrared emission that can be most simply understood as arising from a shell of large particles, >>10 μm in radius, located approximately 85 A.U. from the star and heated to 85 K by absorbed starlight (Aumann, Gillett et al. 1984).

Photometric Measurements:

Vega plus a number of other bright standard stars were used to check the

IRAS photometry for systematic errors such as spectral leaks and cracked filters. Table 2 shows the results of a series of observations designed to give the best possible photometric accuracy. The upper part of the table gives the star name, Bright Star number, observed flux densities, spectral type and effective photospheric temperature. The second column of the lower part of the table gives the ratio of the IRAS 12 μm measurement for each star to that obtained from ground-based telescopes. The agreement is better that 5 %. The next three columns give the ratio of the IRAS measured flux density to that predicted from a blackbody extrapolation from the 12 μm measurement based on the known photospheric temperature. For five stars the blackbody spectrum provides a good fit to the stellar energy distribution, within the estimated accuracy of the preliminary IRAS calibration, approximately 25 %. Vega is the notable exception and shows excesses of factors of 1.3, 7.4 and 16 at 25, 60 and 100 μm, respectively. The conclusions to be drawn from Table 2 are that the overall quality of the IRAS photometry and absolute calibration is good and that something quite odd is happening near Vega.

TABLE 2

PHOTOMETRIC OBSERVATIONS

Observed Flux Density (Jy)*

SOURCE	BS#	12 μm	25 μm	60 μm	100 μm	Sp Type[a]	Teff[b](K)
α Lyr	7001	28.6	8.7	8.9	7.0	A0V	9,850
α Leo	3982	6.	1.52	0.3e	<0.4e	B7V	12,200
α CMa	2491	102	25	4.0	2.0f	A1V	9,400
α Car	2326	106	23.4	4.1	1.5	F0II	7,500
α Boo	5340	500	110	19.7	6.8	K2III	4,460
β Gru	8636	630	147	28.0	10.2	M5III	2,950

	$F_\nu(obs)/F_\nu(grd)$	$F_\nu(obs)/B_\nu(Teff)$ normalized to 12 μm			Ref.
	12 μm	25 μm	60 μm	100 μm	
α Lyr	0.97	1.28	7.4	16.1	(c)
α Leo	1.04	0.98	1.10e	<4e	(c)
α CMa	1.00	1.03	0.93	1.29f	(c)
α Car	0.99	0.92	0.90	0.91	(d)
α Boo	0.95	0.88	0.87	0.83	(c)
β Gru	1.00	0.89	0.91	0.90	(d)

References--(a) Hoffleit,1964, (b) Johnson,1966, (c) Gehrz, et al, 1974, (d) Thomas et al., 1973

(e) low signal-to-noise measurement---based on inspection of raw data
(f) source confusion with position dependent background.
* Uncertainties are in almost cases dominated by the 25 % uncertainty in the absolute calibration of the IRAS photometry.

Spatial Measurements:

Observations to maximize the positional accuracy and spatial resolution were obtained to measure or set limits to the size of the far-infrared source and to insure that the far-infrared source was truly associated with Vega and not due to an unrelated object along the same line of sight such as a distant, infrared bright, spiral galaxy. Scans made of Vega, β Gru and α Boo led Aumann, Gillett et al. to conclude that for Vega the centroids of the emission at 25, 60 and 100 μm coincide with that of the 12 μm source to within <3''. The probability of the chance coincidence of an unrelated, far-infrared source along the line of sight to Vega is vanishingly small.

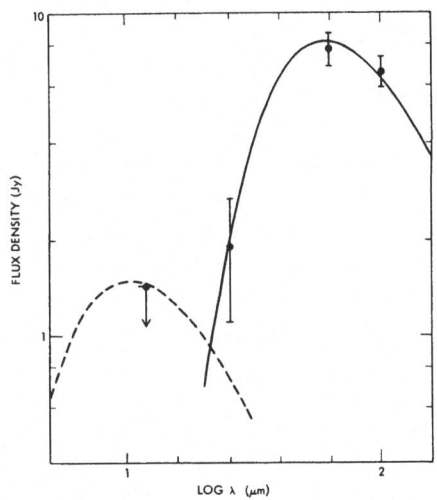

Fig. 6. Energy distribution of excess emission from α Lyrae over that expected from 10,000 K blackbody (solid line). The dashed line shows the maximum emission from a 500 K blackbody that is consistent with the limit to the excess at 12 μm.

The observations also showed that at 60 μm the size of the emitting region is approximately $15\pm4''$ and $23\pm6''$ in two orthogonal directions, or, roughly 20''. There is no statistically significant evidence for asymmetry. Vega is 8.1 pc from earth so that the linear radius of the shell is approximately 85 A. U.

Figure 6 shows the spectrum of the excess obtained by subtracting from the data the spectrum of a 10,000 K blackbody normalized to the observed 2.2 μm flux density of Vega. The shape of the excess is inconsistent with that of optically thick or thin free-free emission from a stellar wind such has been observed around early type stars (Gehrz et al. 1974).

Emission from an 85 K blackbody accounts for the observations very well, considerably better than emission from lower temperature grains with an emissivity proportional to frequency. The size of the emitting grains must exceed $\lambda/2\pi$ where λ is the longest wavelength at which the spectrum is

Planckian. Thus, the grains in the shell must be larger than 10 µm in radius. The luminosity of the infrared shell is 2.5×10^{-5} L_O, implying a mean visual optical depth of 2.5×10^{-5} if one interprets the emission as arising from an optically thin shell of material around the star.

A check on the consistency of this picture comes from comparing the measured angular size of the shell with the value deduced from the luminosity of the star and the temperature of the shell. From simple energy balance considerations Aumann, Gillett et al. derive

$$\theta_d/\theta_* = 0.5(\epsilon_{vis}/\epsilon_{ir})^{1/2}(T_*/T_d)^2$$

where θ_d, θ_*, T_d and T_* are the angular sizes and temperatures of the shell and star and $\epsilon_{vis,ir}$ are the emissivities of the grains at visual and infrared wavelengths. Using known quantities for the star (Drelling and Bell 1980) one can show that

$$\theta_d = 20.7(\epsilon_{vis}/\epsilon_{ir})^{1/2} \text{ arcsec,}$$

which is in excellent agreement with the observations if the ratio of the emissivities is roughly unity as is expected for large particles. Models incorporating typical interstellar grains with infrared emissivities <<1 would result in a shell much larger than the observed 20''.

It should be pointed out that Vega differs in two ways from other stars with infrared excesses. First, the size the of grains around most stars with shells is very small, typically <0.1 µm, while the grains around Vega are larger than 10 µm. Second, and most importantly, other stars with dust shells, such as late type OH/IR stars, are known to be losing mass. Vega is in a quiescent part of the main sequence and observations of UV resonance lines set a stringent limit of 10^{-12} M_O yr^{-1} to the mass loss rate from Vega (Lamers et al. 1978; Lamers and Waters 1983).

An important physical property of the shell around Vega is its total mass. Unfortunately the IRAS observations set only the most crude limits on this quantity. IRAS measures the surface area of the emitting material, but the mass depends on the volume of material. The ratio of the two depends on the poorly known size of the particles. The mass of the shell can be written as

$$M = 4/3 \, A \, \rho \, \langle a^3 \rangle / \langle a^2 \rangle / \epsilon_{ir}$$

where A is the projected physical area of the shell; ρ the density of the shell material; taken to be 2 gm cm^{-3}; and a, the particle radius; ϵ_{ir} is assumed to be unity. Aumann, Gillett et al. invoked the Poynting-Robertson effect to estimate the minimum size for the particles in the shell. The effects of radiation pressure over

10^8 yr would have either blown away from or dragged into Vega any grains smaller than 0.6 mm in radius. If all grains are this minimum size, then the shell mass is 1.2×10^{-2} $M_\oplus$. If one assumes a power law distribution, $dN(a) \propto a^{-3.5} da$, characteristic of material in the solar system up to the largest asteroids, then $\langle a^3 \rangle / \langle a^2 \rangle = (a_{min} \cdot a_{max})^{1/2}$. Leaving $a_{min} = 0.6$ mm as before and arbitrarily adopting a_{max} to be the radius of Ceres, 5×10^7 cm, yields a shell mass of 300 $M_\oplus$.

Aumann, Gillett et al. suggest that this material is the remnant of the nebular material out of which Vega formed. They point out that the grains around Vega are much larger than typical interstellar grains so that there must have been a significant growth in particle size during the life of the star. Finally, they note that the main sequence lifetime of an A0 V star, 3×10^8 yr, is within the timescale expected for active planet formation, 10^7–10^9 yr (Safronov 1980).

IRAS OBSERVATIONS OF OH/IR STARS

Many oxygen rich M supergiant and Mira variable stars undergoing mass loss are known to be strong emitters of infrared radiation and OH maser emission at 1612 MHz. In recent years sensitive searches along the galactic plane in the OH lines have yielded some 300 OH maser sources (Baud et al. 1981, Herman 1983). An infrared counterpart is usually found if the radio positional uncertainty is small and the wavelength of search is sufficiently long (e.g. Jones et al. 1982)

Stellar OH maser emission is pumped by 35 μm photons. Theory predicts (Elitzur et al. 1976) and observations confirm (Herman 1983) that the ratio of radio to infrared flux density is approximately 0.25. As the limiting flux density of the radio searches is about 1 Jy, the infrared counterparts of these OH sources should be stronger than 4 Jy at 25 μm–60 μm, considerably above than the IRAS survey limit. In unconfused parts of the sky IRAS should find all OH/IR stars known from radio searches and discover new, fainter objects as well.

Olnon et al. (1984) used the IRAS survey to locate the infrared counterparts of 40 OH masers brighter than 4 Jy at 1612 MHz and located in the first quadrant of the galaxy (Baud et al. 1981). Figure 7 shows a color-color diagram for OH/IR stars including optical Miras with 2000 K photospheres, IRC and AFGL sources with dust temperatures around 400 K and extremely red objects with color temperatures less than 100 K. The range in redness is attributed to a large variation in mass loss rate, from less than 10^{-5} M_0 yr^{-1} to in excess of 10^{-4} M_0 yr^{-1}. In the figure, stars with large radio variations are shown as filled circles, those with small variations (<0.3 mag) as open circles. There is a definite break in the distribution for a 12 to 25 μm flux ratio around 1 with redder stars showing less variation. This result can be understood, Olnon et al. suggest, if, at the end of the mass loss phase, stars pulsate less than at earlier times (Herman 1983) and lose mass at the greatest rate (Baud and Habing 1983).

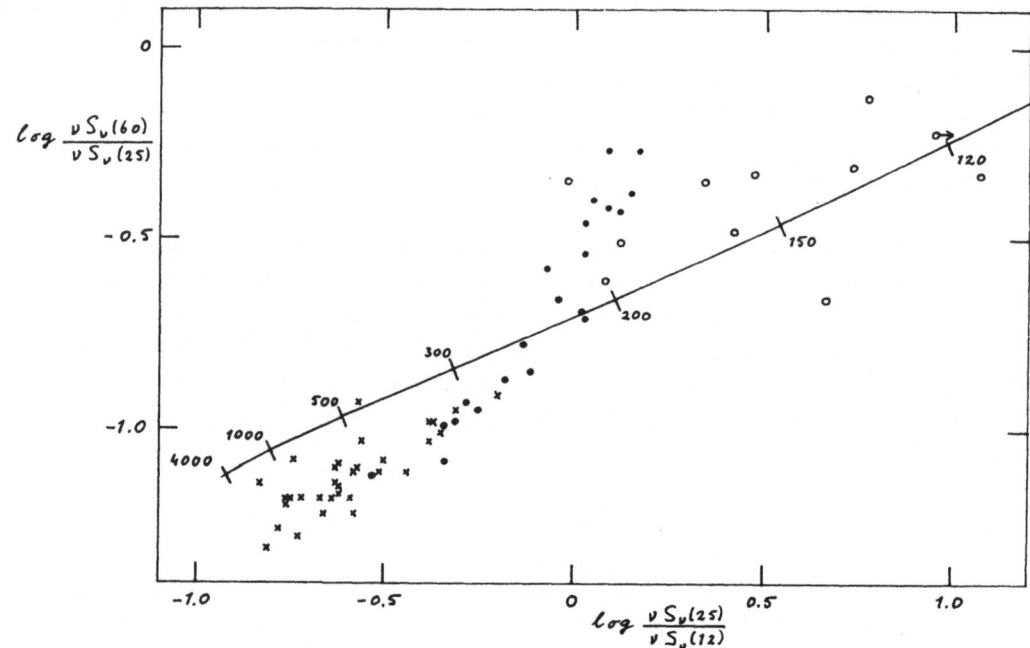

Fig 7. Color-color diagram for OH/IR stars as discussed in the text. The line shows the locus of blackbodies at various temperatures. The open circles indicate OH/IR stars with peak-to-peak OH amplitudes less than 0.3 mag and the filled circles those with lager amplitudes. Crosses indicate optical Miras, IRC or AFGL sources associated with OH emission.

Acknowledgements

    We thank all of the hardware and software engineers and the operations personnel on both sides of the Atlantic whose skills and dedication made IRAS such a great success. CAB would like to thank Dr. A. Dupree for making it possible to present this talk.

References

Baud, B., Habing, H. J., Matthews, H. E., and Winnberg, A. 1981,
    Astr. Ap., 95, 156
Baud et al. 1984, Ap J. (Letters), in press.
Beichman, C. A. and Harris, S. H. 1981, Ap. J., 245, 589.
Beichman et al. 1984, Ap J. (Letters), in press.
Benson, P. J. 1983, unpublished Ph. D. Dissertation, M. I. T.
Benson, P. J., Myers, P. C. and Wright, E. L. 1984, in prparation.

Cohen M. 1983, Ap J. (Letters), 270, L69

Drelling, L. A., and Bell, R. A. 1980, Ap. J., 241, 736.

Elitzur, M., Goldreich, P. and Scoville, N. Z.
1976, Ap. J., 205, 384.

Freidlund, C. V. M., Nordh, H. L., Van Duinen, R. J.,
Alders, J. W.G., Sargent, A. I. 1980, Astr. Ap. 91, L1.

Gehrz, R. D., Hackwell, J. A. and Jones, J. 1974., Ap.J.,191,675.

Harvey, P. M., Thronson, H. A., and Gatley, I. 1979, Ap. J., 231, 115.

Harvey, P. M., and Wilking, B. A. 1982 Pub. A. S. P., 94, 285.

Herman, J. 1983, Ph. D. Thesis, Leiden University.

Hildebrand, R. H. 1983, Quart J. R. A. S., in press.

Hoffleit, D. 1964, Yale Catalog of Bright Stars, Third Rev.
Yale Univ. Press., New Haven, Conn.

Houck et al. 1984, Ap J. (Letters), in press.

Hyland, A. R., Jones, T. J., and Mitchell, R. M. 1982,
M.N.R.A.S., 201, 1095.

Johnson, H. L. 1966, Ann. Rev. Astr. Ap., 4, 193.

Jones, T. J., Hyland, A. R., Wood, P. R., and Gatley, I. 1982, Ap. J., 253, 208.

Keene, J., Davidson, J., Harper, D. A., Hildebrand, R. H., Jaffe, D. T.,
Loewenstein, R. F., Low, F. J. and Perenic, R. 1983, Ap J., in press.

Lamers, H., Stalio, R., and Kongo, Y. 1978, Ap. J., 223, 207.

Lamers, H., and Waters, R. 1983, private communication.

Low et al. 1984, Ap J. (Letters), in press.

Myers, P. C., Linke, R. A., and Benson, P. J. 1983,
Ap. J., 264,517.

Neugebauer et al. 1984, Ap J. (Letters), in press.

Rowan-Robinson et al. 1984, Ap J. (Letters), in press.

Safronov, V. S. 1980, Accumulation of the Protoplanetary Bodies. Early Solar
System Processes and the Present Solar System: Proceedings of the International
School of Physics, "Enrico Fermi", Course LXXIII.,
North Holland Publishing Company, p. 58.

Soifer et al. 1984, Ap J. (Letters), in press.

Stahler, S. W., Shu, F. H., and Taam, R. E. 1980. Ap. J., 241, 637.

Thomas, F. A., Hyland, A. R., Robinson, G. 1973.,
M. N. R. A. S.,165,201.

Young, J. S., Goldsmith, P. F., Langer, W. D., Wilson, R. W.,
and Carlson, E. R. 1982, Ap. J., 261, 513.

Young et al. 1984, Ap J. (Letters), in press.

RS CVn BINARY SYSTEMS

Jeffrey L. Linsky[*]

Joint Institute for Laboratory Astrophysics, National Bureau of Standards and
University of Colorado, Boulder, Colorado 80309

ABSTRACT

In this review I will attempt to place in context the vast amount of data obtained in the last few years as a result of X-ray, ultraviolet, optical, and microwave observations of RS CVn and similar spectroscopic binary systems. Since this topic is now very broad, I will concentrate on the RS CVn systems and their long-period analogs, and restrict the scope by attempting to answer on the basis of the recent data and theory the following questions: (1) Are the original defining characteristics still valid and still adequate? (2) What is the evidence for discrete active regions? (3) Have we derived any meaningful physical properties for the atmospheres of RS CVn systems? (4) What are the flare observations telling us about magnetic fields in RS CVn systems? (5) Is there evidence for systematic trends in RS CVn systems with spectral type?

I. INTRODUCTION

While the study of close binary systems that we now consider members of the RS Canum Venaticorum class goes back many decades in time, the recognition that these systems constitute a well defined class of objects with common characteristics began with Hall's (1976) review paper published only seven years ago. Subsequently, ultraviolet and X-ray observations with the Copernicus, IUE, HEAO-1, and HEAO-2 spacecraft, together with microwave and optical observations from the ground have revealed a vast range of fascinating phenomena that were not anticipated even seven years ago. Later reviews by Catalano, Frisina, and Rodono (1980), Dupree (1981), Hall (1981), Rodono (1983), Bopp (1983), and Catalano (1983) have summarized much of this complex phenomenology and have presented the generally accepted starspot model in which magnetic fields and their interactions with the ambient plasma are presumed to be responsible for the dark starspots in the photosphere and the greatly exaggerated but solarlike activity occurring in the chromosphere and corona.

In this review I propose to take a somewhat different approach. Instead of reviewing phenomena per se, I will ask a number of basic questions concerning the RS CVn systems and then attempt to answer these questions by describing the relevant

---

[*]Staff Member, Quantum Physics Division, National Bureau of Standards.

244

data and theoretical work on the topic. The purpose is to attempt to place in context the vast amount of data obtained in the past few years from the X-ray to the microwave regions of the spectrum. Of particular concern are the atmospheric structures of these stellar systems and the roles played by magnetic fields. I will also emphasize very recent observations, many of which are not yet published.

II. ARE THE ORIGINAL DEFINING CHARACTERISTICS STILL VALID AND STILL ADEQUATE?

a) Defining Characteristics

Close binary stars display a vast array of phenomena depending on the spectral types, mass ratios, and evolutionary status of the component stars, whether one or both stars fill their Roche lobes, and the proximity of the two stars. In order to bring some order out of this chaos and to define a relatively homogeneous class of systems with similar characteristics and phenomena observed in the visible, Hall (1976) proposed a working definition of RS CVn systems consisting of the three properties listed below. Are these three defining characteristics still valid?

1. Binaries with orbital periods of 1-14 days. Hall chose these upper and lower limits rather arbitrarily on the basis that for the 24 stars, which he felt have similar characteristics and thus should be members of the RS CVn class, the distribution of periods was such that there were no suspected candidates with periods of 0.9-1.9 days and 11-17 days. Since the list of known RS CVn binaries keeps growing (Hall [1981] lists 69 members), we should question this original definition and search for a definition that rests on a physical basis. Probably the main characteristic separating the more evolved (and in general more active) components in these systems from single stars of the same spectral type is rapid rotation, a direct consequence of tidally-induced synchronism of rotation and orbital periods.

Zahn (1977) has shown that for stars with convective envelopes, the synchronization time is

$$t_{synch} \approx 10^4 \left[\frac{(1+q)}{2q} \right]^2 P^4 \text{ yr} \quad , \tag{1}$$

where q is the mass ratio, which is generally close to unity for these systems, and P is the orbital periods in days. Thus $t_{synch} = 10^8$ yr for P = 10 days and 1.6 × 10^9 yr for P = 20 days when q ≈ 1. Evolutionary considerations then suggest that subgiants with orbital periods less than 20 days should be tidally synchronous, and thus rapid rotators and active stars as is indeed the case. A well-known nonsynchronous system, λ And (G8 III-IV + ?), with P_{orb} = 20.5 days and P_{rot} = 54 days confirms this argument. Thus a natural division between RS CVn systems and the so-called long period RS CVn systems is about 20 days, not 14 days as originally suggested. The proper short period cutoff to the RS CVn class is unclear at this time, but it seems reasonable to exclude contact binaries such as the W UMa systems (Dupree 1983) on the basis that they exhibit somewhat different properties, such as a common coronal envelope.

2. <u>Strong Ca II H and K line emission outside of eclipse.</u> Bright H and K
emission historically has been used to identify new members of the RS CVn class because these emission features stand out against the relatively weak photospheric
continuum and photospheric Ca II absorption line wings even at relatively low dispersion. Bopp (1983) has tabulated Ca II surface fluxes for 19 RS CVn systems,
which are typically 2-20 times that of the quiet Sun. The Hα line is a pure emission feature in only the most active systems (e.g. HR 1099, UX Ari, II Peg, and
IM UMa), and appears as a filled-in absorption feature in the other systems. While
one can subtract the Hα profile of a standard single star to obtain a net Hα emission profile (e.g. Fraquelli 1982), this is not a simple technique; thus filled-in
Hα absorption should not be a defining characteristic for these systems. Enhanced
ultraviolet emission lines (e.g. Dupree 1981) and X-ray luminosity (Charles 1983)
are common properties of these systems and could be used as defining spectral characteristics, but it is not yet feasible to search large numbers of stars for these
features. On this basis, it seems reasonable to continue to use bright H and K line
emission as a defining characteristic, but we should recognize that the strong Ca II
line emission is only one indicator of the enhanced nonradiative heating rate, which
is the physical basis responsible for this defining characteristic.

3. <u>The hotter star is of spectral type F or G and luminosity class V or IV.</u>
If we adopt P_{orb} = 20 days as defining the long period cutoff of the RS CVn group,
the only systems that are inconsistent with this definition (Hall 1981) are single
line spectroscopic binaries for which the other component is unknown. Three are
luminous systems with periods of 17-20 days (σ Gem, ζ And, and V350 Lac) and the
other is II Peg (P_{orb} = 6.7 days). What are the secondary components in these systems? Are they less evolved stars close to the main sequence as is generally the
case for RS CVn systems, or could they be white dwarfs as in V471 Tau, but cooler?

b) <u>Nondefining Characteristics</u>

Hall (1976) also lists 15 additional characteristics that appeared to be valid
for most but not all of the systems for which he had data at that time, and thus
were not suitable as defining characteristics. I would like to comment on several
of these.

1. <u>The Ca II H and K emission is from the cooler star (or both).</u> This is generally true for the RS CVn systems, but it is important to recognize that the hotter
star is often also quite active, and when the orbital velocity separation of a spectrum is large the contribution of the hotter star is often apparent in high resolution spectra. For example, Simon and Linsky (1980) detected Mg II h and k emission
from the G5 V star in UX Ari about 1/5 as strong as that of the K0 IV star. But
since the radius of the K0 IV star is about twice that of the G5 V star, the Mg II
surface fluxes of the two stars, indicative of the general activity level, are similar. The long period Capella system (G6 III + F9 III, P_{orb} = 104 days) is a clear
exception in that the earlier type star is the predominant emitter (Ayres and Linsky

1980; Ayres, Schiffer, and Linsky 1983), presumably because the F9 III star is a
rapid rotator whereas the G6 III star is not. In some sense this exception proves
the rule since for synchronous systems the more evolved (i.e. cooler) star has a
larger radius and is thus the more rapid rotator.

2. <u>A wave-like distortion in the optical light curve is detected outside of
eclipse</u>. This characteristic is now deemed fundamental as it indicates the presence
of dark starspots (e.g. Eaton and Hall 1979) that migrate in phase (Catalano and
Rodono 1974), are cool (Vogt 1979; Ramsey and Nations 1980), and are presumed to be
magnetic in character by analogy with sunspots. This last point is critical because
strong magnetic fields presumably underlie all of the interesting activity seen in
RS CVn systems, yet rapid rotation makes it difficult to measure the magnetic fields
directly by the Zeeman effect. To my knowledge the only direct measurement of a
magnetic field is the measurement of a field of 1290 ± 320 Gauss covering half the
visible surface of λ And (Giampapa, Golub, and Worden 1983). It is important that
this work be extended.

3. <u>RS CVn systems are detached binaries with mass ratios close to unity</u>. To
my knowledge the only exceptions to this statement are RT Lac and SZ Psc, which are
semidetached systems, and the single line spectroscopic systems for which the mass
ratios are unknown. Popper and Ulrich (1977) have called attention to the interesting evolutionary status of the RS CVn systems. They point out that binaries develop
RS CVn characteristics when one or both stars enter the Hertzsprung gap and develop
convective envelopes. The tidal synchronism mechanism will halt the rapid loss of
rotational velocity when $P_{rot} = P_{orb}$, so that the K0 IV and K0 III cool stars in
these systems have equatorial rotational velocities of 40-60 km s^{-1} instead of 2-5
km s^{-1} (Gray 1982). This combination of rapid rotation and convection presumably is
responsible for the efficient generation of strong magnetic fields by dynamo processes that results in the RS CVn phenomena. For most systems there is no evidence
for streams, tidal distortions, or reflection effects, consistent with the detached
geometry.

III. WHAT IS THE EVIDENCE FOR DISCRETE ACTIVE REGIONS?

a) Chromospheres and Transition Regions

RS CVn systems have been monitored extensively in the Hα line (see Bopp 1983
for a detailed review), but these data exhibit no evidence for active regions correlated with starspots. For example, Ramsey and Nations (1980) observed stronger Hα
emission in HR 1099 at a phase near spot maximum (photometric minimum) than at spot
minimum, but they did not monitor for flares at that time. In an extensive study of
HR 1099 during 1977-79, Dorren <u>et al</u>. (1981) monitored the Hα line with narrow band
photometry. They detected strong net emission in Hα with large nightly variations,
but no correlation with photometric phase even though the photometric wave showed
a large amplitude at that time. They concluded that the variable Hα net flux indi-

cated flaring rather than the presence of active regions. Similarly, Fraquelli (1982) also detected variable Hα net flux from HR 1099 that correlated with the microwave radio flux, a good indicator of flaring. When she removed from the data set those observations taken during flares, she also found no correlation of Hα emission with photometric phase.

There are fewer observations of RS CVn systems in the Ca II H and K lines, but these data do provide some evidence for active regions correlated with the star spots. For example, Weiler (1978) observed six systems in the Ca II lines and Hα. The data are sparse, but the Ca II emission equivalent widths appear to strengthen at photometric minimum for RS CVn itself, consistent with the hypothesis that chromospheric active regions cluster above starspots. Weiler's observations of UX Ari and Z Her are also marginally consistent with the above hypothesis.

The connection between active regions and starspots became clear only with observations by IUE. The first such evidence by Rodono, Romeo, and Strazzula (1980) is based on low dispersion Mg II fluxes of II Peg (K2-3 V-IV + ?). Subsequently, Baliunas and Dupree (1982) observed λ And at photometric maximum and minimum with IUE. They detected transition region line (e.g. C II, C IV, Si IV) emission 30–50% brighter, and the Ca II lines brighter, at photometric minimum than at maximum, but for some unexplained reason the Mg II line fluxes were nearly unchanged. Also the Ca II lines have different asymmetries, suggesting that the flows are different on the two hemispheres. Walter, Gibson, and Basri (1983) observed Ar Lac (G2 IV + KO IV) with IUE at egress (phase 0.053) from primary eclipse (KO IV star in front) and at quadrature (phase 0.256). These data (see Figs. 1 and 2) show the Mg II flux

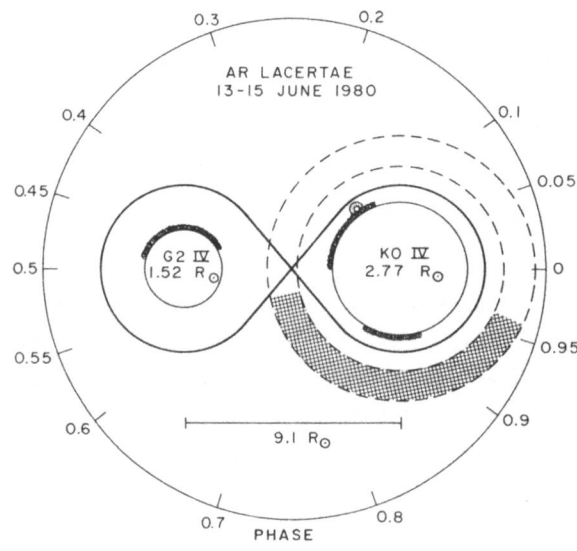

Fig. 1. A scale drawing of the AR Lacertae system. The line of sight at a given phase is found by lining up the phase indicated on the outer circle with the center of mass. The solid line is the Roche surface; the dashed lines surrounding the K star indicate the inner and outer radii (1.5 and 2.0 R_k) of the extended component of the K star corona. Crudely indicated are the location and extent of the observed chromospheres and coronae. Note that the extended component of the K star corona exceeds the Roche radius (from Walter, Gibson and Basri 1983).

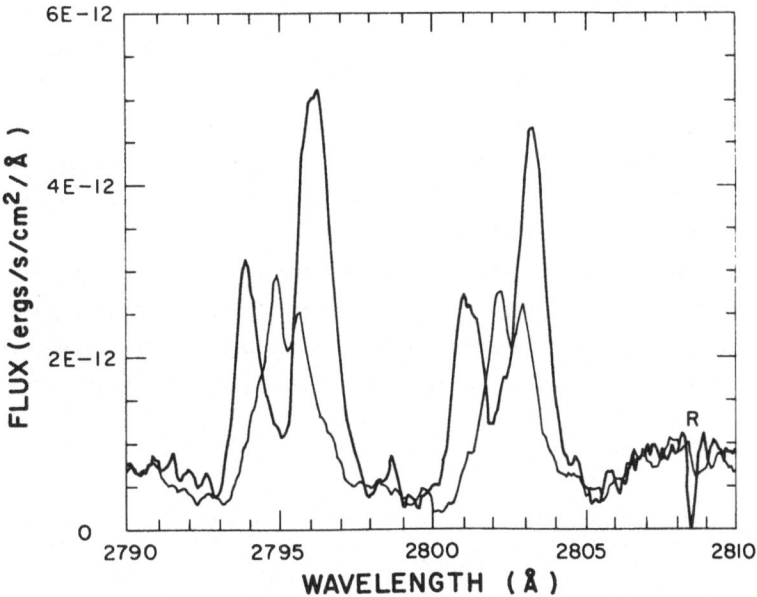

Fig. 2. High-resolution observations of the Mg II resonance lines for AR Lac during egress from primary minimum (thin lines) and at quadrature (thick lines). Approximately half of the G star is visible at this stage of egress. Note that although the G star is the one eclipsed, it produces the approximately constant blueward component while the K star shows a dramatic change (having remained unocculted). The flux ratio of K to G star is 2:1 at quadrature, indicating the G star actually has stronger surface flux here (from Walter, Gibson and Basri 1983).

attributable to the G2 IV star unchanged, but the Mg II flux attributable to the KO IV star a factor of two larger at quadrature. They argued that the data are consistent with an active region (visible in both observations) above the spot group on the G2 IV star producing the photometric minimum at phase 0.25, and an active region on the KO IV star in view only during the quadrature observation.

Two sets of IUE high resolution spectra with the short wavelength camera indicate the presence of compact active regions by changes in the line integrated fluxes and centroid velocities. In the first data set, Ayres and Linsky (1982) observed HR 1099 (G5 IV + K1 IV) at opposite quadratures (phases 0.21 and 0.76). They found that at phase 0.76 the transition region lines (e.g. C II 1336 Å and C IV 1548 Å) are brighter and displaced +40 km s^{-1} relative to the K1 IV star velocity, indicating an active region near the receding limb of the KO IV star at this phase but on the back side of the star at phase 0.21. This would put the central meridian passage of the active region at phase 0.6 in June 1980, but photometric minimum in 1979 occurred near phase 0.95. Thus the connection between spot and active region is not clear for this observation.

A better example is the high resolution spectra of σ Gem (K1 III + ?) obtained at phases 0.53 and 0.58 by Ayres, Simon, and Linsky (1984). They found that the transition region lines were both stronger and blue-shifted at phase 0.53 compared to phase 0.58, consistent with an active region on the receding limb at phase 0.53 but over the limb at phase 0.58. Since starspot group B (see Fried et al. 1983) was also on the receding limb at phase 0.53, the spatial connection of the active region and a spot group is indicated by these data.

The clearest example yet of the spot-active region connection is the October 1-7, 1981 IUE monitoring of II Peg. In this program, Marstad et al. (1982) found that the ultraviolet emission line flux is well correlated with photometric minimum (see Fig. 3). In particular, all the emission lines rise sharply at phase 0.45 and

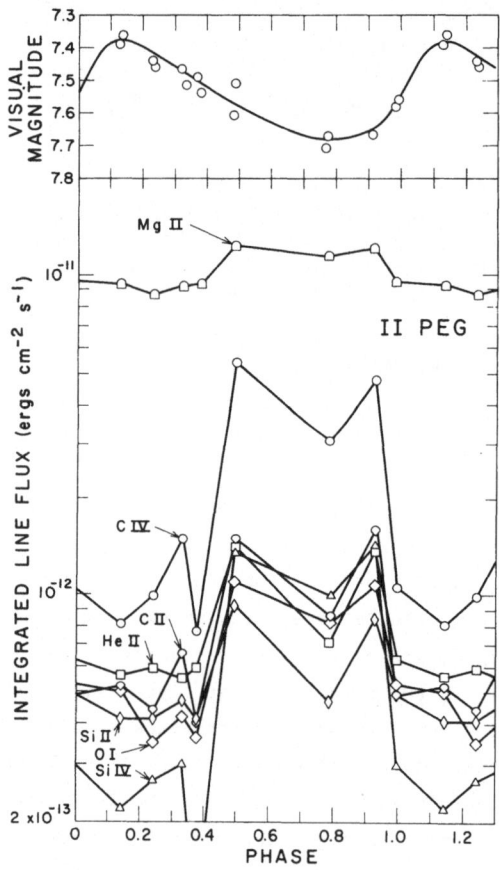

Fig. 3. Lower Panel: integrated emission line fluxes for II Peg obtained in October 1981 by Marstad et al. (1982). Note the rapid rise in flux near phase 0.45 and rapid fall near phase 0.95 indicating the rotational modulation of a compact active region across the disk. Upper Panel: photometric variation obtained with the FES simultaneously with the IUE spectra.

fall sharply at phase 0.95, indicating a rather compact active region. The much larger rise of the transition region lines (a factor of five) compared to the chromospheric lines (less than a factor of two) is consistent with the difference between the spectra of solar active and quiescent regions. Marstad (1983) used these data to locate the active region on the stellar surface and compared this active region with the location of the two spot groups he derived from the IUE optical light curve, which is similar to the 1976 light curve and spot group positions derived by Bopp and Noah (1980). Marstad placed the active region near the leading edge of the larger spot group (see Fig. 4), and concluded that its area is no larger than 6% of the visible hemisphere, compared to the spot area of 25-30% of the visible hemisphere.

This result is remarkable to say the least. Whereas solar active regions are much larger than the photospheric spots they overlie, the situation is reversed for II Peg. Second, the small active region size implies that the active region surface fluxes for the transition region lines must be very large as indicated by Table 1. For an assumed surface covering factor of 0.06, the maximum value indicated by Marstad, the C IV surface flux is 4200 times the quiet Sun value and the C IV 1550 Å lines carry 0.1% of the total stellar luminosity per unit area. These extreme

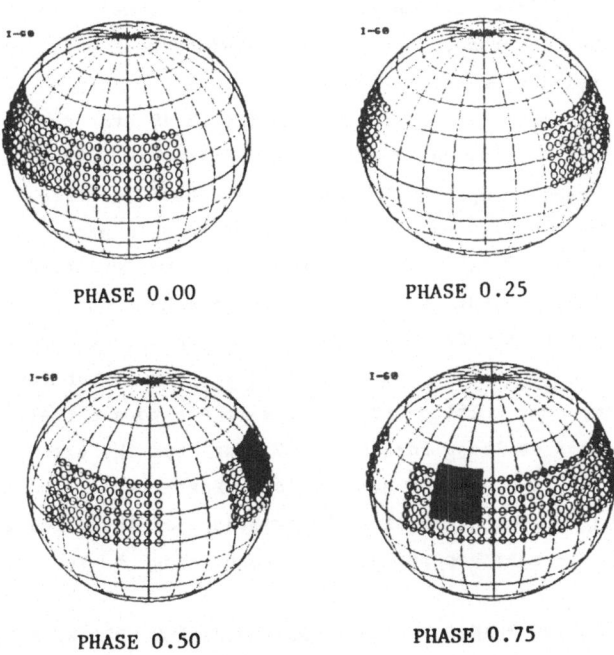

PHASE 0.00   PHASE 0.25

PHASE 0.50   PHASE 0.75

Fig. 4.   The location of the two spot groups (small circles) and the active region (solid black) derived by Marstad (1983) from the optical photometry and emission line flux versus phase observations of II Peg in October 1981. Note that the active region overlies a small portion of the larger spot group.

Table 1. Active Region Surface Fluxes

Assumed Surface Covering Factor	$F_{C\ IV}$(active region) (ergs cm^{-2} s^{-1})	$\dfrac{F_{C\ IV}\text{(active region)}}{F_{C\ IV}\text{(quiet Sun)}}$	$\dfrac{F_{C\ IV}\text{(active region)}}{\ell_{bol}}$
1.00	1.7×10^6	2.6×10^2	7.3×10^{-5}
0.06	2.8×10^7	4.2×10^3	1.2×10^{-3}
0.02	8.3×10^7	1.3×10^4	3.6×10^{-3}

values point out the necessity for deriving active region areas and the folly of describing RS CVn chromospheres and transition regions by one-component theoretical models.

b) Coronae

I now consider the evidence for discrete active regions in the coronae of RS CVn systems based on observations with the Einstein X-ray observatory. The most important set of observations has been obtained by Walter, Gibson, and Basri (1983) of AR Lac during a total primary eclipse (KO IV star in front at phase 0.0) and a annular secondary eclipse (G2 IV star in front). The geometry for these eclipses is shown in Figure 1. Eclipses are powerful probes of coronal X-ray brightness distributions and thus the location of active regions since each position in the corona is covered and uncovered during each eclipse.

From these data Walter et al. found that the G2 IV star contributes 40% of the total X-ray flux. Its corona has a small scale height (~0.02 R_*) and is located primarily over the spot group, which is near phase 0.25 (see Fig. 1). The KO IV star exhibits a more complicated coronal structure with a geometrically thin component (scale height ~0.02 R_*) located primarily at two longitudes and with an extended component (scale height ~R_*) located over one hemisphere. In addition, the IPC spectral height distribution indicates a two temperature coronal plasma, as does the SSS data (Swank et al. 1981), and Walter et al. argue that the hotter plasma is in the extended component. This data set thus points to a correlation of spots and coronal active regions at least for the G2 IV star, but the extended (hot?) component is not connected to any known spot region. Furthermore, the location of hot plasma far from the KO IV star implies the existence of large loops extending beyond the Roche lobe of this star and perhaps interconnecting the two stars in the system.

IV. HAVE WE DERIVED ANY MEANINGFUL PHYSICAL PROPERTIES FOR THE ATMOSPHERES OF RS CVn SYSTEMS?

a) Chromospheres and Transition Regions

In the past few years there have been at least four major studies of RS CVn systems that purport to derive the run of temperature, pressures, and density with height using different spectroscopic diagnostics. In each case the authors solved the radiative transfer and statistical equilibrium equations for an assumed homogeneous, one component atmosphere in hydrostatic equilibrium so as to match computed and observed line fluxes and in some cases line profiles. For example, Baliunas et al. (1979) derived chromospheric models representative of λ And and Capella to match the observed Ca II, Mg II, and Hα profiles. Their best fit models are characterized by top pressures, appropriate for the base of the transition region, of 1-1.7 dynes cm^{-2}. Subsequently, Simon and Linsky (1980) derived models for HR 1099 and UX Ari to match the observed Mg II line profiles, fluxes of C II, Si II, and Si III lines, and three density sensitive line ratios. Their best fit models have a top pressure of roughly 0.5 dynes cm^{-2} and are inconsistent with a transition region that is conductively heated. On the basis of models constructed to match the Hα profile of λ And, Mullan and Cram (1982) derived transition region pressures of 0.06 or 0.4 dynes cm^{-2} depending on the assumed macroturbulence. Finally, Baliunas and Dupree (1982) proposed transition region pressures of 1.3 dynes cm^{-2} for the unspotted hemisphere and 1.9 dynes cm^{-2} for the spotted hemisphere transition region of λ And, based on observed C I and C II line fluxes.

These models, based on a range of spectroscopic diagnostics, different RS CVn systems, and computed by different groups, are in reasonable agreement with each other, but are they realistic representations of the mean atmospheric properties of RS CVn systems? It seems to me that they are not, because they ignore what is likely a fundamental property of the atmospheres of these systems -- extreme inhomogeneity. If the results for II Peg described above are representative, then most of the observed chromospheric and transition region emission originates in one or a few active regions covering a small portion of the observed hemisphere of the star. Thus the active region line surface fluxes could be 10-100 times that of the spatial average, and the transition region pressures and densities for the active region correspondingly larger. I suspect, but cannot prove, that the same is true for those systems that show little rotational modulation of the UV emission line fluxes, except that in these systems there are several active regions widely distributed in longitude observed at all phases. It seems vitally important, therefore, that future studies concentrate on determining the active region filling factor and then on computing models of the active and quiescent regions separately. One way of deriving active region filling factors is by rotational modulation studies of systems with simple photometric waves of large amplitude, as was the case for II Peg in October 1981. The second way involves deriving the densities of active regions from density-sensitive line ratios when the active regions dominate the observed flux.

b) Coronae

Numerous RS CVn systems were detected as X-ray sources by HEAO-1 because they are intrinsically bright ($29.4 \leq \log L_x \leq 31.5$) and are quite numerous. Walter et al. (1980) listed 15 sources, and Charles (1983) listed 45 sources detected by HEAO-1 and Einstein and reviewed their properties. The HEAO-1 data suggested coronal temperatures of 10^7 K. Assuming this temperature and the Rosner, Tucker and Vaiana (1978) scaling law for static loops smaller than a pressure scale height,

$$T = 1400 \ (PL)^{1/3} \quad , \tag{2}$$

Walter et al. (1980) estimated values of loop heights (L), fractional filling factors (f), and number of loops (N) for different stars. Clearly these quantities depend on the pressure (P) in the loops, but assuming that for Capella P = 1.5 dynes cm^{-2}, the mean transition region pressure, then f = 0.2, N = 100, and L = 2 × 10^{11} cm ≈ 0.3 R_*. These loops are large but still only about 0.05 of the coronal pressure scale height.

Since the RS CVn systems are bright X-ray sources, Swank et al. (1981) were able to observe seven systems and Algol with the Einstein Solid State Spectrometer. These low resolution spectra indicate that all of the systems have coronae characterized by at least two temperatures: a low temperature component with T = 4-8 × 10^6 K and log L_x = 30-31, and a high temperature component with T = 20-200 × 10^6 K and log L_x = 29.1-31.1. Furthermore, the high temperature components appear to be more variable than the low temperature components, and the most widely separated system (Capella) has the smallest ratio of high to low temperature component luminosity. I will return to this important point later. They also applied the Rosner-Tucker-Vaiana scaling law [Eq. (2)] with the conclusion that if P $\lesssim$ 10 dynes cm^{-2}, then L/R_* for the hot coronal component is similar to the binary separation and L/R_* for the cool coronal component is $\lesssim$1 R_*. However, if P $\gtrsim$ 100 dynes cm^{-2} then both components are relatively compact geometrically. They could not decide between these two scenarios.

The Einstein IPC observations of AR Lac during eclipses provided Walter et al. (1983) with the critical data on the coronal emitting region volumes and locations needed to derive the loop pressures directly from Eq. (2). Their only assumption was that the extended component of the KO IV star corona (see Fig. 1) consists primarily of the hot gas detected in earlier SSS observations. Their results, summarized in Table 2, indicate that the coronal loop pressures are large (25-140 dynes cm^{-2}), two orders of magnitude larger than the average transition region pressures previously discussed. This result provides further evidence for the inadequacies of one component models. It is interesting that the II Peg data imply that the active region surface fluxes are 10-100 times larger than quiescent, which may be consistent with the AR Lac coronal loop pressures if pressures scale proportionally to the surface flux. This point needs further consideration.

Table 2. Coronal Loop Parameters for AR Lac

Parameter	G2 IV Star	KO IV Star (extended component)	KO IV Star (inner component)	Flaring Sun
P	100	25	70	140
L/R_*	0.02	2	0.01	0.01
N	10^6	10	10^6	10^6

V. WHAT ARE THE FLARE OBSERVATIONS TELLING US ABOUT MAGNETIC FIELDS IN
RS CVn SYSTEMS?

Flares are highly energetic events in RS CVn systems with time scales of hours
to weeks, much longer than for flares on M dwarf stars like UV Ceti. Reviews of
flare phenomena include discussions of the Hα data (Bopp 1983), X-ray data (Charles
1983), and radio observations (Gibson 1980, Feldman 1983). Also there are a number
of important papers on the very long-lived flare on HR 1099 in February-March 1978
that are included in the December 1978 issue of the Astronomical Journal. From this
wealth of data, I would like to call attention to those data that provide information on the geometry, flows, and magnetic field topology of the flaring plasma.

First, Bopp (1983) pointed out that while the Hα line brightens significantly
during flares, this emission is not modulated at the orbital or rotational period.
In other words, the emitting volume is either large compared to a stellar radius or
the emission occurs in the binary system well away from either star and perhaps between the two stars. The VLBI microwave observations (cf. Feldman 1983) also point
to emission from a large volume, several times the binary separation for the specific case of the April 1981 flare on HR 5110.

Second, there is evidence for mass flows during flares. Bopp (1983) pointed out
that during flares the Hα line becomes very broad ($\sim$400 km s^{-1} during the February-
March 1978 flare on HR 1099) with occasional redward asymmetries. A very important
observation in this regard is a high resolution spectrum of the Mg II lines obtained
by Simon, Linsky and Schiffer (1980) during the New Year's Day 1979 flare on UX Ari.
They found the Mg II lines to be very asymmetric with wings extending out to +475 km
s^{-1}, roughly the escape velocity from either star, and interpreted these profiles as
indicating a mass flow from the K0 IV to the G5 V star that could occur if the large
flux tubes of the two stars interact. This hypothesis suggests that RS CVn flares
are powered by magnetic field annihilation of interacting flux tubes as in solar
flares, except that the scale is vastly larger because it is the flux tubes of two
separate stars that are responsible. The long time scales of RS CVn flares could
be a consequence of the large geometrical scales. Furthermore, the large circular
polarization of flare microwave emission (e.g., Brown and Crane 1978) indicates that
the emission process is magnetic in character.

Third, Uchida and Sakurai (1983), in their theoretical calculations, have shown
that the magnetic flux tubes of active regions on both stars in RS CVn systems will
interact and often interconnect the two stars. In particular, the strong coronal
heating and flares could result from magnetic reconnection as individual starspots
drift across the active longitudes of both stars. They interpreted the low temperature coronal component as plasma confined in small loops and the high temperature
component as plasma confined to loops interconnecting the two stars. This picture
is consistent with that proposed by Walter et al. (1983) for AR Lac on purely observational grounds.

VI. IS THERE EVIDENCE FOR SYSTEMATIC TRENDS IN RS CVn SYSTEMS WITH SPECTRAL TYPE?

I would like to conclude this rather selective review by calling attention to what appears to be a significant difference between the F9 III active star in Capella and the K0 IV active stars in typical RS CVn systems like HR 1099, UX Ari, and II Peg. Ayres, Schiffer and Linsky (1983) obtained high dispersion IUE short wavelength spectra of Capella at three quadratures and one conjunction. These data confirm earlier work (Ayres and Linsky 1980) that the F9 III component has transition region surface fluxes about 25 times brighter than the G6 III primary, presumably because the F9 III star is a rapid rotator ($P_{rot} \approx 8$ days) whereas the G6 III star is a slow rotator. Ayres et al. (1983) also found that Capella is a remarkably steady ultraviolet emission line source (to the few percent accuracy of the IUE photometry) on time scales of hours to 9 months. In a further study involving 22 observations over half an orbital period, Ayres (1983) also found no ultraviolet flux changes at a sensitivity level of <5%.

This remarkably steady flux from the F9 III star in Capella implies either uniform emission across the stellar surface or, more likely, a large number of active regions well distributed in longitude. By contrast, the K0 IV stars in typical RS CVn systems exhibit highly variable emission line flux indicative of only a few active regions or only one active region in the previously discussed case of II Peg in October 1981. I summarize these differences in Table 3.

I believe that these very different properties are not due to different rotational velocities since the rotational period of the F9 III star in Capella is comparable to the periods of the K0 IV synchronously rotating stars. Instead, I believe that these differences can be explained by two effects:

Table 3. Comparison of Active Stars in Capella and Shorter Period RS CVn Systems

Property	Capella	HR 1099, UX Ari, II Peg
Active star	F9 III	K0 IV - K2 IV
Orbital period	104^d	$2.8-6.7^d$
Rotational period	8^d	$2.8-6.7^d$
UV flux variations	<5%	large
Number of active regions	large	few
Radio emission (6 cm)	<0.2 mJy	up to 1000 mJy
$L_x(hot)/L_x(cool)$	0.1	1-3
a/R_{active}	23	4-7
Flaring	never detected	common

(1) The Capella F9 III star has a shallower convective zone and thus a much larger number of convective cells. Also the dynamo and convective zone velocity field together appear to generate many rather than few magnetic active regions.

(2) The separation of the stars in the Capella system is much larger than in the shorter period systems. Thus it is difficult for the magnetic fields of the two Capella stars to interact. According to the previous discussion there should be few flares, weak radio emission, and little hot plasma in the corona, as is observed. The difference between Capella and the short period systems thus strengthens our conclusion that interacting magnetic fields are fundamental to explaining much of the fascinating phenomenology of the RS CVn systems.

I would like to acknowledge partial support of the National Aeronautics and Space Administration through grants NGL-06-003-057 and NAG5-82 to the University of Colorado. It is a pleasure to thank my colleagues Drs. T. Ayres, A. Brown, K. Carpenter, S. Drake, D. Gibson, T. Simon, and F. Walter for stimulating discussions that clarified my thinking on RS CVn systems.

REFERENCES

Ayres, T. R. 1983, in preparation.
Ayres, T. R. and Linsky, J. L. 1980, Astrophys. J., 241, 279.
Ayres, T. R. and Linsky, J. L. 1982, Astrophys. J., 254, 168.
Ayres, T. R., Schiffer, F. H. III, and Linsky, J. L. 1983, Astrophys. J., 272, 223.
Ayres, T. R., Simon, T., and Linsky, J. L. 1984, Astrophys. J., in press.
Ayres, T. R., Stencel, R. E., Linsky, J. L., Simon, T., Jordan, C., Brown, A., and Engvold, O. 1983, Astrophys. J., in press.
Baliunas, S. L., Avrett, E. H., Hartmann, L., and Dupree, A. K. 1979, Astrophys. J. (Letters), 233, L129.
Baliunas, S. L. and Dupree, A. K. 1982, Astrophys. J., 252, 668.
Bopp, B. W. 1983, in Activity in Red Dwarf Stars (IAU Colloq. No. 71), ed. P. B. Byrne and M. Rodono (Dordrecht: Reidel), p. 363.
Bopp, B. W. and Noah, P. V. 1980, Publ. Astron. Soc. Pac., 92, 717.
Brown, R. L. and Crane, P. C. 1978, Astron. J., 83, 1504.
Catalano, S. 1983, in Activity in Red Dwarf Stars (IAU Colloq. No. 71), ed. P. B. Byrne and M. Rodono (Dordrecht: Reidel), p. 343.
Catalano, S., Frisina, A., and Rodono, M. 1980, in Close Binary Stars: Observations and Interpretation (IAU Colloq. No. 88), ed. M. J. Plavec, D. M. Popper, and R. K. Ulrich (Dordrecht: Reidel), p. 405.
Catalano, S. and Rodono, M. 1974, Publ. Astron. Soc. Pac., 86, 390.
Charles, P. A. 1983, in Activity in Red Dwarf Stars (IAU Colloq. No. 71), ed. P. B. Byrne and M. Rodono (Dordrecht: Reidel), p. 415.
Dorren, J. D., Siah, M. J., Guinan, E. F., and McCook, G. P. 1981, Astron. J., 86, 572.
Dupree, A. K. 1981, in Solar Phenomena in Stars and Stellar Systems, ed. R. M. Bonnet and A. K. Dupree (Dordrecht: Reidel), p. 407.
Dupree, A. K. 1983, in Activity in Red Dwarf Stars (IAU Colloq. No. 71), ed. P. B. Byrne and M. Rodono (Dordrecht: Reidel), p. 447.
Eaton, J. A. and Hall, D. S. 1979, Astrophys. J., 227, 907.
Feldman, P. A. 1983, in Activity in Red Dwarf Stars (IAU Colloq. No. 71), ed. P. B. Byrne and N. Rodono (Dordrecht: Reidel), p. 429.
Fraquelli, D. 1982, Astrophys. J. (Letters), 254, L41.
Fried, R. E. et al. 1983, Astrophys. Space Sci., 93, 305.

Giampapa, M. S., Golub, L., and Worden, S. P. 1983, Astrophys. J. (Letters), in press.

Gibson, D. M. 1980, in Close Binary Stars: Observations and Interpretation (IAU Colloq. No. 88), ed. M. J. Plavec, D. M. Popper and R. K. Ulrich (Dordrecht: Reidel), p. 405.

Gray, D. F. 1982, Astrophys. J., 262, 682.

Hall, D. S. 1976, in Multiple Periodic Variable Stars (IAU Colloq. No. 29), ed. W. S. Fitch (Dordrecht: Reidel), p. 287.

Hall, D. S. 1981, in Solar Phenomena in Stars and Stellar Systems, ed. R. M. Bonnet and A. K. Dupree (Dordrecht: Reidel), p. 431.

Marstad, N. 1983, unpublished Masters Thesis, University of Colorado.

Marstad, N., Linsky, J. L., Simon, T., Rodono, M., Blanco, C., Catalano, S., Marilli, E., Andrews, A. D., Butler, C. J., and Byrne, P. B. 1982, in Advances in Ultraviolet Astronomy: Four Years of IUE Research, NASA Conf. Publ. No. 2238, p. 554.

Mullan, D. J. and Cram, L. E. 1982, Astron. Astrophys., 108, 251.

Popper, D. M. and Ulrich, R. K. 1977, Astrophys. J. (Letters), 212, L131.

Ramsey, L. W. and Nations, H. L. 1980, Astrophys. J. (Letters), 239, L121.

Rodono, M. 1983, in Advances in Space Research, 2, No. 9, p. 225.

Rodono, M., Romeo, G., and Strazzula, G. 1980, Proc. Second European IUE Conference, ESA-SP 157, p. 55.

Rosner, R., Tucker, W. H., and Vaiana, G. S. 1978, Astrophys. J., 220, 643.

Simon, T. and Linsky, J. L. 1980, Astrophys. J., 241, 759.

Simon, T., Linsky, J. L., and Schiffer, F. H. III, 1980, Astrophys. J., 239, 911.

Swank, J. H., White, N. E., Holt, S. S., and Becker, R. H. 1981, Astrophys. J., 246, 208.

Uchida, Y. and Sakurai, T. 1983, preprint.

Vogt, S. 1979, Publ. Astron. Soc. Pac., 91, 616.

Walter, F. M., Cash, W., Charles, P. A. and Bowyer, C. S. 1980, Astrophys. J., 236, 212.

Walter, F. M., Gibson, D. M., and Basri, G. S. 1983, Astrophys. J., 267, 665.

Weiler, E. J. 1978, Monthly Notices Roy. Astron. Soc., 182, 77.

Zahn, J. P. 1977, Astron. Astrophys., 57, 383.

STARSPOTS, DIFFERENTIAL ROTATION AND A POSSIBLE SIX-YEAR SPOT CYCLE ON LAMBDA ANDROMEDAE

J.D. Dorren
National Central University
Chung-li, Taiwan 320
Republic of China

E.F. Guinan
Dept. of Astronomy
Villanova University
Villanova, PA 19085

I. Introduction

Lambda Andromedae (HR 8961) is a bright ($\langle V \rangle$= +3.88), nearby single-line spectroscopic binary with an orbital period of 20.5, in which the visible component is a G8III-IV star with variable chromospheric emission lines enhanced by factors of 10-100 over corresponding lines in the Sun (Baliunas and Dupree 1982). The optical variability of Lambda And was discovered by Calder (1938) who found a quasisinuosiodal light variation modulated with a mean period of ~54 days and a range in brightness (at λ4600) of about 0.3 mag. Subsequent photometry by Archer (1960), Landis et al. (1978), Dorren, Guinan, and Paczkowski (1982), and Boyd et al. (1983) show, that cycle-to-cycle variations in the amplitude, mean brightness and shape of the light curves occur. Bopp and Noah (1980) and Dorren, Guinan, and Paczkowski (1982) have modeled selected light curves and find strong evidence to support the presence of large subluminous regions on the star's surface. From the spectral type, the presence of strong CaII H and K emission, and the photometric behavior, Hall (1976) has classified Lambda And as a long period RS CVn star. Unlike in the case of the shorter period RS CVn systems, the assumed rotational period of ~54^d is not synchronous with the orbital period of 20.5 days.

II. The Observations

Photoelectric photometry of Lambda And was carried out on 210 nights from 1977 July up to the present time (1983 October). The 1977 observations were obtained at Biruni Observatory (Shiraz, Iran) with the 51-cm reflector by M.J. Siah and the authors. A pair of intermediate-and narrow-band intermediate band interference filters centered near the rest wavelength of Hα line at λ6565 and a Strömgren u filter were used. The Hα observations were continued at Villanova University with a matched Hα filter set starting in 1978. Additional observations were obtained at Villanova with intermediate-band filters centered at λ4530 and λ7790. The filter characteristics are given elsewhere (Dorren et al. 1984). We note that the Hα intermediate-bandpass filter has a bandwidth broad enough (FWHM=280A) that the included line feature does not significantly contribute to measure. Thus, the intermediate bandpass measure is essentially that of the continuum centered at λ6600. The recent publication of V-band photometry of Lambda And by Boyd et al. (1982), obtained from 1976 through 1981 chiefly at private observatories, greatly adds to the corpus of data and extends the continuous photometric coverage to seven years. Additional V- observations made during 1982-83 have been generously made available by D.S.Hall prior to publication. Since all of the above photometry was made relative to the same comparison star (ψ And), which has nearly identical colors with the variable star, it is possible to combine the differential V and λ6600 observations with little loss of accuracy. Figure 1 shows the differential V and λ6600 magnitudes for Lambda And, obtained between 1976 to early 1983, plotted against Julian Date and year. As shown in the figure, the light curve undergoes systematic cycle-to-cycle changes in amplitude, mean brightness, and shape. Noteworthy is the development of the light curve from a large amplitude, quasisinsoidal form as observed during 1976 and again during late 1978 to a light curve showing two maxima and two minima with the ~54^d period - as observed during mid-1977 and during 1980-81.

III. Analysis

We have investigated the light curves using the starspot model developed previously by us (see Dorren et al. 1981; Dorren and Guinan 1982). To model the star we have assumed circular spots of uniform brightness on a spherical star of 4900 K where the tmeperature was obtained from its G8III-IV spectral classification and colors. Since the spectrum of the secondary star is invisible at ultraviolet to near-infrared wavelengths, we have assumed that it contributes a neglible fraction of flux to the total systematic brightness. We have adopted an intermediate value of the inclination of the star's rotational axis (measured relative to the line of sight) of i = 25°. This value was obtained by assuming coplanarity between the star's rotational equator and its orbital plane, where the orbital inclination was inferred to lie between $10° \lesssim i \lesssim 40°$ from its low mass function of f(m) = 0.0006 M⊙ (Walker 1944; Gratton 1950) when the primary star is assumed to have a mass of 1.9 M⊙. The mass of the primary star was estimated from its position in the theoretical H-R diagram of Iben (1967a,b). A change of ±15° in the adopted value of the rotational pole inclination will not, however, significantly alter the results of the analysis. An inspection of Fig.1 shows that the light curves are at times asymmetric showing two distinct maxima and minima during the star's rotation period. These asymmetries cannot be produced by a single spot region and we introduce the next simplest case of two spots. In addition, the observed changes in the form of the light curve indicate that the longitudinal separation of the two spots changes with time. To accommodate such changes we permit the longitudinal separation of the spot to change with time. The ratio of the radiant flux from the spot to the flux from the rest of the star is taken from the spectrophotometric flux tables of Straizys and Sviderskiene (1972). This is a more accurate procedure than the assumption of black body flux distributions that is often assumed. Thus, the adjustable parameters in fitting the light curves are temperatures and radii of the spots, their location on the surface of the star (i.e. stellar longitude and latitude) and the rate of which the two spots move in stellar longitude.

IV. Results and Conclusions

In the following we outline the chief results and conclusions derived from the analysis of the light curves of Lambda And. Some of the conclusions are still tentative and further, more detailed analyses of the light curves are planned. Observations of Lambda And are continuing and we plan a more complete treatment of the data and a more thorough discussion of results in the near future.

A). Determination of the Starspot Temperature

The temperature of the spot region relative to the star was determined chiefly from the wavelength dependence of the light variation. Analysis of the multi-bandpass data indicates that spots are about 800 ± 100K cooler than the adopted photospheric temperature, or $T_{spot} \cong$ 4100K. From the available data, no significant change in the spot temperature appears to have occurred over the six years for which observations are available in more than one wavelength. We also investigated the possibility of bright spots (i.e. spots hotter than the photosphere) and found that they do not correctly describe the morphology and wavelength dependence of the light curves.

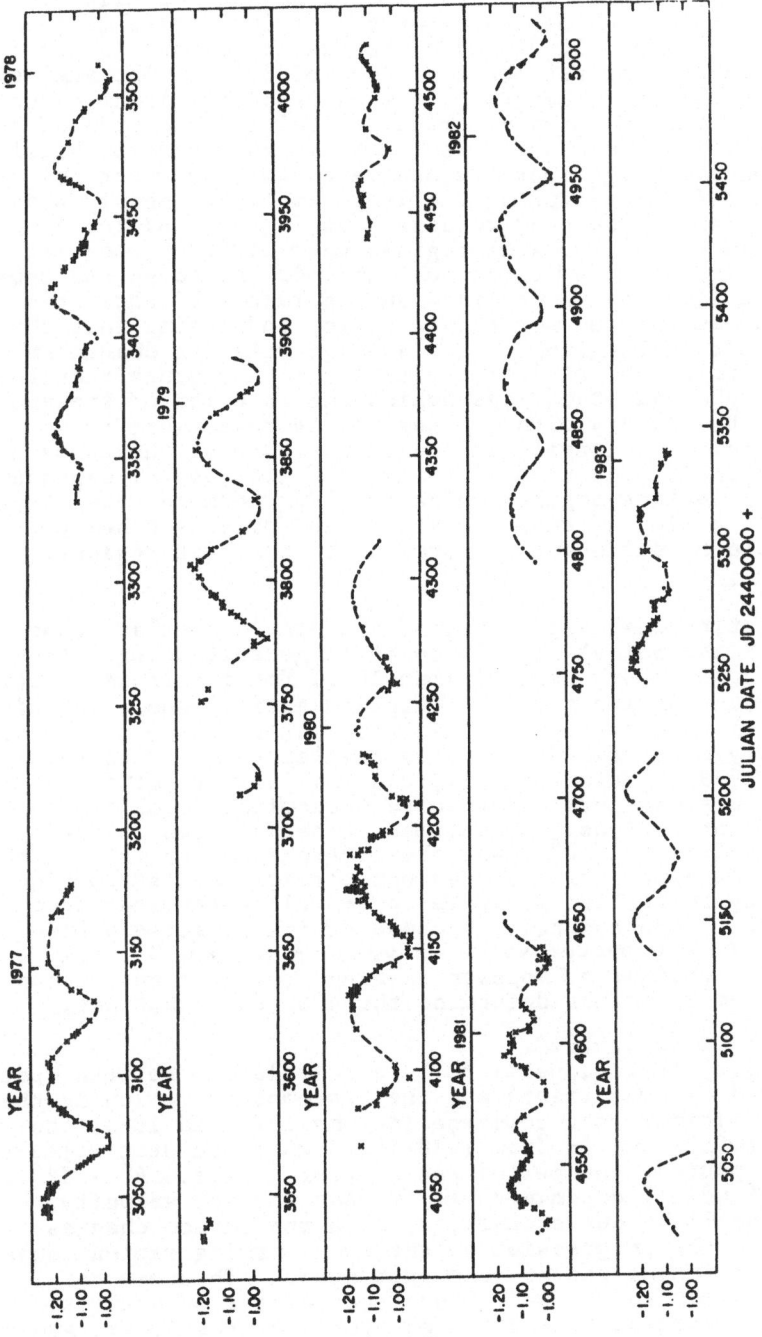

Fig. 1. The differential intermediate-band λ6600 and V magnitudes of Lambda And obtained from 1976-1983 are plotted. The intermediate-band λ6600 observations were obtained at Biruni and Villanova Observatories and are represented by circles. The V measures were obtained by Boyd et al. and are represented by crosses.

261

B). Evidence for Differential Rotation

The analysis of the light curves with the non-static starspot model indicates the presence of two large spot regions of slightly different areas in which the smaller spot typically appears to be at a higher stellar latitude ($\sim 50°-70°$) than the faster moving, larger spot region, which is located at stellar latitudes of $30°-50°$. The analysis indicates that the observed changes in the shape of the light curve are explicable from the relative movement in stellar longitude of one spotted region relative the other. For example, referring to Fig.1 during 1978-79 the two large spot regions were close together in stellar longitude ($\Delta l \lesssim 60°$) producing the large amplitude nearly sinusoidal light variation as the star rotates. During 1980-81 the longitudinal separation of the two spot regions was $\sim 150°-180°$ producing the two minima during one $\sim 54^d$ rotational period. Although not shown in the figure, the light curve obtained during late 1983 shows two maxima and minimum during one rotational cycle, indicating that the spots are $\sim 180°$ apart in longitude. The average relative change in longitude of the two spots is $\sim 16°$ per rotation in the sense that the spot complex at higher latitude lags behind the rotation of the spot at mid-latitude. By analogy with the Sun, it is reasonable to suppose that the relative longitudinal displacements of the spots are produced by differential rotation of the star - i.e. by the dependence on latitude of the surface angular velocity. Furthermore, the changes in the photometric period as shown in Fig.2 (and discussed below), are also explicable by differential rotation if the spot regions change latitude.

The inferred latitudinal dependence of the star's angular velocity (decreasing toward the poles) appears to be in general accord with what is expected from theory (see Gillman 1979) for a star with weak-to-intermediate rotation and a fairly deep convective zone. In this case the equatorial acceleration profile with latitude is expected to be broad with a nearly monotonic decrease in angular velocity to the poles (Gillman 1979). Unfortunately, the analysis of the disk-integrated light does not permit a precise determination of the latitudes of the spot regions. We estimate uncertainties in the spot latitudes of the order of $\sim \pm 20°$, thus, the inherent uncertainty in the determination of the spot latitude does not permit the distribution of angular velocity in stellar latitude to be well determined from the starspot model. Unfortunately, because of its relatively slow rotational period (the expected value of $v \sin i \sim 2-3 \ kms^{-1}$), the recently employed technique of Doppler imaging (see Vogt and Penrod 1983) cannot be used to better determine the spot distributions.

C). Evidence for a Spot Cycle

As shown in Fig.2, there appears to be a correlation between spot area (as determined by mean light) and the photometric period in the sense that during minimum spot coverage (during 1977 and 1983) the period is longer ($\sim 55^d$), while from 1978-1982, when the star appeared more extensively spotted, the period was shorter. (i.e. $\sim 54^d - 52^d$). If there is a latitudinal dependence of surface angular velocity, as suggested by the light curve analysis, then the period changes could be related to the progression of the spot forming region toward the stellar equator during the ~ 6 year spot cycle. This may be similar to the behavior in the Sun, where the centers of activity progress from high latitude toward the equator over the 11 yr. sunspot cycle. The abrupt change in period observed during mid-1981 could indicate the latitude of the spot forming region rapidly shifted poleward. As shown above, the latitude of the spot regions can only be approximately determined from modeling the light curves.

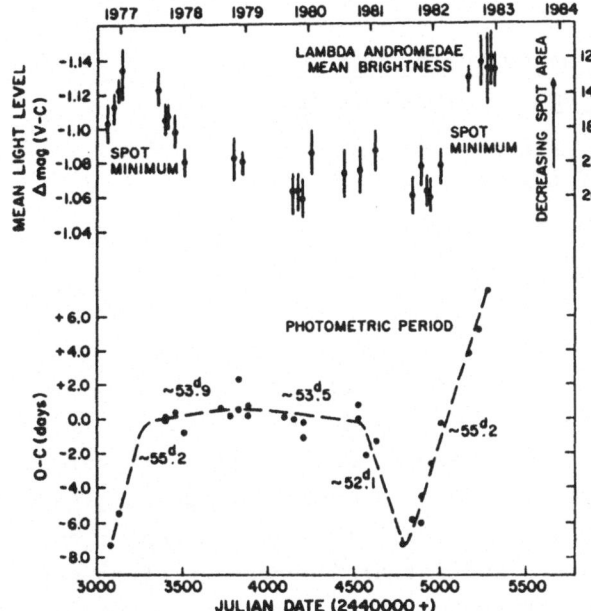

Fig. 2 The top panel shows the change in the mean brightness of Lambda And over a ~6 year interval. The approximate total spot area is shown where the spot minima occur during early 1977 and 1983. The lower panel shows the change in the photometric period which may arise from latitudinal variations of the spot regions.

Acknowledgements:

We wish to thank Brian Paczkowski of J.P.L. for making observations during early 1981, while a student at Villanova University. We are grateful to Mrs. Hildred Nason for preparing the manuscript. EFG also wishes to acknowledge support in part by the Langley-Abbot program of the Smithsonian Institution.

References:

Archer, S. 1960, J. British Astron. Assoc. 70, 95.
Baliunas, S.L. and Dupree, A.K. 1982, Ap.J. 252, 668.
Bopp, B.W. and Noah, P.V. 1980, P.A.S.P. 92, 717.
Boyd, R.W. et al. 1983, Astrophys. Sp. Sci. 90, 197.
Calder, W.A. 1938, Harvard Obs. Bull. 907, 20.
Dorren, J.D. and Guinan, E.F. 1982, Ap.J., 252, 296.
Dorren, J.D., Guinan, E.F., and McCook, G.P. 1984, P.A.S.P. (in press).
Dorren, J.D., Siah, M.J., Guinan, E.F., and McCook, G.P. 1981,
    A.J. 86, 572.
Dorren, J.D., Guinan, E.F. and Paczkowski, B. 1982, Bull. Am. Astron.
    Soc. 14, 634.
Gillman, P.A. 1979, in Stellar Turbulence, eds. D.F. Gray and J.L.
    Linsky (Springer-Verlag) No. 114, p. 19.
Gratton, L. 1950, Ap.J. 111, 31.
Hall, D.H. 1976, in "Multiple Periodic Phenomena in Variable Stars",
    ed. W.S. Fitch (Dordrecht: Reidel) p. 287.
Iben, I. 1967a, Ap.J. 147, 624.
Iben, I. 1967b, Ap.J. 147, 650.
Landis, H.J., Lovell, L.P., Hall, D.S., Henry, G.W. and Renner, T.R.
    1978, Ap.J. 83, 176.
Straizys, V. and Sviderskiene, Z. 1972, Bull. Vilnius Astron. Obs.,
    Nov. 35.
Vogt, S.S. and Penrod, G.D., 1983, P.A.S.P. 95, 565.
Walker, E.C. 1944, J. Roy. Astro. Soc. Canada, 38, 249.

VISUAL AND INFRARED PHOTOMETRY OF THE
LONG-PERIOD RS CVN SYSTEM HR 7275

M. Zeilik, R. Elston, and P. Smith
University of New Mexico
Albuquerque, NM   U.S.A.   87131

D. S. Hall
Dyer Observatory
Vanderbilt University
Nashville, TN   U.S.A.   37235

HR 7275 consists of an early KIV star and an unclassified companion; its orbital period is 28.6 days, making it a member of the long-period group of RS CVn stars, as classified by Hall (1976). We report on V-band photometry spanning 1981-1982 and infrared photometry from early 1981. These data show that the starspots have dramatically evolved over a 1-year period: The full amplitude of the distortion wave dwindled to 0.05 mag (at V) from over 0.3 mag.

The visual (V-band) observations were carried through 1981 and 1982 (ending in December) at 11 different observatories. All observations used the comparison star HR 7229, and the magnitude differences are (HR 7275 - HR 7229). Statistical errors in each point are generally better than 0.005 mag. Figures 1-5 show the results, with the magnitude differences transformed to the standard UBV system.

We can now examine the activity of this system from 1978 through 1982 (using data from Fried et al. (1982). We note: (1) the maximum distortion wave amplitudes occurred in late 1980 (0.25 mag) and middle 1981 (0.30 mag); (2) by the middle of 1982, the amplitude has dwindled to at most 0.05 mag; (3) the low-amplitude distortion wave continued through 1982; (4) the most similar prior episode occurred at the end of 1978 into the beginning of 1979; and (5) as the distortion wave died out in 1982, the average magnitude declined -- implying that the starspots evolved to a roughly uniform distribution around the star.

For infrared observations, we used the 1.3-meter telescope at Kitt Peak National Observatory with an InSb photometer ("Otto") to carry out the observations during the daytime (Zeilik et al., 1983). Magnitudes were calculated relative to Alpha Lyr, whose infrared magnitudes were taken to be 0.0. The observations spanned December 1980 to February 1981; most were clustered in January 1981. The magnitude differences (Alpha Lyr - HR 7275) at JHKL were found as a function of phase, using the same ephemeris as Fried et al. (1982):

$$HJD = 2431043.57 + 28.59E.$$

(The period is the spectroscopic one based on the observations of Young, 1944.) The infrared distortion wave, has a full amplitude of 0.11 mag at J, 0.09 at H, 0.09 at K, and 0.06 at L. (Statistical errors in each datum are no more than 0.01 mag.) The peak of the wave occurs at about phase 0.1, as can be seen in Figure 6 for the J-band observations.

Combining the visual and infrared data, we have developed a preliminary starspot model for the HR 7275 system. The combined observations provide a stronger constraint on the spot temperature than do the visual data alone. Our ultimate goal is to understand the evolution of the active regions on the K star. We have started with a very simple model in which we assume: (1) one spot on the surface (penumbra only); (2) photospheric temperature from spectral type; (3) surface clean except for spot; (4) limb darkening; (5) blackbody output for photosphere and spot. We then manipulate the temperature and spot area to maximize the light to the photometry, from the visual through the infrared for January 1981. Our best model to date requires a spot with T = 3700 K covering 90% of the spotted hemisphere--clearly an unrealistic result. A major problem is that we must model the light from the secondary star, but we do not yet know its spectral type.

References

Fried, R.E., Eaton, J.A., Hall, D.S., Henry, G.W., Lovell, L.P., Krisciunas, K., Chambliss, C.R., Detterline, P.K., Landis, H.J., Louth, H., and Skillman, D.R., 1982, Astrophysics and Space Sci., vol. 82, p. 181.
Hall, D.S., 1976, IAU Colloquium 29, p. 137.
Young, R.K., 1944, J. Roy. Astron. Soc. Canada, vol. 38, p. 366.
Zeilik, M., Elston, R., Henson, G., and Smith, P., 1983, Information Bull. on Variable Stars #2333.

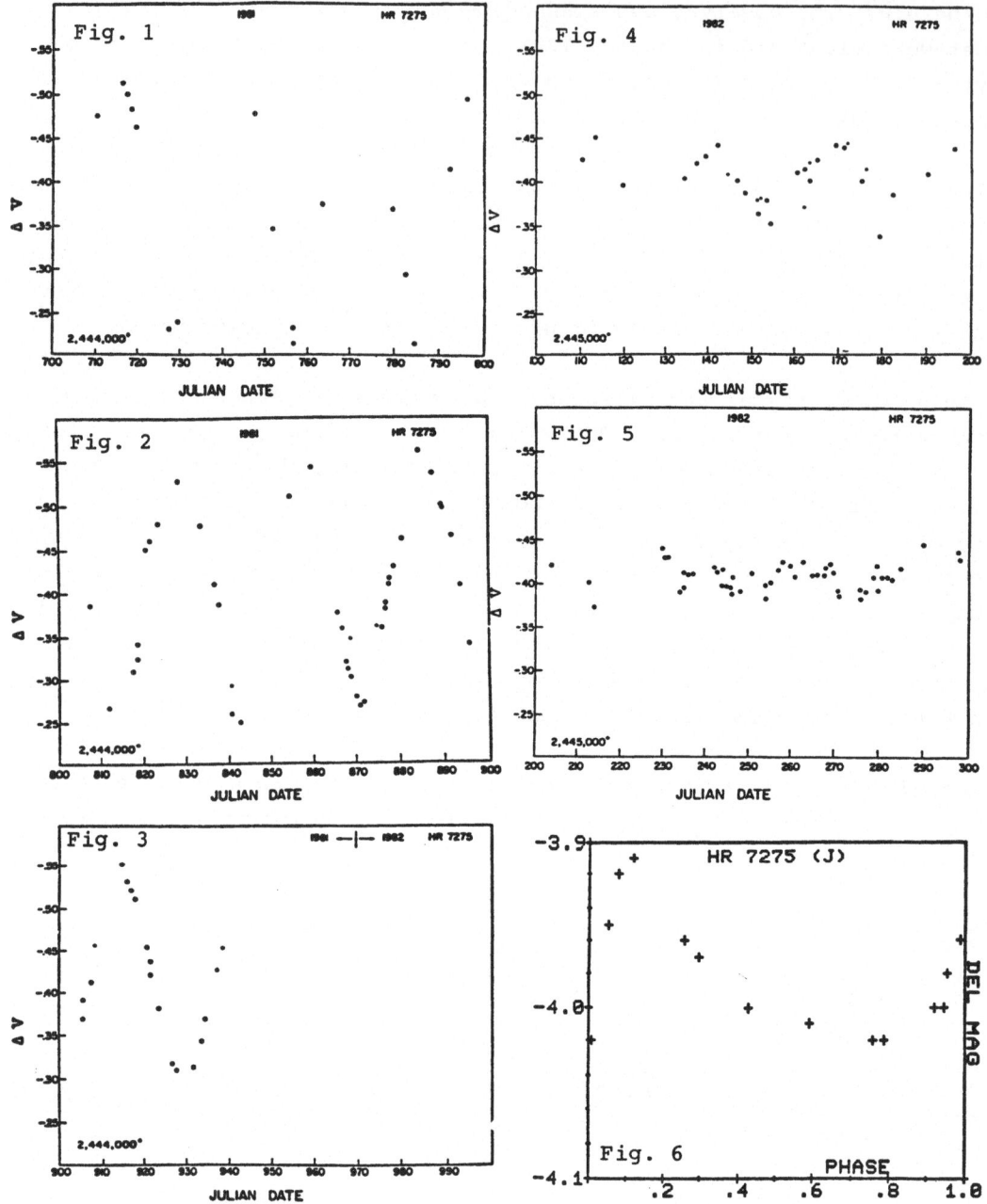

Figures 1-6   Light curves for HR 7275, plotted as differential
magnitudes.   Figures 1-3 are 1981 V-band data;
Figures 4-5 1982 V-band data; Figure 6 is J-band
in early 1981.

Hα OBSERVATIONS OF THE RS CVn STAR, RT LAC: EVIDENCE FOR GAS STREAMING: PRELIMINARY RESULTS

David Huenemoerder
Pennsylvania State University
Department of Astronomy
University Park, PA 16802

RT Lacertae is a 5.07 day period, totally eclipsing RS CVn system. Hα spectra obtained with the Fiber-coupled Spectrograph at Black Moshannon Observatory for two seasons are presented. Observations are continuing, so only a preliminary report is given here. Hα shows some very unusual but generally repeatable behavior.

Excess Hα emission (or absorption) profiles were derived by subtracting composite comparison spectra synthesized from spectra of normal stars with the same spectral types as the individual components of RT Lac (G9 IV and K1 IV) using the orbital solution and photometry of Eaton and Hall (1979) and Milone (1977), and the radial velocity curve of Joy (1931). This technique has proven useful in the analysis of other double-lined eclipsing RS CVn systems (Huenemoerder and Ramsey, (1983)).

The excess profiles (Fig. 1) for primary eclipse and quadratures show that the fainter K component is active, consistently displaying excess Hα emission. The G star usually has little, if any, Hα emission. It does, however, occassionally show emission comparable to that of the K star. The K star's emission measure, using a distance of 47 pc (Walter and Bowyer, 1981), is about 10^{55} cm^{-3}. Its Hα luminosity is about 10^{31} erg s^{-1}.

This origin of the activity is consistent with the Ca II H and K observations of Oliver (1974) and the photometry of Milone (1977), which both show that the K star to be the active component. The eclipse photometry of Eaton and Hall (1979) suggests that there is also some intermittent surface activity on the G star as well.

Near secondary eclipse, however, the behavior of Hα is most unusual. The absorption becomes very strong (see the 1983 spectra in Fig. 1). It is, in fact, much deeper than in normal stars. In one instance in 1982, not only was there strong absorption, but it was at -300 km/s from the velocity of the G star!

The G star is nearly filling its Roche surface, according to eclipse photometry. Eaton and Hall (1979) argue that it actually underwent expansion to completely fill it. Milone (1977) also proposes Roche overflow from the G9 and a stream fed envelope about the K1 star to explain some peculiarities in the colors (the cooler component is bluer!).

After the initial secondary eclipse observation of 1982, it was assumed that the high velocity absorption was due to cool gas seen in projection against the K star, flowing between the two stars. Order of magnitude estimates of the physical conditions were made under some simplifying assumptions. If the gas is optically thin and at about 10^4K, then the column density of hydrogen is about 10^{21} cm^{-2}. This also assumes, however, that there is no velocity gradient along the stream. Imposing a velocity structure dictated by the gravitational potential between the stars and applying the Sobolev method results in a thin gas if the total hydrogen density is below 10^9 cm^{-3}, yielding a column density of about 10^{21} cm^{-2}.

This extremely simplistic analysis has several shortcomings. First, the absorption component position and width are much more than gentle Roche lobe overflow can account for. Secondly, the current season's data indicate that the strong absorption is seen even when most (>80%) of the fainter K star is occulted by the G star. The high velocity gas may not have been flowing between the stars. The strong absorption is from gas seen in projection against the G star. This is most curious indeed, since the excess absorption is not seen at other phases. The G star apparently has a cool extended atmosphere on one side only; it may be material flowing out through the outer Lagrangian point (L2).

A more rigorous approach will be investigated incorporating the current season's data and utilizing the shapes of the Hα profiles as well as the depths.

That RT Lac shows such strong indications of circumstellar matter and gas flow is very interesting. According to the binary gas dynamics theory of Lubow and Shu (1975), the system parameters permit only a weak stream which impacts the companion and does not form a disk. Other binaries in this regime have only shown weak evidence for gas streams (Polidan and Peters, 1982).

REFERENCES

Eaton, J.E., and Hall, D.S. (1979). Ap.J., 227, 907.
Huenemoerder, D.P. and Ramsey, L.W. (1983). A.J., submitted.
Joy, A.H. (1931). Ap.J., 74, 101.
Milone, E.F. (1977). A.J., 82, 998.
Oliver, J.P. (1974). Ph.D. Thesis, UCLA.
Polidan, R.S. and Peters, G.J. (1982). in Four Years of IUE, Kondo, Y., et al., eds., 534.
Walter, F. and Bowyer, S. (1931). Ap.J., 245, 671.

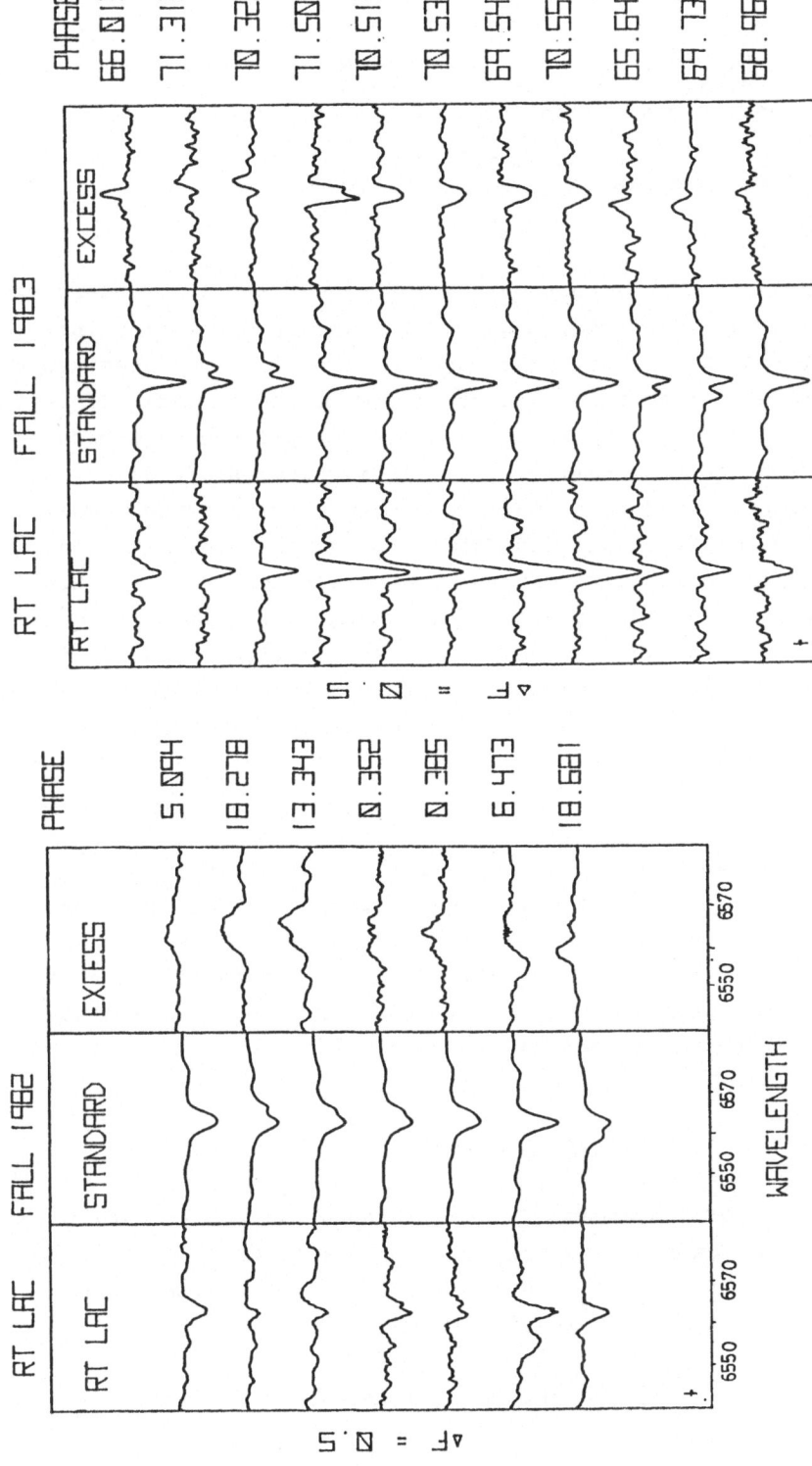

Figure 1. Data for two seasons are shown. The leftmost panels are the Hα spectra of RT Lac. The center panels are synthesized composite comparison spectra. The rightmost panels are their normalized difference. Cumulative phases (orbit no. + phase) are given at the right of each set of spectra. Spectra are aligned on the brighter G9 component.

NEAR INFRARED PHOTOMETRY OF SOME RS CVn SYSTEMS

R. P. Verma, S. K. Ghosh, K. V. K. Iyengar, T. N. Rengarajan,
S. N. Tandon, and R. R. Daniel
Tata Institute of Fundamental Research, Bombay

N. B. Sanwal
CASA, Osmania University, Hyderabad

We observed 11 RS CVn systems during 1981-82 at the 1.2m telescope
of Osmania University, Hyderabad, and the 1m telescope of Indian Institute of Astrophysics, Bangalore. Data on the RS CVn systems observed by
us as well as those available in literature are listed in Table I. The
spectral type and $(\Delta M_v)_{max}$, the maximum observed amplitude of optical
wave (col. 4) are taken from Hall (1981). IR magnitudes are given as
excess over those predicted based on a knowledge of visual magnitude,
spectral type, luminosity class, and Johnson's standard color tables.
For those systems for which information on a single member of the binary only is known, the excess is calculated assuming that member to be
the dominant one. If the unknown companion is of earlier type it will
lead to an underestimate of IR excess; if it is a later type, the excess
will be overestimated. The errors in IR excess are in the region of
0.1 to 0.2 magnitude. It is seen that there are 10 systems with excess
> 0.2 mag.

In Fig. 1, we plot IR excesses ΔJ, ΔH, ΔK against $(\Delta M_v)_{max}$. In
all three bands a good correlation is seen between the two. This shows
that the excess is related to the activity of the system. Fig. 2 shows
the spectrum of excess radiation normalised to the J band, for 4 systems for which L band data are available. The curves shown are blackbody spectra for different temperatures.

Origin of Excess

1. Reddening: For 4 systems (UX Ari, HR1099, AR Lac, RT Lac) with
large IR excess, the observed B-V colors are close to the expected
ones, thus setting a limit of A_v < 0.3 mag for any extinction. The
corresponding IR excess is much less than observed.

2. Cool spots: UX Ari, and HR1099 have excess of 0.8 mag in K,
whereas the optical wave amplitude is < 0.15 mag. RT Lac exhibits excess

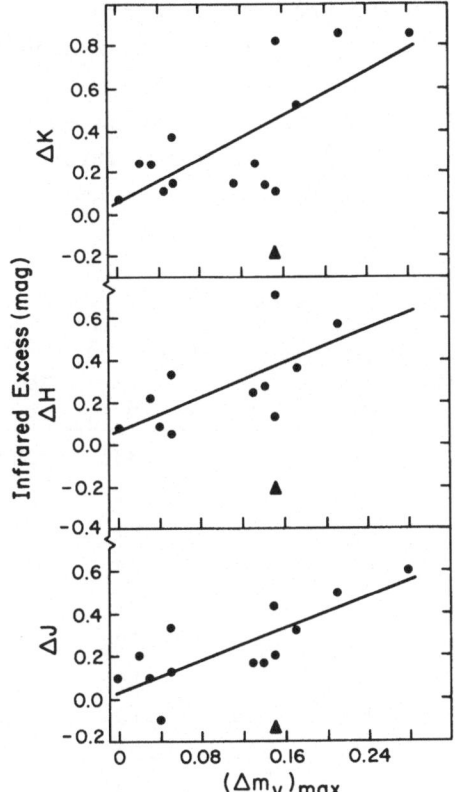

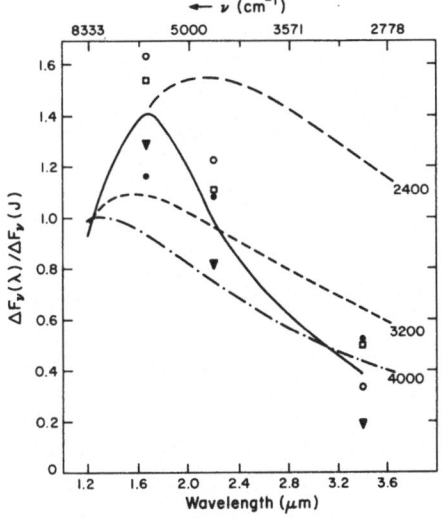

Fig. 1 Infrared Excess (mag) in J, H and K bands as a function of the maximum amplitude of optical wave, $(\Delta m_v)_{max}$ for the RS CVn systems listed in Table I. The points ▲, refer to α Aur which deviates from the general trend. The lines through the data points are the best fits.

Fig. 2 Wavelength dependence of the infrared excess emission $\Delta F_\nu(\lambda)$ normalised at 1.25 μm. The symbols o, •, ▼, □ refer to data corresponding to the systems UX Ari, HR 1099, AR Lac and RT Lac, respectively. The different dashed lines refer to blackbody emission at the temperatures (°K) indicated beside them.

at all binary phases (Milone, 1976). Therefore, cool spots cannot explain excess > 0.2 magnitude.

3. Circumstellar grains: Color temperatures of excess radiation are > 2000 K, too high for grains.

4. Circumstellar free-free emission: The spectral shape does not

fit either the flat spectrum expected for optically thin free-free emission or the $\nu^{2/3}$ powerlaw expected for a central source undergoing mass loss at a constant velocity.

No correlations are seen between IR excess and any other parameter like period, X-ray activity, period change, etc. Futher observations up to 10 microns will be helpful in understanding the origin of the IR excess radiation.

TABLE I   Infrared Excess of RS CVn Systems

System	Sp. Type and Lum. Class	P (Day)	$(\Delta m_v)_{max}$	IR Excess (mag)			Ref.
				J	H	K	
UX ARI	G5 V+K0 IV	6.44	0.15	0.44	.72	.83	P,H
TY Pyx	G5 + G5	3.20	.04	-.10	.09	.11	P
HR 1099	G5 IV+K1 V	2.84	.21	.50	.58	.87	P,A
HR 5110	F2 IV+K1 V	2.61	.0	.01	.08	.07	P
σ CrB	F8 V	1.14	.05	.33	.33	.37	P
σ Gem	K1 III	19.60	.15	.20	.14	.12	P
12 Cam	K0 III	80.17	.14	.17	.28	.14	P
54 Cam	F8 V	11.08	.05	.13	.05	.15	P
93 Leo	A+G5 III-IV	71.7	.03	.09	.22	.23	P
λ And	G8 IV-III	20.5	.28	.61	-	.87	J
ξ And	K1 II	17.77	.02	.20	-	.25	J
α Aur	G5 III+G0 III	104.02	.15	-.14	-.21	-.19	J
RT Lac	G9 IV+K1 IV	5.07	.17	.41	.51	.61	M
CG Cyg	G9.5 V+K3 V	0.63	.11	-	-	∿.15	M'
AR Lac	G2 IV+K0 IV	1.98	.13	.17	.25	.25	M

$(\Delta m_v)_{max}$ is the maximum amplitude of the optical wave. Spectral types are from Hall (1981). References for J, H and K magnitudes are: P - present study; J - Johnson et al., 1966; M - Milone, 1976; M' - Milone and Naftilan 1980; N - Needham et al., 1980; H - Hall et al. 1975; A - Antonopoulou and Williams 1980.

Antonopoulou, E.: 1983. Astron. Astrophys., 120, 85.
Hall, D. S.: 1981. Solar Phenomena in Stars and Stellar Systems, 431-447, eds. Bonnet, R. M., Dupree, A. K., Reidel D. Publishing Company, Dordrecht, Holland.
Hall, D. S., Montle, R. E. and Atkins, H. L.: 1975. Acta Astr., 25, 125.
Johnson, H. L.: 1966. A. Rev. Astron. Astrophys., 4, 193.
Milone, E. F.: 1976. Astrophys. J. Suppl., 31, 93.
Milone, E. F. and Naftilan, S. A.: 1980. Close Binary Stars: Observations and Interpretation, IAU Symposium No. 88, 419-422, eds. Plavec, M. J., et al., Reidel D. Publishing Company, Dordrecht, Holland.
Needham, J. D., Phillips, J. P., Selby, M. J., and Sanchez Magro, C.: 1980. Astron. Astrophys., 83, 370.

THE PUZZLE OF THE UV CONTINUA OF THE HYADES GIANTS

Erika Bohm-Vitense
University of Washington
Seattle, Washington 98195

We are convinced that the Hyades giants have the same chemical composition. They should have the same age and hence, the same mass and gravity. Their continuum energy distribution should therefore be a function of T_{eff} only. So should be their B-V color. They are all of spectral type K0 and therefore should have nearly the same UV continua. On Figure 1, however, we see that in the wavelength range $1800 < \lambda > 2000A$ their fluxes differ by large factors. (In Figure 1 the energy distributions are normalized to $m_v = 0$, so we measure the flux relative to the visual flux). At 1950A the flux varies by a factor of 6. For ϵ Tau the flux is slightly variable.

In Figure 1 we also see that the flux at 1950A is correlated with the CIV (1550A) emission line intensity. Stars with $f(1950A) > 10^{-12} erg/cm^2/sec/A$ show the CIV emission. In fact, on one image ϵ Tau shows a larger flux and seems to show a weak CIV emission line, while on the other image it has less flux at 1950A and no CIV emission is seen.

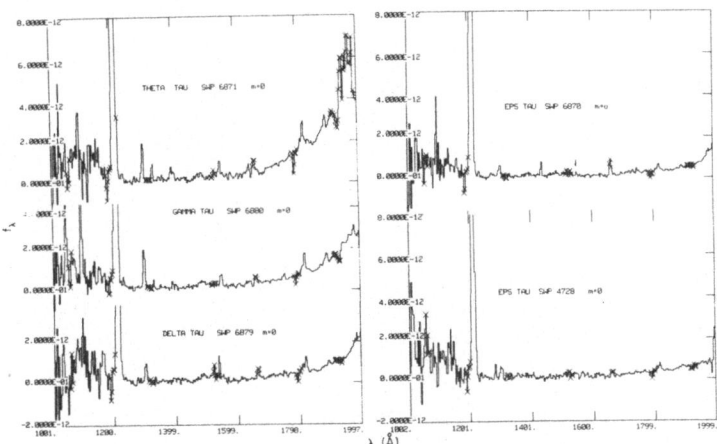

Figure 1: The radiative fluxes of the Hyades giants are shown as a function of wavelengths. Around 1950A they are very different. The intensities of the CIV emission lines at 1550A change in accordance with the continuum fluxes around 1950A. All fluxes were normalized to $m_v = 0$ and measured in $erg/cm^2/sec/A$.

_If_ the flux around 1950A would be a unique function of the effective temperature, then we would have to conclude that the CIV intensity is an extremely steep function of T_{eff}. In Figure 2, however, we show the relation between the flux at 1950A and B-V for field giants. For these giants, the flux at 1950A changes only slightly and not systematically with B-V. Therefore, the change in flux at 1950, as observed for the Taurus giants, is not due to a difference in T_{eff}, as was also concluded by Baliunas, et al. (1983). _For a given Z, T_{eff}, and gravity g, we observe different fluxes in the continuum around 1900A._

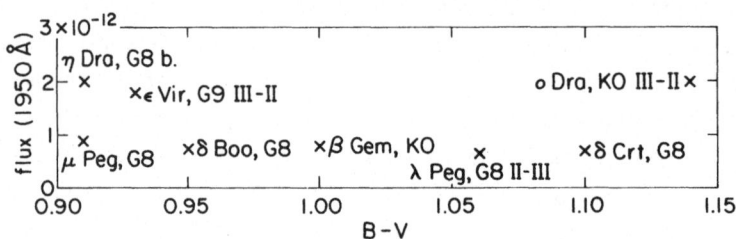

Figure 2: For non-Hyades giants, the flux at 1950A in erg/cm²/sec/A (normalized to $m_V=0$) is shown as a function of B-V. It is only slightly dependent on B-V. It is not a steep function of T_{eff}.

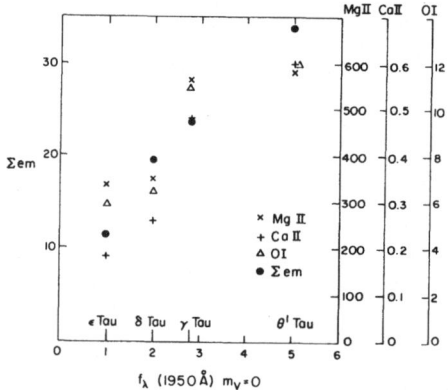

Figure 3: The emission line fluxes of the Hyades giants for the MgII, CaII and OI lines (scale on the right-hand side) and for the sum of the other emission lines (scale on the left-hand side) are shown as a function of the continuum fluxes around 1950A (for $m_V=0$). The data were taken from Baliunas, et al. and are given in their units.

274

In Figure 3 we have plotted the emission line intensities of the MgII (h+k), the CaII K, and the OI lines, as well as the sum of the other emission lines for the Hyades giants (according to Baliunas, et al.) as a function of f_λ (1950A). The correlation is obvious for all lines.

Baliunas, et al. point out that θ^1, γ and δ Tau have companions with orbital periods between 1.5 and possibly 20 years. Tidal effects should not be important, but the fluxes at 1950A may be partly due to the companion. For θ^1 Tau, the companion was measured by Peterson to be 3.5 magnitudes fainter than the giant. The spectral type was determined from its colors to be between G0 and G5V. Such a star would contribute very little to the fluxes at 1950A. It would have to be of earlier spectral type. The larger the flux at 1950A, the earlier the spectral type of the companion would have to be.

If indeed the companions would be responsible for the excess flux around 1950A, then the observed correlation between emission line intensities and flux at 1950A (Figure 3) would mean a correlation between emission line intensities and spectral type of the companion. Such a correlation would be very interesting but hard to understand unless the emission lines would be due to the companion. Baliunas, et al. give strong arguments against the latter hypothesis. We therefore think that the excess flux at 1950A is intrinsic to the Hyades giants. We may speculate that, for those cool giants which generate large amounts of mechanical energy, like the Hyades giants, a considerable amount of this mechanical energy is either dissipated in the photospheric layers or a large flux is conducted back from the transition layer and chromosphere into the high photosphere leading to a temperature increase in the region of the temperature minimum. For larger mechanical energy generation, we would then expect a higher radiation in the continuum as well as an increase in emission line intensities.

Another possibility would be a reduction of the UV absorption coefficient due to overionization of the metals by excess UV radiation. The metals are mainly absorbing in the UV. We might then in the UV look deeper into the atmosphere where the temperature is higher.

In any case, it is interesting to note this correlation between photospheric temperature stratification and transition layer activity. We may see here that the chromosphere reacts back on the photosphere changing its temperature stratification.

References:

Baliunas, S.L., Hartmann, L., and Dupree, A.K. 1983: Ap.J. 271, 672.
Peterson, D., et al. 1981: A.J. 86, 280.

MULTICOLOR PHOTOMETRY OF THE ACTIVE G GIANT FK COM

Jon A. Holtzman and Harold L. Nations
Van Vleck Observatory
Wesleyan University
Middletown, CT 06457

I. INTRODUCTION

The star HD 117555 (FK Com) has been recognized as peculiar for several decades. Merrill (1948) observed the star spectroscopically and found it to be a G giant with broad absorption lines indicating a v sin i of about 100 km/sec. Since then, emission features have been discovered at Ca II H & K and at Hα (Bidelman 1954; Herbig 1958). The Hα line is very broad (approximately 20 angstroms full width) and has a variable profile (Ramsey, Nations, and Barden 1981; Walter and Basri 1982). The star also shows strong chromospheric and transition region emission lines in the ultraviolet (Bopp and Stencel 1981). In this paper we describe and analyze UBVRI and Hα photometry of FK Com obtained at Kitt Peak National Observatory in February 1983. A more complete account of these results has been submitted to the A.J.

II. OBSERVATIONS AND DISCUSSION

The photometry was obtained using the #4 16-inch telescope at Kitt Peak on 11 nights from 10-21 February 1983. Observations were made with a single channel photometer using a Ga-As photomultiplier tube and the UBVRI filters of the Cousins/Bessell system. In addition to the UBVRI photometry, observations were obtained through two Hα filters with FWHM of 37 and 176 angstroms.

The light curves for FK Com at all wavelengths are presented in Figure 1. The data for the flare event observed on 21/22 February is shown in Figure 2. All phases have been using the ephemeris of Chugainov (1976), which defines phase 0.0 to be at the photometric minimum: JD(min) = 2442192.345 + 2.400E. Several features are evident from these graphs. They are:

1. The light curve is asymmetric, with a relatively flat minimum around phase 0.0 which gets increasingly flatter at longer wavelengths.

2. There are well correlated color changes, in the sense that the star is redder when fainter at all wavelengths. This has been noted previously by both Chugainov (1976) and Rucinski (1981).

3. There is evidence for a flare-like event on 21/22 February; this is most evident in the U light curve and at Hα. Using the absolute visual magnitude of a G8 giant (Allen 1973), we calculate the lower limits to the energy of the flare to have been approximately 5.3×10^{36} ergs in U and 6.0×10^{36} ergs in B.

It seems likely that the photometric variability of FK Com can be attributed to the presence of starspots. Such an interpretation is supported by other characteristics of the star (rapid rotation and an active chromosphere) which are similar to those observed on other spotted stars.

In order to better determine the starspot parameters for FK Com, we have chosen to model the light curves in detail using a procedure developed by Budding (1976). Thus, using a grid search least-squares minimizing routine, we have simultaneously fit the UBVRI data to yield values of i, the inclination of the rotational axis, the longitude and latitude of the center of each of two spots, the angular radius of each spot, and the unspotted light level in intensity units (the observed max is normalized to 1.0) for each band, U_λ. These best fit (in the least squares sense) parameter values along with their estimated errors are given in Table 1 while the light curves generated with these parameters are compared with the observations in Figure 1.

Table 1. Starspot Parameters for FK Com

i	= $55° + 3°$	Long(2)	= $275° + 5°$
Rad(1)	= $32° + 3°$	U_U	= 1.15
Lat(1)	= $52° + 5°$	U_B	= 1.12
Long(1)	= $28° + 5°$	U_V	= 1.06
Rad(2)	= $24° + 3°$	U_R	= 1.04
Lat(2)	= $21° + 4°$	U_I	= 1.02

References

Allen, C.W. (1973) in Astrophysical Quantities (Athlone, London).

Bidelman, W.P. (1954) Astrophys. J. Suppl. 1, 175.

Bopp, B.W., and Stencel, R.E. (1981) Astrophys. J. Lett. 247, L131.

Budding, E. (1976) Astrophys. Space Sci. 48, 207.

Chugainov, P.F. (1976) Izv. Krymskoj. Astrof. Obs. 54, 89.

Herbig, G.W. (1958) Astrophys. J. 128, 259.

Merrill, P.W. (1948) Publ. Astron. Soc. Pac. 60, 382.

Ramsey, L.W., Nations, H.L., and Barden, S. (1981) Astrophys. J. Lett. 251, L101.

Rucinski, S.M. (1981) Astron. and Astrophys. 104, 260.

Walter, F.M., and Basri, G.S. (1982) Astrophys. J. 260, 735.

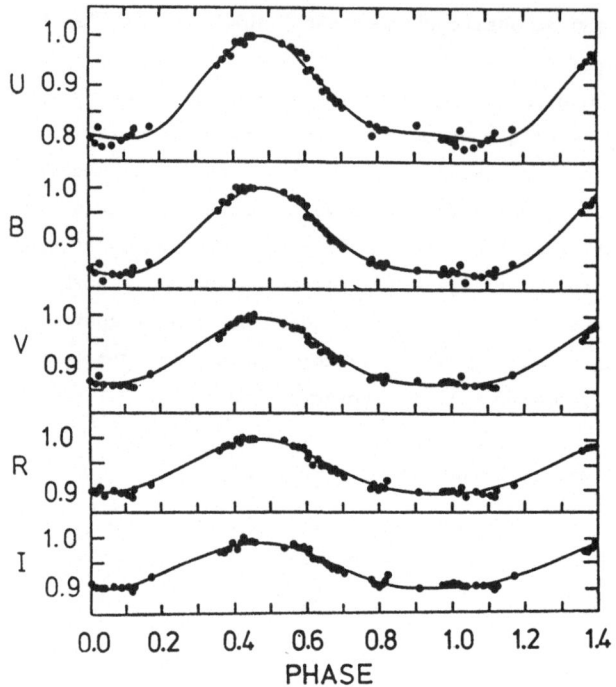

Figure 1: Comparison of best
fit starspot model light curves
with the observations. The ordinate is in intensity units,
normalized to 1.0 at max in
each band. (The flare points
have been omitted.)

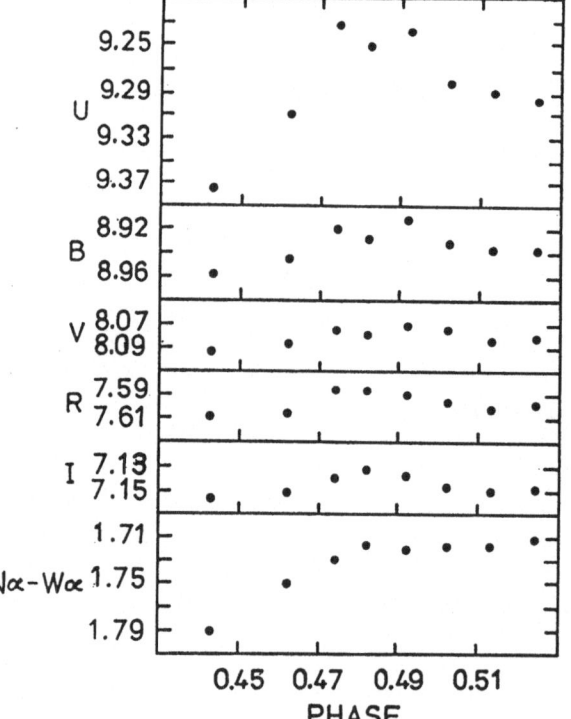

Figure 2: Observations in
UBVRI & Hα of a flare on
21/22 February 1983. Note
that Hα remains enhanced
after the flare has started
to decline in the other bands.

ROTATIONAL MODULATION AND HIGH SPEED STREAMS IN FK COMAE BERENICES:
EVIDENCE FOR A MASSIVE, HIGHLY-EVOLVED BINARY SYSTEM

F. M. Walter and J. E. Neff: Joint Institute for Laboratory Astrophysics, University
of Colorado and National Bureau of Standards, Boulder, Colorado 80309
B. W. Bopp: Dept. of Physics and Astronomy, University of Toledo, Toledo, Ohio
R. E. Stencel: NASA

INTRODUCTION

FK Comae Berenices is a very peculiar star. First noted by Merrill (1948) because of its diffuse absorption lines and a broad Hα emission, it has become more
and more enigmatic with each new observation. Herbig (unpublished) noted a possible
$4^{d}8$ periodicity in the V/R ratio of the double peaked Hα line, whereas Chugainov
(1974) found a $2^{d}4$ photometric period. Walter (1981) found FK Com to be a strong
soft X-ray source, while Bopp and Stencel (1981) reported extremely strong UV emission lines. Walter and Basri (WB, 1982) and Ramsey et al. (1982) then found a periodic variation in the shape of the Hα line modulated at the photometric period.
WB showed that the Hα variations could be attributed to a brightening of the chromosphere over the bright hemisphere of the star, and that the chromosphere rotated
about the star with V sin i = 110 km s^{-1}. Recently Morris and Milone (1983) have
refined the photospheric rotation period, and have argued that the color variations
cannot be due to normal dark star spots, although Holtzman and Nations (1983) arrive
at the opposite conclusion.

Bopp and Stencel (1981) proposed that FK Com is a coalesced W UMa system, since
it was a rapidly rotating, apparently single early G giant. McCarthy (1982) has found
that any variations in the K velocity are less than ~5 km s^{-1}, which places strong
constraints upon any binary system. Nevertheless, WB considered the star to be a
high mass ratio, mass transferring system -- a terminally evolving Algol system.

In either case, the star is unusual, and represents a short-lived state of the
evolution of a close binary system. We present here new observations in the hope of
elucidating the nature of this star.

OBSERVATIONS

We obtained high resolution spectroscopic observations of FK Com in the spring
of 1983 using the IUE and the $2^{m}1$ coudé at KPNO. The emission lines are broad
(Fig. 1), and unlike any active late-type stars we have seen. Comparison of the
photospheric absorption lines to those of an artificially spun up star of similar
spectral type show that V sin i $\gtrsim$ 200 km s^{-1}, which is in excess of the V sin i for
the Hα emitting region. This seems to indicate non-synchronous rotation of the photosphere and chromosphere.

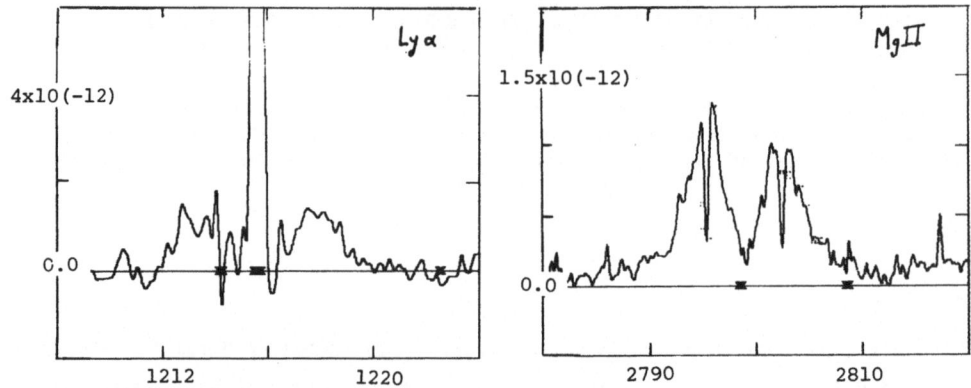

Fig. 1. Emission line profiles for Lyman α, and Mg II. Widths of Mg II lines correspond to ~500 km s^{-1} FWHM; the width of Lyman α is twice that. The spike in Lyman α is geocoronal emission. ×'s indicate bad data. The profiles have been smoothed with Gaussians of indicated width.

The IUE spectra show that C II, C IV, and the Mg II and Ca II resonance lines have velocity widths of ~500 km s^{-1} FWHM, as has been noted for Hα (WB). Ly α has double this width, perhaps due to opacity broadening. There is a deep, unresolved, presumably circumstellar absorption feature in Mg II which is stationary in wavelength. We have used the eight LWR-HI spectra obtained over the past three years to study phase dependent asymmetries of the Mg II h & k lines. Because of the breadth of the lines, h & k are blended so we assume similar line shapes and fold the red wing of this h line over the blue wing of the k line. There is a marginally convincing phase dependent asymmetry, which could be due to material streaming towards the observer at velocities of up to ~400 km s^{-1} near phase 0.1 (see Fig. 2). Also, the far UV emission line fluxes apparently modulated by about 30% with the stellar rotation period, with the brightest emission occurring near phase 0.25.

We have conducted an emission measure (EM) analysis using IUE SWP-LO data and an X-ray observation from the Einstein IPC. The UV fluxes are enhanced by two orders of magnitude over an extrapolation of the EM back from the X-ray point, assuming thermal

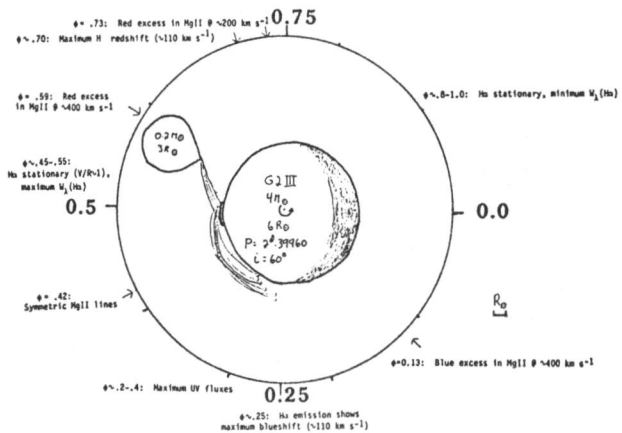

Fig. 2. A scale model of FK Com. The accretion stream and splashing material are indicated, as is the dark hemisphere of the G giant. Significant spectroscopic phases are indicated.

conductive losses ($EM \propto T^{3/2}$) as in solar-like loops. In this regard, the atmospheric structure of FK Com appears similar to that of coronal supergiants (e.g., β Dra), with a large fraction of its emission regions having maximum temperatures $\lesssim 10^6$ K.

INTERPRETATION

It is not easy to attribute the bizarre line shapes and profile changes to a single star, and the observation that the angular velocity of the photosphere (V sin i ~ 200) exceeds that of the Hα emitting region (V sin i ~ 110) by roughly a factor of two makes it especially difficult. One might put a hot spot in a disk as in a CV, but this would presumably have to be maintained by an external driver, since it could not be maintained by magnetic loops from the G star, as envisioned by Ramsey et al. (1981), because the Hα emitting region and the photosphere do not corotate. The existence of an external driver immediately implies a binary system.

Our revised model (Fig. 2) is a modified version of that presented by WB. FK Com is a high mass ratio, mass transferring binary wherein the mass ratio has reversed due to Roche lobe overflow. The mass loser, having lost >90% of its mass, is now the less massive star, while the mass gainer has ~doubled its mass. No accretion disk can form because the G giant is too large. We assume that the system is detached, the mass loser fills its Roche lobe, and that the observed emission line widths are comparable to the escape velocity of the system. The distance is ~200 pc. A constraint on the system masses is obtained from requiring that the core of the secondary not be degenerate. We arrive at a terminally evolving Algol system, of ~5-6 $M_\odot$ original mass, now consisting of stars of masses 4 and 0.2 $M_\odot$, with radii of 6 and 2.8 $R_\odot$. The semi-major axis is 12.5 $R_\odot$. The system is drawn to scale in Fig. 2, and significant spectroscopic and photometric phases are indicated. We predict that the photometric variations are due primarily to heating of a large fraction of the G star atmosphere by the accretion stream, which can mimic small dark spots for an appropriate geometry. Note that photospheric spots should rotate in ~$1^d.2$ if V sin i ~ 200, whereas atmospheric heating will be tied to the $2^d.4$ binary period.

There may be difficulties with the evolutionary scenario, as the mass loser should now be the helium core of a once massive star, and it is not clear what this would like like or whether it could fill its Roche lobe. Clearly much more work must be done before this model should be accepted. We shall present a detailed analysis of these data, and full details of the model, elsewhere.

REFERENCES
Bopp, B. W., and Stencel, R. E., 1981, Ap. J. (Letters), 247, L131.
Chugainov, P. F., 1974, Izv. Krymskoj Astrof. Obs., 54, 89.
Holtzman, J. A., and Nations, H. L., 1983, these proceedings.
McCarthy, J. K., 1982, B.S. Thesis, Pennsylvania State University.
Merrill, P. W., 1948, PASP, 60, 382.
Morris, S., and Milone, E., 1983, PASP, 85, 376.
Ramsey, L., Nations, H., and Barden, S., 1981, Ap. J. (Letters) 251, L101.
Walter, F. M., 1981, Ap. J., 245, 677.
Walter, F. M., and Basri, G. S., 1982, Ap. J., 260, 735 (WB).

OUTER ATMOSPHERES OF GIANT AND SUPERGIANT STARS

Alexander Brown
Joint Institute for Laboratory Astrophysics, University of Colorado
and National Bureau of Standards, Boulder, Colorado 80309

ABSTRACT

The properties of the chromospheres, transition regions and coronae of cool evolved stars are reviewed based primarily on recent ultraviolet and X-ray studies. Recent determinations of mass loss rates using new observational techniques in the ultraviolet and radio spectral regions are discussed and observations indicating general atmospheric motions are considered. The techniques available for the quantitative modeling of these atmospheres are outlined and recent results discussed. Finally our current rudimentary understanding of the evolution of these outer atmospheres and its causes are considered.

I. INTRODUCTION

The range of astrophysical research implied by the title of this review is so large that the subject matter must be restricted to fit both the time and space available to me. The review is therefore restricted to:

a) single, cool (spectral-type F-M), evolved stars,

b) the chromospheres, transition regions and coronae of these stars (i.e. that portion of the stellar atmosphere outside the photosphere), and

c) primarily to advances made during the last two or three years using the International Ultraviolet Explorer (IUE) satellite in the ultraviolet spectral region (1200-3200 Å) and the Einstein satellite in the X-ray spectral region (3-60 Å).

In recent years several excellent reviews have been written on the general topic of stellar chromospheres and coronae and I refer the reader to Linsky (1980, 1982), Dupree (1981a) and Jordan (1983) for a wider coverage of this research area, which has flourished with the general availability of ultraviolet and X-ray data. Many new and exciting results are emerging from this research allowing for a far deeper understanding of these outer atmospheres than has ever been possible before. Detailed studies of many chromospheric and transition region phenomena require observations at the limits of current instrumental capabilities in terms of both length of observation and required accuracy. The work currently being done is creating the basis of a new research area and laying the foundation for studies involving future space observatories.

II. OBSERVED PROPERTIES OF GIANTS

Observations of giant stars made during the first year of IUE operations clearly showed that two distinct forms of ultraviolet spectrum are seen from these stars (Linsky and Haisch 1979; Brown, Jordan and Wilson 1979; Dupree et al. 1979). Giants with spectral types earlier than K1 have ultraviolet spectra similar to those of dwarf stars in which most of the lines are formed by collisional excitation in an atmosphere containing material at temperatures from ~6000 K (Mg II resonance doublet) to over 10^5 K (C IV and N V resonance lines). X-ray observations by the Einstein satellite showed that these stars also possess coronal regions with temperatures up to ~10^7 K (Vaiana et al. 1981; Ayres et al. 1981; Haisch and Simon 1982). On the other hand, giants of spectral type K1 and later show no ultraviolet emission lines formed at temperatures much greater than 10^4 K nor do these stars show X-ray emission. Many of the emission lines seen from these stars are formed by fluorescence, e.g., the S I 1295, 1296 Å lines pumped by the O I resonance lines (Brown and Jordan 1980), and other radiative processes rather than collisional processes. A fuller review of radiative processes in the atmospheres of such stars is given by Jordan and Judge (1983). A typical example of each type of spectrum is shown in Figure 1.

Observations of the profiles of the Mg II resonance doublet at 2796 and 2803 Å have shown that a systematic change also occurs in the asymmetry of these optically thick, self-absorbed line profiles (Stencel and Mullan 1980). Stars showing coronal emission tend to have a stronger blue wing, while the stars showing only cool emission lines have a stronger red wing and this is interpreted as the presence of an accelerating outflowing stellar wind from the stars with strong red asymmetries (cf. Hummer and Rybicki, 1968). However, when considering individual stars, especially distant high luminosity stars, the influence of interstellar absorption is severe and can totally alter the observed line profile (Bohm-Vitense 1981). Stencel et al. (1981) showed that stars with outflowing winds have extended atmospheres while models of coronal stars indicate an essentially solar-like structure.

Much controversy has ensued concerning the nature and cause of the division between the two types of atmospheric structure implied by the IUE observations. Linsky and Haisch (1979) first proposed the presence of a _sharp_ dividing line in the HR diagram between stars showing coronal and non-coronal structure, i.e., through the presence of C IV emission. Various authors proposed that the change was more gradual in terms of line strengths and other atmospheric properties (Jordan and Brown 1981; Reimers 1981; Hartmann, Dupree and Raymond 1982) while others have sought to strengthen the argument for a sharp division (e.g. Simon, Linsky and Stencel 1982). Baliunas, Hartmann and Dupree (1983) showed that the C IV emission line fluxes of the four Hyades K0 giants were not equal but differed by significant amounts even when differing distances were taken into account. The present situation for giants is that while a large range in terms of line strength is seen near

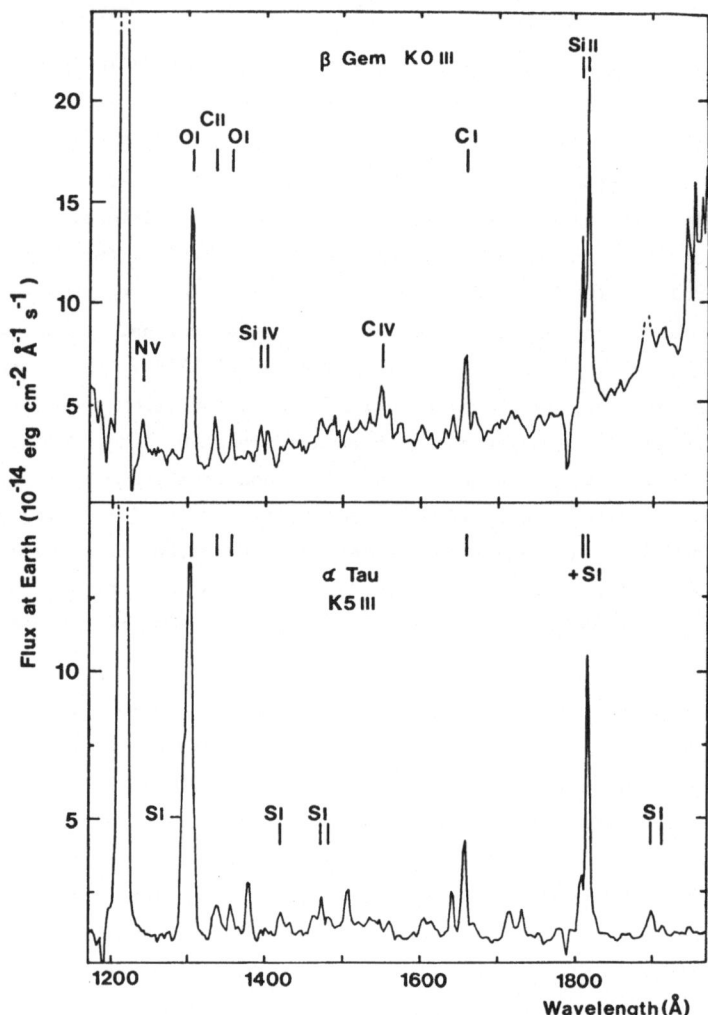

Fig. 1. The low dispersion IUE spectra of β Gem, a coronal giant, and α Tau, a star with only a cool extended chromosphere. The β Gem spectrum is from a single 120 minute exposure while the α Tau spectrum is the summation of a series of spectra, the longest having an exposure time of 150 minutes. The upper limit on the C IV emission is ~2.5 × 10^{-14} ergs cm^{-2} s^{-1} with the noise level on the combined spectrum being slightly less than but of the same order as this value. These spectra were reduced using the Oxford University ICL 2980 computer and the methods described by Brown and Jordan (1981).

the dividing line, it seems that all giant stars, as they evolve up the red giant branch, suffer a fundamental change in outer atmospheric structure and this change occurs near spectral type K1. Therefore the position of a giant star in the HR diagram, i.e., its luminosity and temperature, are major factors in determining its outer atmospheric structure but the actual level of emission is severely affected by an as yet ill-defined third parameter.

III. OBSERVED PROPERTIES OF BRIGHT GIANTS AND SUPERGIANTS

While the coronal/non-coronal dividing line of Linsky and Haisch still seems
relevant for giants, this is not true for more luminous stars where the situation
is far more complicated. Hartmann, Dupree and Raymond (1981) showed that the star
α TrA (K4 II) has C IV emission although it lies considerably to the right of the
Linsky-Haisch dividing line. Subsequently, Reimers (1982) identified three further
K bright giants which showed C IV emission (ι Aur, θ Her and δ TrA) and Hartmann et
al. (1983), in addition to confirming the findings of Reimers, showed that γ Aql has
similar properties. These stars are known as hybrid stars and are identified by the
presence of C IV emission and high velocity blue-shifted absorption components in
the Mg II and Ca II resonance doublets presumed to be caused by an outflowing wind.
It is the simultaneous presence of transition region material and a stellar wind
which makes these objects so interesting and important in understanding the transition between coronal-type structure and the cool, extended chromospheres of cooler
stars. The early G supergiants, α and β Aqr, also show the properties of hybrid
stars and represent the supergiant equivalents of the K bright giant hybrid stars
(Hartmann, Dupree and Raymond 1980). (Early K supergiants show only cool extended
chromospheres.) No known hybrid star has yet been detected as an X-ray source.

The hybrid stars are variable both in TR line strength and in the velocity
and mass flux of their winds as indicated by the Mg II absorption components. The
variability of C IV line strength may be responsible in part for earlier arguments
as to whether or not particular stars are hybrid stars. Hartmann et al. (1983)
have shown that over a year the velocity of the α TrA Mg II absorption component
increased from −84 to −180 km s^{-1}. Also, Drake, Brown and Linsky (1983) find that
the high velocity blue-shifted Mg II components of hybrid stars have radial velocities that show a greater scatter than those of the narrower low velocity absorption
components. The variability and breadth of the high velocity features suggest that
they are formed in a high-velocity rather turbulent stellar wind. The low-velocity
absorption features, on the other hand, are most probably formed in the interstellar
medium.

In the past, there has been much discussion as to the nature of the atmospheric
structure of hybrid stars. Hartmann, Dupree and Raymond (1981) proposed that the TR
emission lines were formed in the outflowing wind and that the heating required for
this might be derived from deposition of energy by Alfvén waves. Linsky (1982), on
the other hand, proposed that the structure had two components, namely, magnetically
confined TR material and a cooler outflowing stellar wind. The present situation
(see for instance, Hartmann et al. 1983) seems to be that the observational evidence
suggests that the C IV resonance and C III and Si III intersystem lines are not
formed in the stellar wind and that the temperature of the wind is no more than a
few ×10^4 K.

IV. MASS LOSS AND MASS LOSS RATES

Mass loss is an important process among cool giants and supergiants since the onset of mass loss is intimately related to the significant changes in atmospheric structure already noted. Accurate determination of mass loss rates is important in establishing what effect the observed mass loss will have on the evolution of these stars. Drake and Linsky (1983c) discuss in some detail new methods for the determination of mass loss rates and reference to those areas shall therefore be brief here. Recent reviews by Castor (1981), Dupree (1981b), Reimers (1981) and Linsky (1981) discuss in depth the nature of and possible mechanisms causing mass loss in cool stars; here only new methods and subsequent results will be discussed.

A. Ultraviolet Observations

Observations of stellar winds in the ultraviolet are superior to optical studies because lines of ions which are dominant species can be studied. This removes many of the uncertain assumptions necessary in the optical derivation of mass loss rates.

The first observations which definitely showed that mass was lost from stars were those of the G giant companion of α Her by Deutsch (1956) where absorption lines from the wind of the M5 bright giant primary were seen in the spectrum of the companion. Such observations allow the use of the companion as a probe of the stellar wind structure and this technique has been elegantly used by Che, Hempe and Reimers (1983) in the ultraviolet to study the winds of the K supergiant primaries of ξ Aurigae-type binary systems. Che et al. modeled the equivalent widths and line profiles of Fe II, Si II and S II lines seen in the B dwarf companion's spectra using a non-spherically symmetric, three-dimensional radiative transfer code. Mass loss rates and wind velocities were determined for ξ Aur, 32 Cyg and 31 Cyg. The mass loss rates fell in the range 0.6-2.8 (-8) $M_\odot$ yr^{-1} with wind velocities of 30-80 km s^{-1}.

Mass loss rates can also be determined by modeling the Mg II and, less importantly, Ca II resonance doublets. The Mg II lines have the advantages of being formed higher in the chromosphere than the Ca II lines due to their greater optical depth and of being seen against a lower photospheric background. The derivation of mass loss rates from observations of these lines is not a simple calculation nor can unique results be guaranteed. Drake and Linsky (1983a) have developed a co-moving frame, spherical symmetry, partial redistribution (PRD) radiative transfer code which they have used to model the Mg II lines of the K0 giant α Boo. The mass loss rate determined is 1(-10) $M_\odot$ yr^{-1} with a velocity of 40 km s^{-1}.

B. Radio Observations

The Very Large Array (VLA) radio telescope has the ability to detect cool giant and supergiant stars due to the free-free emission originating from their partially or fully ionized stellar winds. Drake and Linsky (1983c) present a table of derived mass loss rates for the sample of cool stars so far studied with the VLA. Of particular importance is the probable detection of emission from K giants for the first time (α Boo and β Gem). To illustrate the type of data which the VLA can provide, a 6 cm VLA map of the region around α Her (from Drake and Linsky 1983b) is shown in Figure 2.

The most significant fact to be noted from Table 1 of Drake and Linsky (1983c) is how small the newly derived mass loss rates are compared with previous results. For the rate derived from radio observations there is always the possibility that the wind is predominently neutral, in which case the radio observations place severe upper limits on the fractional ionization. The upper limit shown for the mass loss rate of the hybrid star ι Aur is smaller than previously quoted values which were used to compute Alfvén-wave driven wind models for such stars (Hartmann, Dupree and Raymond 1981).

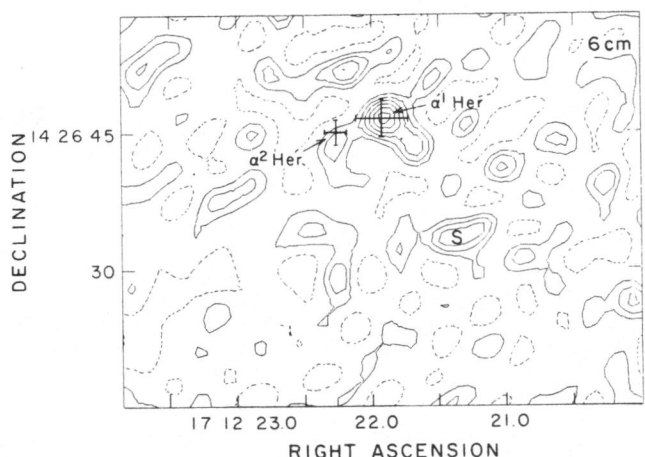

Fig. 2. A VLA 6 cm (4885 MHz) map of the region surrounding the binary star α Her. The base contour level (the first solid contour) is equivalent to the r.m.s. noise level. The optical positions of α^1 Her and α^2 Her are indicated, but note that the uncertainties in these positions are much less than the crosses shown. A serendipitous source (S) is also indicated. α^2 Her has a flux density of 1.2 ± 0.2 mJy (from Drake and Linsky 1983b).

V. ATOMSPHERIC MOTIONS

In addition to the information that IUE spectra provide concerning the outflowing winds of cool stars, other effects of atmospheric motions are detectable.
Ayres et al. (1983) and Brown et al. (1983) have shown that in the IUE spectra of
a wide variety of coronal stars, ranging from dwarf stars to the supergiant β Dra,
systematic wavelength shifts are seen between transition region and chromospheric
emission lines. The shift is in the sense that the transition region lines are red
shifted with respect to the chromospheric lines.

This phenomenon is best illustrated by observations of α Aur (Capella) and
β Dra. Ayres (1983) obtained a series of wavelength calibrated high dispersion IUE
spectra close to conjunction of the Capella binary system. These spectra were then
co-added to give the results shown in Figure 3. The zero velocity represents the
rest velocity of the system and therefore at conjunction the rest velocities of both
stellar photospheres. The striking feature is that all the emission lines, even the
chromospheric lines are red shifted. Before examining the other features of this
diagram it is best to consider β Dra, which Brown et al. (1983) have studied in detail. From the observed line widths, profiles and ratios of the C IV and Si IV
resonance doublets, these lines are seen to be optically thick and asymmetric to

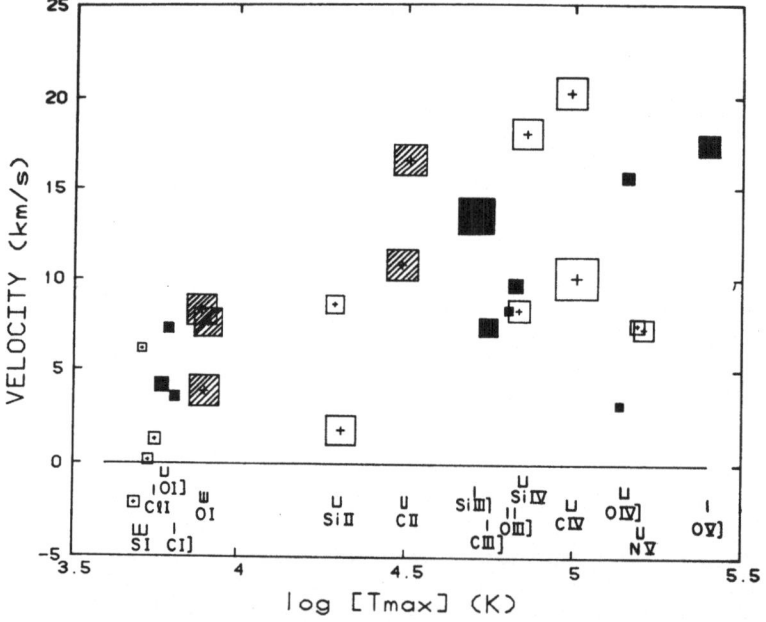

Fig. 3. The observed line centroid velocities for a range of emission lines from
the Capella binary system at conjunction. Open symbols indicate lines
which have uncertain optical thickness, solid symbols lines which are
definitely optically thin and hatched symbols lines which are definitely
optically thick (from Ayres 1983).

the red, resulting in the measured redshifts. The observed shifts for these lines are therefore most likely due to optically thick line formation in a turbulent medium containing accelerating upflows and/or decelerating downflows. This conclusion also applies to other optically thick lines such as the C II and O I resonance lines. Hence, simple interpretation of the line shift data is best restricted to the optically thin intersystem lines which are redshifted by $+8 \pm 4$ (1σ) km s^{-1} in β Dra and also by a similar amount in α Aur as can be seen from Figure 3. These values are near the limits possible using IUE and clearly confirmation of such results by future higher-quality measurements is desirable. However at face value these shifts indicate a net excess of emission from downward moving material over that from material moving upward.

VI. ATMOSPHERIC MODELING

A. Modeling Techniques

As might be expected from the very different spectra of coronal and noncoronal stars, the techniques used to calculate atmospheric models are different for the two groups of stars.

For coronal stars the two techniques which have been widely used are emission measure analysis and radiative transfer modeling of the Mg II and Ca II resonance lines. Emission measure analysis can be used to model the portion of the outer atmosphere at temperatures $\gtrsim 10^4$ K where hydrogen is predominantly ionized. This type of analysis was originally developed for solar work (Pottasch 1964; Jordan and Wilson 1971) and its general application to stellar chromospheres and coronae has been discussed in detail by Jordan and Brown (1981). The emission measure, E_m, is $\int_{\Delta H} N_e^2 dh$ where ΔH is the region of line formation and is related to the total surface flux of a collisionally-excited, effectively-thin emission line, using our current formalism, by

$$F_* = \frac{8.6 \times 10^{-22}}{\lambda(cm)} \frac{\Omega_{12}}{\omega_1} \frac{N_E}{N_H} \frac{N_1}{N_{ion}} g(T_m) \int_{\Delta H} N_e^2 dh \quad .$$

Here Ω_{12} is the averaged collision strength, ω_1 is the statistical weight of the lower level, N_E/N_H is the elemental abundance, N_1/N_{ion} is the population of the lower level and $g(T_m)$ is the value at peak ion population of the temperature-dependent function

$$g(T) = T_e^{-1/2} \exp(-W_{12}/kT_e) \frac{N_{ion}}{N_E} \quad .$$

W_{12} is the excitation energy and all other symbols have their usual meaning.

From the emission measures of individual lines a mean emission measure distribution is derived and then plane-parallel atmospheric models can be computed in hydrostatic equilibrium using

$$dT_e/dh = P_e^2/\lceil 2.0\ E_m(T_e)T_e\rceil$$

and

$$dP_e/dh = -7.14 \times 10^{-9}\ P_e g_*/T_e$$

where P_e is the scaled pressure ($N_e T_e$) and g_* is the stellar surface gravity.

The alternative modeling method for coronal stars is to use the Mg II and, less importantly, Ca II resonance line profiles to determine the chromospheric structure by means of partial-redistribution (PRD) line transfer codes and then to derive the transition region structure by matching the line fluxes of lines such as the Si IV, C IV, and N V resonance lines. This method can also be applied to noncoronal stars with extended atmospheres although in this case the systematic outflow of the stellar wind must be explicitly included in the calculation. In order to produce reasonably unique results from modeling the Mg II resonance line profiles, constraints on the atmospheric pressure must be obtained from density sensitive line ratios. (This is also true for emission measure analysis.)

The estimation of electron densities from line ratios is a critical step in the modeling process. For coronal stars the most useful line ratios are between (i) the intersystem lines of C III 1909 Å and Si III 1892 Å, although in this case the original density calibration of Cook and Nicholas (1979) must be corrected for the new Si III atomic data (Baluja, Burke and Kingston 1980, 1981), and (ii) the members of the O IV intersystem multiplet at 1401, 1405 and 1407 Å (Nussbaumer and Storey 1982). For stars with cool extended chromospheres the C II intersystem multiplet near 2325 Å provides the best density estimates (Stencel et al. 1981).

A further method for calculating atmospheric models of cool extended chromospheres involves the study of the ratios of lines of differing opacity which originate from the same upper level. In the situation where one line is optically thick and the other is not, Jordan (1967) showed that the opacity in the optically thick line can be determined and a mass column density derived. Application of this method to stellar spectra observed with IUE was discussed by Brown, Ferraz and Jordan (1980).

Basically the method is as follows: The fluxes (F) in each pair of lines are related to their branching ratios, b, and probabilities of escape, q, so that

$$F_1/F_2 = \lambda_2 b_1 q_1/\lambda_1 b_2 q_2 \quad .$$

Assuming a Gaussian profile, the probability of escape is related to the opacity at line center, τ_0, by

$$q = 1 - \mathrm{erf}(\ln \tau_0)^{1/2} \quad .$$

Finally, for a Doppler-broadened line formed at temperature T_1, τ_0 is related to the

mass column density, $\int N_H dh$, by

$$\tau_0 = 6.0 \times 10^{-15} \lambda(A) f_{12} M^{1/2} \frac{N_E}{N_H} \int \frac{N_1}{N_{ion}} \frac{N_{ion}}{N_E} N_H T_1^{-1/2} dh$$

where f_{12} and M are the oscillator strength and atomic weight, respectively. This method is applicable to lines of several ions such as C I, S I, O I and Fe II. (For details see Jordan and Judge 1983.)

B. Models of Atmospheric Structure

Four giant and bright giant stars have now been modeled using emission measure techniques. These are ß Dra (G2 Ib-II; Brown et al. 1983), ß Gem (K0 III; Brown and Jordan 1983), α TrA (K4 II) and ι Aur (K3 II), with the latter two models by Hartmann et al. (1983). The general conclusion is that the atmospheres are not extended, i.e. the extent is less than or equal to the stellar radius. Representative models of ß Dra are shown in Figure 4. For this star, density-sensitive line ratios and opacity arguments suggest the 1.1×10^{14} cm^{-3} K model is appropriate for the transition region but a slightly larger scaled pressure (3.5×10^{14} cm^{-3} K) is needed to

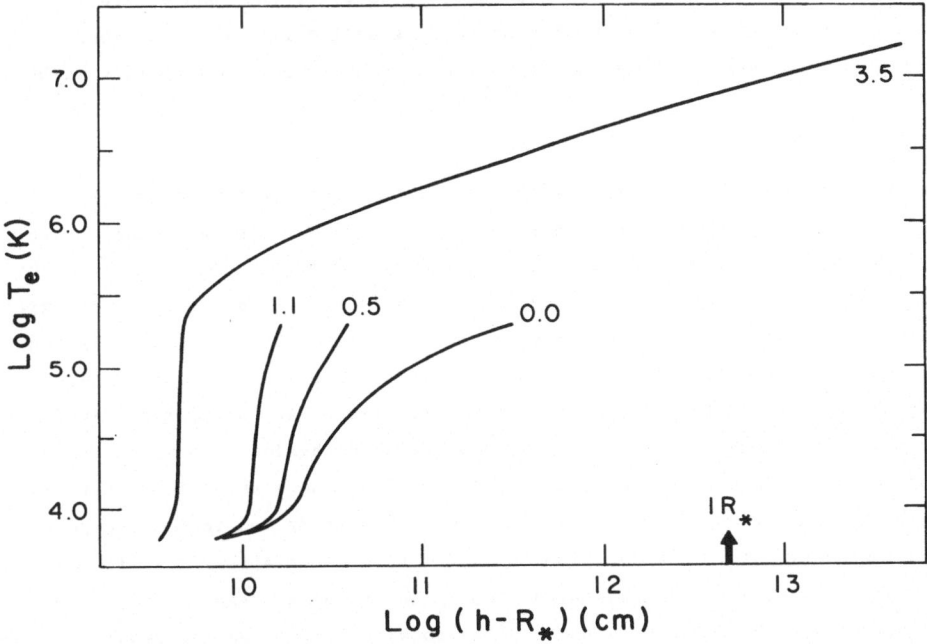

Fig. 4. Simple spherically-symmetric models of temperature versus height for ß Draconis based on emission measure analysis. The scaled pressure for each model at log T_c = 5.3 is indicated in units of 10^{14} cm^{-3} K. Note that the atmospheric extent is less than or of the order of the stellar radius. Only the model at $P_0 = 3.5 \times 10^{14}$ cm^{-3} K can consistently be extended to coronal temperatures without assuming that the X-ray emitting plasma is geometrically confined (from Brown et al. 1983).

match the coronal emission, although this discrepancy can be removed by reducing the fractional area coverage of the transition region plasma. Additional evidence that the atmosphere may have more than one component is that the coronal temperature ($\sim 1.5 \times 10^7$ K) is much greater than the escape velocity of the star ($\sim 8 \times 10^5$ K) and the coronal regions must then be either magnetically confined or heated in an out-flowing wind. Generally, the pressures derived for these stars are lower than those found for dwarf stars of similar spectral type. The density derived by Hartmann et al. (1983) for α TrA from the C II intersystem lines (4×10^8 cm^{-3}) is comparable to those found for extended chromosphere stars and probably reflects the density in the outflowing wind of α TrA.

Other stars have been modeled using radiative transfer codes to match the Mg II resonance line profiles including β Cet (G9.5 III; Eriksson, Linsky and Simon 1983), and β Dra, ε Gem (G8Ib) and α Ori (M2 Iab) by Basri, Linsky and Eriksson (1981). Comoving frame models by α Boo (K0 III) and α Ori have been computed by Drake and Linsky (1983) and Hartmann and Avrett (1983) respectively. Earlier modeling of chromospheric resonance lines was plagued by lack of constraints on the atmospheric pressure. Hartmann and MacGregor (1980, 1982) have investigated the properties of stellar winds heated and driven by Alfvén waves. However, the counter arguments of Holzer, Flå and Leer (1983) cast doubts on this mechanism for high mass loss winds. No detailed models have yet been published based on the escape probability method.

VII. EVOLUTION OF ATMOSPHERIC STRUCTURE

Our understanding of the detailed evolution of the outer atmospheric structure of cool stars is as yet fairly rudimentary. The most comprehensive study of evolutionary changes in the ultraviolet spectra of giant and bright giant stars is that of Simon (1983) in which great care was taken to separate stars ascending the giant branch for the first time from more evolved stars. Figure 5 shows the variation of C IV surface flux normalized to the stellar bolometric flux found for the stars crossing the Herzsprung–Russell gap for the first time. The normalized C IV flux rises steadily to a maximum at G0 III and then declines again. The range of C IV emission among the late G and early K giants including the more evolved stars is large, about a factor of 50, reflecting the results already mentioned concerning the Hyades giants. Additionally, Simon showed that the normalized C IV flux is well correlated with v sin i indicating, that much of the systematic change seen in Figure 5 is related to changes in the stellar rotation rates and the growth of sub-photospheric convection zones.

Although as yet too few quantitative models of atmospheric structure exist to allow the detailed evolution of stellar chromospheres and coronae to be explained, it is possible to speculate on the major factors affecting the outer atmospheres of stars evolving towards the giant branch. Gray (1982) has shown that the surface

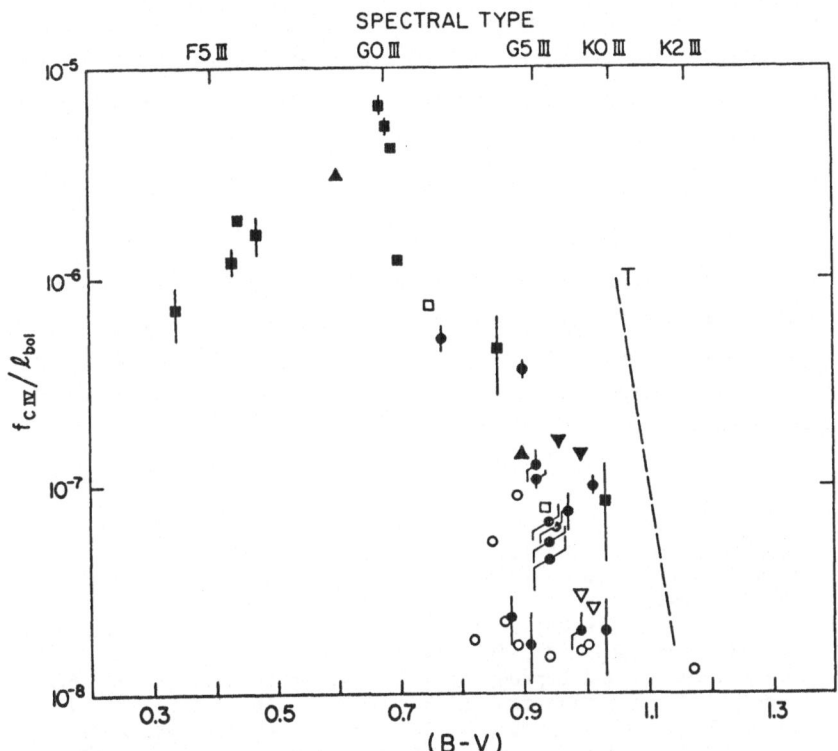

SPECTRAL TYPE

Fig. 5. Normalized C IV fluxes for yellow giants plotted versus B-V. Stars evolving up the giant branch for the first time are shown as squares while more evolved stars are shown as circles. Triangles and inverted triangles indicate the components of the Capella binary and the Hyades K giants respectively. Detections (with 1σ error bars) are plotted as filled symbols and 1σ upper limits as open symbols. The vertical dashed line shows the position of the Linsky-Haisch dividing line (from Simon 1983).

rotation rate of giants decreases dramatically and abruptly at spectral type G5. Rotation is a major factor in the generation of surface magnetic fields by the dynamo mechanism. The coronal activity level of a star that suffers a large decrease in rotation will be greatly reduced and its ability to retain magnetically-confined coronal plasma diminished. Simon, Linsky and Stencel (1982) draw attention to the systematic reduction in the temperature of the critical point of a Parker-type thermally-driven wind and the relation of this change to the cooling and expansion times of the wind. The form of the radiative power loss function of a hot plasma (cf. McWhiter, Thonemann and Wilson 1975) is such that if a plasma has a temperature between $\sim 10^4$ K and $\sim 5 \times 10^5$ K and no heating is supplied, then the plasma will cool rapidly to $\sim 10^4$ K. Simon et al. noted that the critical temperatures for β Gem (coronal) and α Boo (non-coronal) were $\lesssim 2 \times 10^6$ K and $\lesssim 3 \times 10^5$ K respectively and suggested that a lack of confined magnetic regions and the above radiative instability together could account for the lack of transition region and coronal plasma in α Boo. This type of explanation seems reasonable given our current knowledge of the detailed properties of these stars. It is interesting to note that, because of their larger masses, the hybrid bright giants and supergiants have critical temperatures above 5×10^5 K. For example, using the stellar properties adopted by Hartmann et al. (1983) for α TrA (log g = 1.5, R = 79 $R_\odot$), the critical temperature for this star is $\leq 7 \times 10^5$ K.

I would like to acknowledge support by National Aeronautics and Space Administration grant NAG5-82 to the University of Colorado while preparing this review. I would like to thank Drs. T. Ayres, K. Carpenter, S. Drake, L. Hartmann, C. Jordan, J. Linsky and F. Walter for their advice and helpful discussions concerning the subject of this review.

REFERENCES

Ayres, T. R. 1983, Astrophys. J. Suppl., submitted.

Ayres, T. R., Linsky, J. L., Vaiana, G. S., Golub, L. and Rosner, R. 1981, Astrophys. J. 250, 293.

Ayres, T. R., Stencel, R. E., Linsky, J. L., Simon, T., Jordan, C., Brown, A. and Engvold, O. 1983, Astrophys. J., in press.

Baliunas, S. L., Hartmann, L. and Dupree, A. K. 1983, Astrophys. J. 271, 672.

Baluja, K. L., Burke, P. G. and Kingston, A. E. 1980, J. Phys. B 13, L543.

Baluja, K. L., Burke, P. G. and Kingston, A. E. 1981, J. Phys. B 14, 1333.

Basri, G. S., Linsky, J. L. and Eriksson, K. 1981, Astrophys. J. 251, 162.

Bohm-Vitense, E. 1981, Astrophys. J. 244, 504.

Brown, A., Ferraz, M. C. de M. and Jordan, C. 1980, in Proceedings of the Symposium on The Universe at Ultraviolet Wavelengths, Goddard Space Flight Center, NASA-CP2171, p. 297.

Brown, A. and Jordan, C. 1980, Mon. Not. R. astr. Soc. (Short Communication), 191, 37.

Brown, A. and Jordan, C. 1983, in preparation.

Brown, A., Jordan, C., Stencel, R. E., Linsky, J. L. and Ayres, T. R. 1983, Astrophys. J., submitted.

Brown, A., Jordan, C., and Wilson, R. 1979, in The First Year of IUE, A Symposium held at the University College, London, eds. A. J. Willis, p. 232.

Castor, J. 1981, in Physical Processes in Red Giants, eds. I. Iben, Jr. and A. Renzini (Reidel: Dordrecht), p. 285.

Che, A., Hempe, K. and Reimers, D. 1983, Astron. Astrophys. 126, 225.

Cook, J. W. and Nicolas, K. R. 1979, Astrophys. J. 229, 1163

Deutsch, A. J. 1956, Astrophys. J. 123, 210.

Drake, S. A., Brown, A. and Linsky, J. L. 1983, in preparation.

Drake, S. A. and Linsky, J. L. 1983a, Astrophys. J. 273, 299.

Drake, S. A. and Linsky, J. L. 1983b, Astrophys. J. (Letters), 274, in press.

Drake, S. A. and Linsky, J. L. 1983c, in Third Cambridge Workshop on Cool Stars, Stellar Systems and the Sun, eds. S. L. Baliunas and L. Hartmann, in press.

Dupree, A. K. 1981a, in Second Cambridge Workshop on Cool Stars, Stellar Systems, and the Sun, eds. M. S. Giampapa and L. Golub, SAO Special Report 392, Vol. II, p. 3.

Dupree, A. K. 1981b, in Effects of Mass Loss on Stellar Evolution, eds. C. Chiosi and R. Stalio (Reidel: Dordrecht), p. 87.

Dupree, A. K., Black, J. H., Davis, E. J., Hartmann, L. and Raymond, J. C. 1979, in The First Year of IUE, A Symposium held at the University College, London, eds. A. J. Willis, p. 217.

Eriksson, K., Linsky, J. L. and Simon, T. 1983, Astrophys. J. 272, 665.

Haisch, B. M. and Simon, T. 1982, Astrophys. J. 263, 252.

Hartmann, L. and Avrett, A. V. 1983, in Third Cambridge Workshop on Cool Stars,
    Stellar Systems and the Sun, eds. S. L. Baliunas and L. Hartmann, in press.
Hartmann, L., Brown, A., Jordan, C. and Dupree, A. K. 1983, in preparation.
Hartmann, L., Dupree, A. K. and Raymond, J. C. 1980, Astrophys. J. (Letters), 236,
    L143.
Hartmann, L., Dupree, A. K. and Raymond, J. C. 1981, Astrophys. J. 246, 193.
Hartmann, L., Dupree, A. K. and Raymond, J. C. 1982, Astrophys. J. 252, 214.
Hartmann, L. and MacGregor, K. B. 1980, Astrophys. J. 242, 260.
Hartmann, L. and MacGregor, K. B. 1982, Astrophys. J. 257, 264.
Holzer, T. E., Flå, T. and Leer, E. 1983, Astrophys. J. 275, in press.
Hummer, D. G. and Rybicki, G. S. 1968, Astrophys. J. (Letters), 153, L107.
Jordan, C. 1967, Solar Phys. 2, 441.
Jordan, C. 1983, Phys. Reports, to appear.
Jordan, C., and Brown, A. 1981, in Solar Phenomena in Stars and Stellar Systems,
    eds. R. M. Bonnet and A. K. Dupree, NATO ASIC 68, (Reidel: Dordrecht), p. 199.
Jordan C., and Judge, P. 1983, Physica Scripta, in press.
Jordan, C., and Wilson, R. 1971, in Physics of the Solar Corona, ed. C. J. Macris
    (Reidel: Dordrecht), p. 211.
Linsky, J. L. 1980, Ann. Rev. Astron. Astrophys. 18, 439.
Linsky, J. L. 1981, in Effects of Mass Loss on Stellar Evolution, eds. C. Chiosi
    and R. Stalio (Reidel: Dordrecht), p. 187.
Linsky, J. L. 1982, in Advances in Ultraviolet Astronomy: Four Years of IUE
    Research, eds. Y. Kondo, J. M. Mead and R. D. Chapman (NASA-CP2238), p. 17.
Linsky, J. L. and Haisch, B. M. 1979, Astrophys. J. (Letters), 229, L27.
McWhirter, R. W. P., Thoneman, P. C. and Wilson, R. 1975, Astron. Astrophys. 40,
    63.
Nussbaumer, H. and Storey, P. J. 1982, Astron. Astrophys. 115, 205.
Pottasch, S. R. 1964, Space Sci. Rev. 3, 816.
Reimers, D. 1981, in Physical Processes in Red Giants, eds. I. Iben Jr. and A.
    Renzini (Reidel: Dordrecht), p. 269.
Reimers, D. 1982, Astron. Astrophys. 107, 292.
Simon, T. 1983, Astrophys. J., submitted.
Simon, T., Linsky, J. L. and Stencel, R. E. 1982, Astrophys. J. 257, 225.
Stencel, R. E., Linsky, J. L., Brown, A., Jordan, C., Carpenter, K. G., Wing, R. F.
    and Czyzak, S. 1981, Mon. Not. R. astr. Soc. (Short Communication), 196, 47.
Stencel, R. E. and Mullan, D. 1980, Astrophys. J. 238, 221.
Vaiana, G. S., et al. 1981, Astrophys. J. 245, 163.

ENVELOPES OF RED GIANT STARS

P. G. Wannier
Jet Propulsion Labs, Caltech

Abstract

A discussion is presented of the observations of circumstellar envelopes around late-type giant stars. The results are treated in two broad categories. First, there are extensive surveys which provide a few basic parameters of the flows (v_{exp}, dM/dt and/or abundance information). These indicate the importance of rapid mass-loss to stellar evolution and to the enrichment of the interstellar medium in nucleosynthetic products. Second, there are the more detailed observations of structure in the flows, giving information about the mass-loss mechanisms which may be operating, but restricted so far to only a handful of the brightest objects.

1. Introduction

Certainly a key event in post-main-sequence evolution is the rapid loss of material in dense winds. Not only can rapid mass-loss affect the progress of intermediate-mass stars up the asymptotic giant branch branch (AGB), but it may prevent supernova detonation in stars up to $\sim 6-8 M_\odot$ and must certainly enrich the interstellar medium in the great variety of nuclides produced by equilibrium CNO burning, the 3-α process and the s-process. The way in which mass-loss actually occurs is poorly understood, a lack which can be appreciated by our struggles with the Solar wind where observational detail is rich but our understanding still not complete.

Our knowledge at any time is strongly affected by the available observational techniques. For example, studies of the blue-shifted H and K lines lead to a picture of steady mass-loss at a moderate rate throughout much of the post-main-sequence lifetime of giant stars. On the other hand, observations of planetary nebulae (PN) have led to pictures of envelope ejection on a dynamic timescale. Recently, observations at millimeter and IR wavelengths have identified a class of very red long-period variables with super-winds indicating mass-loss rates (dM/dt) of up to 10^{-6} to 10^{-4} $M_\odot$/yr. These massive winds, typically expanding at 10-15 km/s, form the circumstellar envelopes (CSE) which are the subject of this review. Sometimes these CSE's are so dense that the central star is visibly obscured.

2. The Observational Material

From existing observations, we can learn quite a bit about the basic parameters of CSE's. The determination of dM/dt and v_{exp}, especially for a statistically significant sample of late-type giants is obviously important. For example, we may ask whether all weakly bound envelopes develop super-winds or whether there are additional properties which play important roles. Does the formation of a CSE correlate with initial metallicity? Does it correlate with carbon enrichment following deep dredge-up from the carbon core? How significant is the total mass lost in the super-wind stage as opposed to mass-loss in other evolutionary stages? Such questions may be addressed by large surveys of CSE's, even without detailed information about the flow.

However, many other interesting questions can only be addressed when, with enough spatial resolution, we can tell more about the morphology of CSE's than we are currently able to. In terms of radial structure, we may ask if, on a dynamic timescale, the mass-loss is sudden, steady or periodic. Perhaps there are slower variations in either dM/dt or v_{exp}, say on a thermal timescale. Non-radial structure may well indicate either polar or equatorial winds and may point to a role for surface stellar fields or the importance of a close companion for driving winds in a binary system. Much of our present information about structure comes from later stages in the evolution of CSE's when ionization by the central star make visible nebulae.

Planetary nebulae are the oldest known manifestation of stellar mass-loss and continue to yield new and interesting information, especially about the extended morphology of mass-loss. Recent photographic evidence shows the presence of extended halos in PN, probably remnant from dense winds (Jewitt, 1983; Terzian, 1983). In addition, many PN and/or their halos show some symmetric structure indicating that an axis, defined either from stellar rotation or binary orbits, must affect their formation (Calvet and Peimbert, 1983). The correspondence between CSE's and PN is not clear, but recent theoretical investigations indicate that quite acceptable looking PN may form naturally from a CSE by the interaction of the slowly expanding super-wind with a fast, low-density wind streaming off of a newly revealed white dwarf (c.f., Kwok, 1982; Purton, 1983). In that case, the morphology of the PN may find some correspondence in the pre-existing

CSE. We therefore arrive at considering structure in CSE's and much of this paper will be devoted to reviewing the information thus far available and to the observing techniques which show future promise.

2.1 Millimeter-wave Observations

There are several important molecular transitions which have been used to observe CSE's, but our only statistically significant sampling comes from the CO molecule using its J=1-0 and J=2-1 rotational trans-transitions. CO has a low permanent dipole moment and therefore can be collisionally excited at $n(H_2) \sim 1000$ cm^{-3}, corresponding to a radial distance of:

$$r_{max} = sqrt\{dM/dt/[n(H_2)v_{exp}]\}$$

$$= 4.3 \times 10^{17} cm$$

using $dM/dt = 10^{-5} M_\odot$/yr and $v_{exp} = 20$ km/s, applicable to the well-known carbon-rich CSE, IRC+10216. In fact, this massive and fairly close ($\sim$ 300 pc) envelope has been mapped out to such a radius, confirming some of the assumed properties of the envelope (Wannier et al., 1979). Surveys of CO yield reliable values of v_{exp} though there are some discrepancies of factors of two or less with results of maser observations. In addition, values of dM/dt may be derived which are fairly insensitive to assumptions about nuclear abundances or circumstellar chemistry (Knapp et al., 1982), but share with all determinations of dM/dt a quadratic dependance on the assumed distance. To date there have been about 50 objects detected using the CO(1-0) line (Knapp, 1983) falling into the three C/O abundance classes: ~45% C-rich, ~45% O-rich (or "normal") and ~10% S-stars with C/O $\sim$ 1. Values of dM/dt range from 10^{-7} to 10^{-4} $M_\odot$/yr. Values of dM/dt/L do not suffer from uncertainties in the distance determination since both parameters scale with D^2. There is strong evidence from the CO observations that dM/dt/L is systematically larger in C-rich than in O-rich objects.

The presence of C in stellar envelopes has long fascinated students of stellar evolution, implying a dredge-up of material on the AGB from the carbon core (See complete discussions of how this process may work in Iben and Renzini, 1983.). It is usually thought that O-rich Miras are stars which have not yet dredged up carbon but Wood et al. (1983) suggest, from their study of LMC and SMC Miras, that O-rich Miras might sometimes represent former C-rich objects which have recycled carbon back into N by equilibrium CNO processing. Even if a complete reversion of abundance class is not possible, the continued CNO processing on "primary" carbon could well produce a significant amount of "primary" nitrogen for subsequent ejection into the interstellar medium. Millimeter-wave techniques easily lend themselves to observations of isotopic abundances in CSE's and several unusual abundances have been measured in IRC+10216 (Wannier and Linke, 1978). More extensive observations have been made of $^{12}C/^{13}C$ using CO observations and these indicate that $^{12}CO/^{13}CO$ is smaller (4-15) in O-rich stars than in carbon stars. Finally, there is some evidence that C/H, not just C/O, is larger in C-rich objects (Knapp, 1983). All such observations are consistent with, but are not unique to, the enrichment of carbon in carbon stars by a pure component of ^{12}C and in O-rich Miras of material which has been exposed to equilibrium CNO processing. More observations are needed to determine how many of the s-process nuclides may be enhanced in stars having CSE's.

As indicated above, it would be desireable to map out the spatial distribution of CO in order to answer certain questions about the mass-loss mechanism. Spatial maps of millimeter-wave CO emission are severely restricted by the available single-dish resolution (30"). Nonetheless, Wannier et al. (1983) have obtained maps of 6 objects using the Onsala Space Observatory 20 m antenna at the CO (J=1-0) line. In the 5 C-rich and one S-type CSE studied, there is no obvious indication of noncircular symmetry.

Finally, a discussion of millimeter-wave results would not be complete without some mention of the unpublished results of aperture-synthesis observations by the new interferometers at the Hat Creek (Bieging et al., 1983) and Owens Valley radio observatories. To date, only three objects have been observed, carefully chosen for their possible non-symmetric CO distribution. Thus, no meaningful conclusions can be drawn about structure in the majority of CSE's. Maps of IRC+10216 have now been completed using the HCN (J=1-0) (at Hat Creek) and the CO (J=1-0) (at OVRO) transitions, with synthesized resolutions of 10" and 6" respectively. In both maps, the emission appears to be circularly symmetric, presenting an interesting contrast to some of the IR results discussed below. In addition to the observations of IRC+10216, two asymmetric visible nebulae, CRL2688 and NGC7027 have been observed in CO at OVRO and have asymmetric structures consistent with their visible appearances.

2.2 OH Surveys

The OH maser, being brighter than the thermal CO lines, provides a larger sampling of CSE's and a straightforward determination of v_{exp}. However, the estimations of dM/dt are less reliable, involving observed IR brightnesses and assumed properties of the circumstellar dust. Naturally OH surveys cannot provide information about C-rich objects. To date there have been ~300 objects detected (Engels, 1979) with v_{exp} falling in the range 10-50 km/s. A particularly interesting OH survey has been made recently by Bowers et al. (1983), consisting of VLA maps of 1612 MHz OH around 20 giants and supergiants. Interestingly, they find structure consistent with spherical symmetry in all cases except for the super-giant VYCMa, which shows a disc-like structure. To date, there is no more complete survey of the non-radial structure of CSE's. In addition, their results indicate a correlation between the outer OH radius and dM/dt in those cases where dM/dt has been measured from CO observations. That may allow dM/dt to be estimated directly from OH observations in a large additional population of O-rich Miras.

2.3 IR Continuum observations

Most of the well-known CSE's studied to date have initially been picked from candidate objects on IR continuum surveys and it is certain that this process will continue in light of the upcoming, sensitive IRAS results (see Beichman, 1983). However, from IR observations alone, it is difficult to derive much information about the CSE's. There is no velocity information and the IR brightness by itself can only yield a value of dM/dt with knowledge of the dust opacity, of v_{exp} and of the gas/dust ratio. In this regard, a recent 400μ survey by Sopka et al. (1983) may offer some help. By using a long observing wavelength, they have ensured low dust opacity in their observed CSE's and, from CO (J=1-0) observations in the same sources, have concluded that the observed gas/dust ratios are at least consistent with the interstellar value of 100.

Again, as with millimeter-wave observations, spatial interferometry holds great promise for IR continuum observations of CSE's, but the existing results are still somewhat sparse. There have been two approaches to IR interferometry: generating maps with a two-element heterodyne system and using speckle techniques with large, single apertures. The heterodyne technique, capable of higher spatial resolution, was pursued at Berkeley by Sutton and co-workers in the 10μ silicate emission band of five objects. A map of IRC+10216 with a spatial resolution of 0.3" indicates circular symmetry (Sutton et al., 1979), a result consistent with the larger-scale millimeter-wave interferometer results. The remaining CSE's observed with the same equipment were O-rich Miras, and scale-sizes but not two-dimensional brightness distributions were obtained (Sutton et al., 1977). Despite a recent hiatus in the two-element IR interferometry, it is expected that new results will be forthcoming.

Speckle interferometry of CSE's has now been pursued by several workers and with several telescopes (Dyck et al., 1983b; McCarthy, 1979; Ridgway, 1983; Howell, 1980). Of interest are the results reported by McCarthy in which flattened brightness distributions are reported for three objects out of three observed, including IRC+10216 and two oxygen-stars. Ridgway has also reported some N-S elongation in IRC+10216. These observations were carried out at similar wavelengths (5-12µ) and spatial resolutions (several tenths arcsec) to those of Sutton et al. A more extensive 2.2-4.8µ survey by Dyck et al. (1983b) includes 16 carbon and oxygen-stars with limited azimuthal information provided by the placement of a slit at (typically) two different position angles. In this survey all objects were resolved at the ~.2-.3" level but is no evidence for non-circular symmetry from the limited two-dimensional information.

2.4 IR Spectral-Line Observations

As is the case for the millimeter-wave observations, the most extensive information about CSE's results from observations of CO, in this case of the vibrational bands and their overtones. The 4.6µ band is the fundamental CO vibrational transition ($\Delta v=1$) and is further separated into the isotopic variants of CO, into the different initial vibrational states and into the many rotational levels forming the P ($\Delta J=1$) and Q ($\Delta J=-1$) branches for each isotope and upper vibrational level. This wealth of spectral information simultaneously offers lines of vastly different oscillator strengths and provides information about isotopic abundances and rotational and vibrational excitation. The weakness of many IR spectral-line observations is that they only provide line-of-sight information, so that non-symmetric flows cannot be studied and radial information can only be inferred indirectly from the observed rotational excitation temperatures.

A study of 9 CSE's using the 4.6µ line is that of Bernat (1981) in which he reports multiple line-of-sight velocities in 4 objects. However, there is no consistent pattern among the objects studied of velocity versus rotational excitation temperature (radius), so that the meaning of these results is still obscure. The different velocity components appear to be separated by many tens or hundreds of years, short on the CSE lifetimes, but long compared to the stellar pulsational periods (~.5-2 years). The different velocities may result from different initial ejection velocities or from different extents of subsequent acceleration.

Particularly interesting studies are the time sequence observations of R Leo (Hinkle, 1978) and Chi Cyg (Hinkle et al., 1982), showing how the stellar envelopes behave at different phases of the Mira pulsation. Both studies used the CO fundamental bands as well as the $\Delta v=2$ and $\Delta v=3$ overtone bands, the overtone bands providing deeper probes into the stellar envelopes. The observations yield detailed information about the the outwardly propagating shocks which may give rise to the circumstellar mass-loss. A most intriguing result is the appearance, above the shocked regions in both stars, of levitated, quiescent (v_{exp}~0) gas at a radius corresponding to a CO excitation temperature of ~1000 K. Such a temperature is close to that for the condensation of silicate grains and may offer a clue as to how mass-loss is originated (from the Mira shocks) and accelerated (by subsequent acceleration by radiation pressure on grains).

In order make use of the spectral wealth available in the IR, but to avoid the shortcomings of obtaining purely line-of-sight information, two groups have turned to observations of the 4.6μ radiation resonantly scattered in extended CSE's. These studies have detected extended structure on scales both large and small compared to a typical seeing limit of ~2". The larger-scale observations were made by Wannier and Sahai (1983) with a large-area detector and an FTS spectrometer using annular-shaped observing apertures of different sizes, masking the bright central star with its deep blue-shifted, absorption lines while passing emission at the stellar rest velocity. Prominent emission lines have been observed in 7 out of 14 objects observed and radial distributions of CO emission and excitation temperature have been obtained. In a few cases there are plans to extend the technique to include full two-dimensional maps by use of offset circular observing apertures.

The smaller-scale structure of CO in CSE's has been measured by a combination speckle/spectral technique applied thus far to IRC+10216 (Dyck et al., 1983a) and Mira (Beckwith et al., 1983). The initial results of this technique indicate that all observed lines are spatially resolved and larger than the continuum (dust) radiation. Larger radii

are measured for the lower-excitation lines, in keeping with expectations and it should be possible to obtain a run of excitation temperature with radius down to ~0.1-0.2" resolution.

3.0 Discussion

It is useful to draw together the available observational material
on CSE's, both to see where we stand in our understanding of the
phenomena and to see how our knowledge is likely to improve. The IR,
mm-wave and OH surveys provide statistically significant information on
the occurence of super-winds and on their expansion velocities and
mass-loss rates. Of stars with super-winds, there is about an equal
representation of C-rich and O-rich objects, a distribution which might
also apply to PN. Interior models and isotopic determinations support
the assumed operation, in double-shell stars on the AGB, of a dredge-up
process bringing up C-core material, increasing $^{12}C/^{13}C$ and also bringing up material rich in s-process nuclides. The oxygen-stars show low
$^{12}C/^{13}C$ more typical of equilibrium CNO processing. Oxygen-rich supergiants have the largest observed values of dM/dt, but at a given bolometric luminosity carbon-stars seem to have the larger mass-loss rates.

PN often display symmetries which suggest an important role for an
axis of either stellar rotation (and therefore of possible stellar
activity) or of a binary orbit. A study of bipolar nebulae by Morris
(1981) stresses the importance for these objects of binary orbits which
may cause mass-loss in a disc. OH maps indicate that oxygen-stars
eject material mostly into spherically symmetric shells and a suggested
scenario is that mass-loss into discs is a later development after the
dredge-up of carbon. However, the limited mm-wave maps of C-rich
objects show that most (of a handful) are also circularly symmetric.
Also, the ~12 bipolar objects, chosen for study by Morris seem to have
about equal representation of C and O-rich objects, just like the
general statistics of CSE's. It is also instructive to recall that the
incidence of close binaries may be the rule, rather than the exception,
for all intermediate-mass stars (Abt, 1983).

In summary, our general knowledge of CSE's indicates that they
surely play an important role in the late evolution of intermediate-
mass stars. They serve also to enrich the ISM in many of the elements
at one time thought to be produced in massive stars and ejected by
supernovae. However, the interplay of carbon enrichment on the mechanism of mass-loss is still unclear and only more extensive observations
using techniques of high spatial resolution seem capable of addressing
the difficult questions about mass-loss mechanisms.

REFERENCES

Abt, H.A., 1983, Ann. Rev. Astron. Astrophys., 21, 343.

Beckwith, S., Dyck, H.M. and Zuckerman, B., 1983, Ap.J., (submitted).

Beichman, C., 1983, (these proceedings).

Bernat, A.P., 1981, Ap.J., 246, 184.

Bieging, J.H., Chapman, B. and Welch, W.J., 1983, Ap.J., (submitted).

Bowers, P.F., Johnston, K.J. and Spencer, J.H, 1983, Ap.J., (in press).

Calvet, N. and Peimbert, M., 1983, Rev. Mexicana Astron. Astrof., (submitted).

Dyck, H.M., Beckwith, S. and Zuckerman, B., 1983a, Ap.J. (letters), 271, L79.

Dyck, H.M., Zuckerman, B., Beckwith, S. and Leinart, C., 1983b, Ap.J., (submitted).

Engels, D., 1979, Astron. Astrophys. Suppl., 36, 337.

Hinkle, K.H., 1978, Ap.J., 220, 210.

Hinkle, K.H., Hall, D.N.B. and Ridgway, S.T., 1982, Ap.J., 252, 697.

Howell, R.R., 1980, Ph.D. thesis, University of Arizona.

Iben, I.Jr. and Renzini, A., 1983, Ann. Rev. Astron. Astrophys., 21, 271.

Jewitt, D.C., 1983, Ap.J., (submitted).

Knapp, G.R., 1983, (private communication).

Knapp, G.R., Phillips, T.G., Leighton, R.B., Lo, K.Y., Wannier, P.G., Wootten, H.A., and Huggins, P.J., 1982, Ap.J., 252, 616.

Kwok, S., 1982, Ap.J., 258, 280.

McCarthy, D.W., 1979, Proc. IAU Colloq. No. 50, High Angular Resolution Stellar Interferometry, ed J. Davis and W.J. Tango (Sydney: Chatterton Astron. Dept., Sch. of Phys., Univ. of Sydney) p. 18-1.

Morris, M., 1981, Ap.J., 249, 572.

Purton, C.R., 1983, Planetary Nebulae, ed. D.R. Flower, Dordrecht, Holland, Reidel, p. 319.

Ridgway, S.T., 1983 (private communication).

Sopka, R.J., Hildebrand, R.H., Jaffe, D.T., Gatley, I., Roellig, T., Werner, M.W., Jura, M. and Zuckerman, B., 1983, Ap.J., (submitted).

Sutton, E.C., Storey, J.W.V., Betz, A.L., Townes, C.H. and Spears, D.L., 1977, Ap.J. (letters), 217, L97.

Sutton, E.C., Betz, A.L., Storey, J.W.V. and Spears, D.L., 1979, Ap.J. (letters), 230, L105.

Terzian, Y., 1983, Planetary Nebulae, ed. D.R. Flower, Dordrecht, Holland, Reidel, p. 487.

Wannier, P.G. and Linke, R.A., 1978, Ap.J., 225, 130.

Wannier, P.G., Leighton, R.B., Knapp, G.R., Redman, R.O., Phillips, T.G., and Huggins, P.J., 1979, Ap.J., 230, 149.

Wannier, P.G. and Sahai, R., 1983, (in preparation).

Wannier, P.G., Nyman, L.-A., Sahai, R. and Olofsson, H., 1983, (in preparation).

Wood, P.R., Bessell, M.S. and Fox, M.W., 1983, Ap.J., 272, 99.

RADIATIVE TRANSFER IN AN EXPANDING SPHERICAL MEDIUM

W. Kalkofen
Harvard-Smithsonian Center for Astrophysics

R. Wehrse
Institut für Theoretische Astrophysik der Universität Heidelberg

ABSTRACT: We describe a method for solving the transfer equation of a spectral line in statistical equilibrium with a background continuum; the atmosphere may be static or expanding and have plane or spherical symmetry. The method permits the rapid and accurate solution of line transfer problems in media with given gross structure.

The basic approach is to separate the calculation into two parts: the accurate determination of the error with which an assumed, or an iterated, solution satisfies the conservation equations of statistical equilibrium; and the calculation of corrections by means of an approximate operator.

The method combines differential as well as integral equations. It is based on Cannon's (1973, 1983) operator perturbation technique with an approximate matrix operator similar to Scharmer's (1983), and on the solution method of Wehrse and Kalkofen (1984) for the transfer equation in curvilinear coordinates. The detailed features of the approximate matrix operator affect the rate of convergence of the perturbation equations but not the accuracy of the converged solution, which depends only on the precision with which the error made by an approximate solution in the conservation equations is determined.

THE BASIC EQUATIONS

We discuss a fast method for solving transfer problems in complete redistribution for both plane and spherical atmospheres. For convenience of exposition we consider a two-level atom.

The technique is based on Cannon's (1973) perturbation of an integral operator in a formulation similar to that of Scharmer and Nordlund (1983). The method is described in terms of an integral equation technique but, apart from an approximate integral operator, it employs only differential equations, which are used to determine the precision with which an approximate solution satisfies the equations. This error then provides the driving term in an integral equation solution.

The problem is defined by the transfer equation in integral form for the mean integrated intensity, J,

$$J = \Lambda S + K \tilde{B} , \qquad (1)$$

where Λ is the integral operator acting on the line source function, S, and K is the integral operator acting on the source function of the background continuum, $\tilde{B}$. The transfer equation is to be solved subject to the constraint of statistical equilibrium, which can be written in the form of the source function equation,

$$S = (1 - \epsilon)J + \epsilon B , \qquad (2)$$

where ϵ is the well-known collision parameter and B the Planck function at the frequency of the line.

The two equations can be combined into a single integral equation,

$$LS = \varphi \, , \tag{3}$$

where L is the integral operator acting on the line source function and φ the inhomogeneous term.

We solve the integral equation (3) with an approximate integral operator, L, employing Cannon's (1973, 1983) operator perturbation technique in the form proposed by Scharmer and Nordlund (1982). For this purpose we write the transfer problem (3) in a form suitable for an iterative solution,

$$LS^{(n+1)} = \varphi + (L - L)S^{(n)} \, . \tag{4}$$

To obtain a numerical solution we cast the correction equation (4) into a form in which the driving term is given by the error in the conservation equation (2) (cf. Scharmer & Nordlund 1982). Thus, defining the error by

$$e^{(n)} = -S^{(n)} + (1 - \epsilon)J^{(n)} + \epsilon B \, , \tag{5}$$

where $J^{(n)}$ is the solution of the transfer equation (1) for the line source function $S^{(n)}$, the $(n + 1)^{th}$ solution is obtained from

$$L[S^{(n+1)} - S^{(n)}] = e^{(n)} \, . \tag{6}$$

A convenient initial estimate of the source function is $S^{(0)} = 0$, which gives in first-order $S^{(1)} = L^{-1}\varphi$. Note that this would be the exact solution of the transfer problem (3) if L were the exact integral operator. It is, however, much more efficient to solve the equations iteratively by using an appropriate approximate integral operator, L, which must satisfy certain conditions (cf. Kalkofen 1983) in order to insure convergence of the hierarchy of equations (6). We construct this operator with the aid of Scharmer's (1983) quadrature weights for the formal integral of the transfer equation along a ray (cf. Nordlund 1983).

For the calculation of the error (5) of the conservation equation (2) we must determine the mean integrated intensity from the transfer equation, which we write in terms of curvilinear coordinates as

$$\frac{1}{r^2}[\mu\frac{\partial}{\partial r} + \frac{1}{r}\frac{\partial}{\partial \mu}(1 - \mu^2)]r^2 I = k(I - S) + \kappa_c(I - B) \tag{7}$$

(cf. Kalkofen and Wehrse 1983). In the special case of plane geometry, the angle derivative on the left-hand side of the equation is absent; and since the source function $S^{(n)}$ is known, the angle and frequency components of the mean integrated intensity $J^{(n)}$ are not directly coupled. The transfer equation can therefore be solved immediately, either via the formal integral, or via the difference equation according to Feautrier giving second-order accuracy, or Auer (1976, 1983) giving fourth-order accuracy.

In the spherical case, the angle derivative couples the angle components of the mean integrated intensity. This coupling is first removed by diagonalizing the angle operator, assuming it to be depth-independent in the individual layers of the radial grid. The resulting uncoupled equations are then solved using a formal integral (cf. Wehrse & Kalkofen 1984).

DISCUSSION

Savings in computation time relative to the conventional solution by means of integral or differential equations are realized both in the construction of the approximate operator L and in the solution of the transfer equation: The approximate operator L requires much less time than the exact operator L because Scharmer's quadrature weights need few operations and because a single angle point and a few well-chosen frequency points suffice (cf. Kalkofen 1983); and the solution of the transfer equation for known source function $S^{(n)}$ needs less time than the solution of the coupled set of differential equations in the differential equation method because the frequency components of the equation are only indirectly coupled. The reduction in computation time for the transfer equation is therefore of the order of the cube of the number of frequency points.

The necessity of iterations to obtain a solution may appear to be a drawback of the method. However, most transfer problems are non-linear and thus require iterations even without the operator perturbation (for the treatment of non-linear problems by means of the perturbation technique, see Scharmer & Carlsson 1983).

The solution of the transfer problem does not depend on the nature of the approximate operator but only on the error in the conservation equation, which is the driving term in the correction equation. The accuracy of the fully converged solution is that of the (formal) solution of the transfer equation.

This work was supported in part by the NASA grant NAGW 253.

REFERENCES

Auer, L. H., 1976. *J. Quant. Spectr. Rad. Transf.*, **16**, 931.
——————— , 1983. *Methods in Radiative Transfer*, W. Kalkofen, editor; Cambridge University Press, Chapter 9.
Cannon, C. J., 1973. *J. Quant. Spectrosc. Rad. Transfer*, **13**, 627.
——————— 1983. *Methods in Radiative Transfer*, Chapter 6.
Kalkofen, W., 1983. *Methods in Radiative Transfer*, Chapter 15.
——————— & Wehrse, R. 1983. *Methods in Radiative Transfer*, Chapter 11.
Nordlund, Å., 1983. *Methods in Radiative Transfer*, Chapter 8.
Scharmer, G. B., 1983. *Methods in Radiative Transfer*, Chapter 7.
——————— & Nordlund, Å., 1982. *Stockholm Obs. Rep.* **19**.
——————— & Carlsson, 1983. preprint.
Wehrse, R. & Kalkofen, W., 1984, in preparation.

ASYMPTOTIC GIANT BRANCH STARS AS PRODUCERS OF CARBON AND OF

NEUTRON-RICH ISOTOPES[†]

Icko Iben, Jr.
Department of Astronomy
University of Illinois
341 Astronomy Building
1011 W. Springfield Ave.
Urbana, IL 61801   USA

Abstract

Carbon stars are thought to be in the asymptotic giant branch (AGB) phase of evolution, alternately burning hydrogen and helium in shells above an electron-degenerate carbon-oxygen (CO) core. The excess of carbon relative to oxygen at the surfaces of these stars is thought to be due to convective dredge-up which occurs following a thermal pulse. During a thermal pulse, carbon and neutron-rich isotopes are made in a convective helium-burning zone. In model stars of large CO core mass, the source of neutrons for producing the neutron-rich isotopes is the $^{22}Ne(\alpha,n)^{25}Mg$ reaction and the isotopes are produced in the solar system s-process distribution. In models of small core mass, the $^{13}C(\alpha,n)^{16}O$ reaction is thought to be responsible for the release of neutrons, and the resultant distribution of neutron-rich isotopes is expected to vary considerably from one star to the next, with the distribution in isolated instances possibly resembling the solar system distribution of r-process isotopes. After the dredge-up phase following each pulse, the ^{13}C is made by the reactions $^{12}C(p,\gamma)^{13}N(\beta^+\nu)^{13}C$ in a zone of large ^{12}C abundance and small ^1H abundance that has been established by semiconvective mixing during the dredge-up phase. There is qualitative accord between the properties of carbon stars in the Magellanic Clouds and properties of model stars, but considerably more theoretical work is required before a quantitative match is achieved.

The observed paucity of AGB stars more luminous than $M_{BOL} \sim -6$ is interpreted to mean that the AGB lifetime of a star more luminous than this is at least a factor of ten smaller than the AGB lifetime of stars less luminous than this, or, at most 10^5 yr. Since, with current estimates of the $^{22}Ne(\alpha,n)^{25}Mg$ reaction rate R_{22}, only AGB model stars more luminous than $M_{BOL} \sim -6$ can produce s-process isotopes in the

---

[†]Supported in part by the NSF grant AST 81-15325.

solar system distribution, it is inferred that either (1) the current estimates of R_{22} are too small by one to two orders of magnitude, allowing less luminous AGB stars to contribute, (2) the solar system distribution is not equivalent to the average Galactic distribution, being rather the consequence of a unique injection into the protosolar nebula of matter from a massive intermediate-mass AGB star, or (3) the estimates of the temperatures in the convective shell that are given by extant models are too low by, say, 10 or 15 percent.

The absence of carbon stars more luminous than $M_{BOL} \sim -6$ is suggested to be due primarily to the fact that $\sim 10^6$ yr of AGB evolution is necessary to produce surface $C/O > 1$, rather than to be due to the burning of dredged-up carbon into nitrogen at the base of the convective envelope during the interpulse quiescent hydrogen-burning phase. Thus, the positive correlation between the nitrogen and helium abundances in planetary nebulae is perhaps primarily a consequence of the second dredge-up episode rather than a consequence of processes occurring during the thermally pulsing phase.

I. THE OBSERVATIONAL FACTS

A. Galactic Studies

One of the drawbacks in attempts to understand carbon stars from observations of stars in the Galaxy is the fact that very few of them are in clusters for which distances can be estimated. Nevertheless, much can be learned by piecing together all of the available evidence.

Abundances of Carbon, Nitrogen, and Oxygen. From a study of the Sun and of stars, HII regions, and molecular clouds in the vicinity of the Sun, one infers that the CNO elements are produced in abundances such that $O > C > N$. This is not to say that every star which contributes to the synthesis of these elements produces them in the stated order of abundances, but that the net result of the synthesis of these elements by all stars is responsible for the observed order in the interstellar medium and in unevolved stars formed from this medium.

That the observed order of abundances must be the result of an averaging of the contributions from many different types of stars is obvious from the very existence of carbon stars which are, by definition, stars in which $C > O$ at the surface and which are known to be contributing matter (with $C > O$) to the interstellar medium by way of a strong wind. As we shall see, the evidence is overwhelming that the relative abundance of C relative to O in carbon stars is due to synthesis of C in the stars themselves.

Neutron-Rich Isotopes. Two other sets of elements (actually, sets of isotopes) which prove to be of interest in connection with carbon stars are found from studies of neutron-rich isotopes in meteorites. These are the so called s-process and r-process isotopes which are thought to be made by successive neutron captures and β decays beginning with the seed nucleus ^{56}Fe. In one group, adjacent stable isotopes are related to one another by $n_i \sigma_i = n_{i-1} \sigma_{i-1} (1 + \Lambda/\sigma_i)^{-1}$, where n_i is the number abundance of the ith isotope, σ_i is its neutron capture cross section (at an energy of ~ 30 keV), and Λ is a universal constant whose value is of the order of (3-5) millibarns. In theoretical nucleosynthesis experiments, such a distribution can most easily be formed if the flux (or density) of neutrons is sufficiently low that a beta-unstable nucleus will decay before capturing a neutron -- hence the name s(low neutron capture)-process. In contrast, the isotopes in the other group display abundance patterns that can be reproduced theoretically most easily if neutron capture takes precedence over beta decay until neutron number reaches the magic values of 82 and 126. These isotopes are therefore called r(apid neutron capture)-process isotopes.

Over the past three decades, an examination of abundances of neutron-rich isotopes in Galactic S-stars (in which C $\sim$ O) has shown that neutron-rich isotopes are formed in such stars and brought to their surfaces. Of particular interest is the occurrence in many S-stars of Tc. The most stable isotope ^{99}Tc has a half life of only 2×10^5 yr, whereas the progenitor of an S-star must have a main sequence lifetime in excess of a few times 10^7 yr. In the spectrum of Galactic Ba and CH stars, which are not as bright as AGB stars, one finds evidence for neutron capture activity, but recent radial velocity studies have shown that many of these stars are in binaries with a degenerate dwarf companion (McClure, Fletcher and Nemec 1980, McClure 1983a,b). The thought is that the neutron-rich isotopes may have been formed in the AGB precursor of the degenerate dwarf and transmitted by Roche-lobe overflow or by a wind to the star which now shows Ba or CH features. Certainly, the orbital separations of many of the binaries are large enough to have accommodated a star of AGB dimensions. The high degree of variability in the relative abundances of the neutron-rich isotopes from one star to the next suggests that the neutron source is highly variable in strength and duration from one erstwhile AGB companion to the next.

Tc in Miras. Most Mira variables are not at the same time carbon stars. However, approximately half of the brightest fifteen percent of them are (Cahn 1980). Furthermore, except for those of the lowest luminosities, many of them show Tc lines and the probability of finding such lines increases with the brightness of the Mira (Little-Marenin 1983). We infer, first, that perhaps the production of neutron-rich isotopes and certainly the probability of dredging these isotopes to the surface increases with increasing AGB luminosity and, second, that the conversion of the surface C/O ratio from less than 1 to larger than 1 may require several dredge-up episodes.

Frequency of C-Stars as a Function of Metallicity. As shown by Blanco, McCarthy, and Blanco (1978) there is a pronounced gradient in the space frequency of Galactic carbon stars, with none occurring in the Galactic bulge. This is despite the existence of metal rich AGB stars in the bulge (Whitford and Rich 1983, Frogel and Whitford 1983). The simplest interpretation of these facts is that the larger the abundance of Fe/H, the more difficult it is for an AGB star to develop a surface ratio C/O > 1. Whether this means that, with increasing Fe/H, it is more difficult for dredge-up to occur or whether, because of a larger initial abundance of oxygen, that more carbon must be dredged-up in order for C > O at the surface, or both, is not settled by these observations.

In summary, from an examination of stars in our own Galaxy, we have learned that (1) carbon and neutron-rich isotopes are made in cool, bright AGB giants and brought to the surfaces of these giants; (2) the brighter the giant, the greater is the likelihood of producing and bringing to the surface freshly manufactured elements and isotopes; and (3) the larger C and O are to begin with, the harder it is to achieve C > O at the surface.

B. Magellanic Cloud Studies

The development of infrared technology over the past decade has made it possible to take advantage of the fact that all of the stars in each of the Clouds are effectively at the same distance from the earth. This has permitted us (since we can now compare absolute luminosities of real carbon stars with those of model AGB stars) to establish with a certainty that AGB stars bring freshly synthesized carbon and neutron-rich isotopes to their surfaces and that the conversion from O-star, through S-star, to C-star occurs at a definite point along the AGB branch.

Field Stars. Surveys of selected fields in both clouds show that carbon stars are, in general, confined to a very narrow interval in bolometric magnitude: $-6 \lesssim M_{BOL} \lesssim -4$ (survey by Blanco, McCarthy, and Blanco 1980; bolometric magnitudes by Richer 1981, Cohen et al 1981, Frogel et al 1981, and Frogel and Cohen 1982). Extrapolations from the surveyed regions suggest that there are altogether ~ 1.4 x 10^4 carbon stars in both clouds (Blanco and McCarthy 1983) and that there are only a handful of S-stars (Blanco, Frogel, and McCarthy 1981, Lloyd-Evans 1983). This permits us to say that the S-star phase, when C ~ O, is a very transitory one, with the transition from C ~ O x (1-.05) to C ~ O x (1 + 0.05) occurring in perhaps only one dredge-up episode.

The long period variables (LPV's) in the Magellanic Clouds may be assigned to one of two distinct sequences (Wood, Bessel, and Fox 1983) according to whether or not they show evidence for overabundances of the neutron-rich element Zr. Those in which the tracer ZrO appears to be "normal" are in general brighter than M_{BOL} ~

-7.3, which corresponds to the magnitude of a theoretical AGB model whose core has reached the Chandrasekhar mass of 1.4 M_0. Those in which strong ZrO bands appear define a sequence in the M_{BOL}-pulsation period plane that is consistent with theoretical AGB models. The observed sequence that is characterized by strong ZrO bands extends, within the uncertainties, up to the maximum brightness reached by theoretical AGB models. The lowest period representatives of this sequence tend to be carbon stars, but the brightest ones are not carbon stars. The number ($\sim$ 100) of the non-carbon star LPV's with $M_{BOL} \lesssim$ -6 is about a factor of twenty or so less than the number of Cepheids in the Clouds and, since the progenitors of the LPV's are thought to be Cepheids with lifetimes of $\sim 10^6$ yr, we infer that stars which reach the AGB with $M_{BOL} \lesssim$ -6 remain on the AGB for only $\sim 10^5$ yr and this, as it turns out, is only ten percent or so of the AGB lifetime expected if mass loss via a stellar wind were to occur at roughly the Reimers rate (1975).

The frequency of carbon stars in the SMC is about 3 times larger than it is in the LMC (Blanco, McCarthy and Blanco 1978, 1980; Blanco and McCarthy 1983) and, since the metallicity of SMC stars is on average much less (by perhaps a factor of 3-6) than the metallicity of LMC stars, we recover the result of studies of Galactic C-stars that, for whatever reason, AGB stars of low metallicity find it easier to become carbon stars than do AGB stars of higher metallicity.

Stars in Globular Clusters. Searle, Wilkinson, and Bagnuolo (1980) have suggested an ordering of Magellanic Cloud clusters in an approximate age sequence and Cohen (1982) has shown that this sequence is equivalent to an ordering according to metallicity with, naturally enough, the oldest clusters having the lowest metallicity. Frogel and Blanco (1983, this volume) find that, in the oldest ($\sim 10^{10}$ yr) clusters, there are essentially no carbon stars. From this one may infer that, if the initial mass of an AGB star is quite low ($\lesssim$ 0.8 M_0), either dredge-up cannot occur, or the envelope mass of the star is lost by a wind before dredge-up can lead to C > O, or both.

Among stars on the AGB of an intermediate age cluster (0.3 x 10^9 yr $\lesssim$ age $\lesssim$ 3 x 10^9 yr), Frogel and Blanco find that there is a clear separation in absolute luminosity between the dimmer M-stars (O > C) and the brighter C-stars (C > O), with the luminosity at the transition point between M-stars and C-stars increasing with decreasing cluster age. This result has also been obtained independently by Lloyd-Evans (1983) and it is consistent both with the idea that dredge-up occurs on the AGB and with the theoretical result that the luminosity with which a model star reaches the thermally pulsing stage on the AGB increases with increasing mass (decreasing nuclear burning lifetime) of its main sequence progenitor (e.g., Becker and Iben 1979, 1980).

A remarkable result of the Frogel-Blanco survey is that in young clusters which contain Cepheids (age $\lesssim 10^8$ yr) there are no C-stars. And yet, theoretically, AGB stars which have Cepheid progenitors, and hence have masses on the order of $(4 - 6)$ M_0, are expected to dredge-up carbon with great facility (Iben 1975a). Even more astonishing than the absence of C-stars is the total absence of even M-stars brighter than about $M_{BOL} \sim -6$. Since a model AGB star reaches a luminosity given by $M_{BOL} \sim -7.3$ if its core mass reaches $\sim 1.4 \; M_0$, and since the rate at which a theoretical model brightens (in magnitudes per year) is independent of core mass, the inescapable conclusion is that a real AGB star leaves the AGB considerably before its core mass reaches $1.4 \; M_0$. The absence of carbon stars in the field brighter than $M_{BOL} \sim -6$ can now be attributed to the paucity of AGB stars brighter than this rather than to two other possibilities that have been cited in the past: (1) AGB stars brighter than $M_{BOL} \sim -6$ do not dredge-up carbon (contrary to theoretical indications) or (2) such stars convert dredged-up carbon into nitrogen as a consequence of proton captures at the base of the convective envelope during the quiescent, interpulse hydrogen-burning phase.

THE THEORY - OVERVIEW

A. Surface Composition Changes Prior to the AGB Phase

It is important to distinguish the nature of those surface composition changes which may occur prior to the AGB phase from those changes which are due solely to processes occurring during the AGB phase proper. This is not a completely straightforward task, as we cannot deduce from first principles the rates at which rotationally induced forms of mixing actually operate in real stars. Even if we were able to calculate the effectiveness of such forms of mixing, the fact that there is a large spread in rotational velocities among real stars will introduce a large spread in surface composition changes if these forms of mixing are of importance.

Meridional Circulation and Turbulent Diffusion in Main Sequence Stars. In principle, these processes should both increase in effectiveness with increasing rotation rate and might be expected to bring to the surface products of nuclear processing in the interior. In particular, one might anticipate some outward diffusion of products of CN-cycle burning and a reduction in C and an enhancement of N at the stellar surface. However, no evidence has yet been presented by spectroscopists to show that there is a tendency for C/N to decrease in any systematic way from high T_e to low T_e across the main sequence band and there is therefore no compelling evidence that rotationally induced mixing is of importance.

Convective Dredge-Up on the Giant Branch. In contrast, there is good observational evidence to support the theoretical indications that CN-cycle processed material is first brought to the surface as a star begins its initial climb upwards along the (first) red giant branch and the base of the convective envelope extends inward in mass to where C has been converted almost entirely into N (see Iben and

Renzini 1984 for a review). The theoretical calculations provide precise quantitative predictions [surface C down by ~ 1/3 x and N up by ~ 2 x] and these predictions are consistent with the observations.

Meridional Mixing on the Giant Branch in Globular Cluster Stars. Kraft et al (1982) and Carbon et al (1982) find that, among red giant stars in the Galactic globular cluster M92, the abundance of carbon decreases with stellar luminosity. This decrease continues to luminosities much larger than the luminosity at which the first convective dredge-up is expected to have run its course. The best explanation for the observed effect is the one given by Sweigart and Mengel (1979) who point out that, in giants, the distance between (a) the region above the main hydrogen-burning shell where C has been converted into N, and (b) the base of the convective envelope, is quite small and argue that meridional circulation may carry C-depleted, N-enhanced material into the base of the convective envelope. Convective mixing will then carry some of this processed material to the surface. The phenomenon appears to be confined to stars in only a subset of clusters and does not appear to occur in stars in the field (Kraft 1979). One infers that close stellar encounters among stars in clusters of the largest spatial concentration may lead to (probably temporary or reversible) spin-ups that initiate the requisite meridional currents.

Convective Dredgeup After Central Helium Exhaustion. In model stars which, as a consequence of hydrogen-burning either before or during the core helium-burning phase, develop a hydrogen-exhausted core larger than about 1 M_0 by the time that the central helium abundance goes to zero, the base of the convective envelope extends briefly inward in mass into the region through which the hydrogen-burning shell has passed during the preceding core helium-burning phase. Therefore, as the model star once again climbs along the (second) red giant branch and the base of the convective envelope moves inward in mass, fresh ^{4}He and ^{14}N (produced at the expense of both ^{12}C and ^{16}O) are brought to the surface. Thus, the surface abundances of both ^{4}He and ^{14}N increase, whereas the surface abundances of ^{12}C and ^{16}O decrease. This second dredge-up phenomenon occurs only in models with initial main sequence masses larger than ~ (4-5) M_0 and less than ~ (8-9) M_0. Within this narrow range of masses, the calculations show that ^{14}N and ^{4}He enhancements as large as a factor of two occur, with the degree of enhancement increasing with increasing mass within this range (Becker and Iben 1979).

B. Surface Composition Changes During the Thermally Pulsing (TP)-AGB Phase

We have seen that, prior to the AGB phase, enhancements in the surface abundances of N and He and depletions in the surface abundances of C and O (the depletion of C being much more substantial than that of O) occur as a consequence of the convective dredge-up of matter that has experienced hydrogen-burning in varying degrees. The fact that many AGB stars exhibit a surface abundance of oxygen larger

than the surface abundance of carbon, in spite of the fact that surface C is depleted when hydrogen-burning products are dredged to the surface, provides compelling
evidence that such stars must somehow bring to their surfaces matter that has
experienced helium-burning in the interior.

Abundance Changes in Nuclear Burning Regions. During about ninety percent of
the AGB lifetime of a model star, helium-burning proceeds at a very low level, with
effectively all of the energy which flows to the surface coming from hydrogen-burn-
ing in a thin shell (whose mass varies from $\sim 10^{-4}$ M_0 to 10^{-7} M_0 as the mass M_{CO} of
the electron-degenerate CO core is varied from ~ 0.5 M_0 to ~ 1.4 M_0). If the mass
of the stellar envelope is large enough and if convective mixing is efficient
enough, some burning may take place at the base of the convective envelope and, if M
$>$ (4-5) M_0, it is even possible that the abundance of ^{12}C decreases and the abundance of ^{14}N increases in the envelope between pulses (Iben 1975a, Renzini and Voli
1981). When the mass of the He-N zone laid down by the hydrogen-burning shell
reaches about 100 times the mass of the burning shell, densities and temperatures in
this zone become large enough to excite a helium-burning thermonuclear runaway. In
the course of this runaway, the rate of energy generation by helium-burning can
reach as high as $(10^7 - 10^8)$ L_0 and temperatures at the base of the runaway zone can
reach as high as (300-400) x 10^6K. During the thermal flash, nuclear energy is
injected so rapidly that it is converted locally into thermal energy and thence into
local expansion energy before it has a chance to leak out from the burning zone by
radiative or convective "diffusion." The runaway is quenched as a consequence of
the expansion and subsequent cooling.

At the peak of a thermal pulse (or helium shell flash, as it is often called),
the entire region between the base of the helium-burning zone and the hydrogen-
helium discontinuity is unstable to convection. After the pulse subsides and helium
begins to burn quiescently, the abundance of ^{12}C left behind just below the hydro-
gen-helium discontinuity is about 20 percent by mass. Because burning in the
convective zone is far from complete, the amount of ^{16}O left behind is much less
than the amount of ^{12}C. In the course of the quiescent helium-burning phase, which
lasts for roughly ten percent of the interpulse lifetime, the total amount of helium
converted into carbon and oxygen equals the amount of helium produced during the
quiescent hydrogen-burning phase.

Of particular interest is the neutron-capture nucleosynthesis that occurs in
the convective helium-burning shell during pulse peak. The nature of this nucleosynthesis is a function of, among other things, the source of neutrons. If the mass
of the CO core is larger than a critical value M_{CO}^{crit}, temperatures in the convective
zone become large enough for a long enough time that ^{14}N is converted completely
into ^{22}Ne and a substantial fraction of this ^{22}Ne is converted into ^{25}Mg and a
neutron. Most of the neutrons released are captured by ^{22}Ne, ^{25}Mg and the neutron

capture progeny of these isotopes, but the number of neutrons left to be captured by ^{56}Fe and its progeny is precisely what is needed to produce heavy s-process isotopes in the solar-system distribution (Iben 1975b, Truran and Iben 1977). An essential aspect of the environment that produces this result is the overlap in mass between successive convective shells; this overlap ensures that, in any given shell, the fraction of matter which has experienced N neutron exposures is an exponentially declining function of N. The universal parameter Λ characterising the final distribution of s-process isotopes is essentially the average neutron capture cross section of the light elements from ^{22}Ne to, say, ^{27}Al.

Analysis of early studies (Iben and Truran 1978) suggested that $M_{CO}^{crit} \sim 0.95$ M_0, but more recent calculations (Becker 1983) give hope that M_{CO}^{crit} may be as small as (0.75-0.80) M_0. It is worthwhile remarking that a full understanding of the theoretical properties of thermally pulsing AGB models will require substantial amounts of computer time on what are, even today, considered to be "supercomputers."

For smaller CO core masses, temperatures in the convective shell do not become large enough for more than a percent or so of ^{22}Ne to be converted into ^{25}Mg and a neutron (Becker 1980, Iben 1982). However, if the metallicity is low enough, ^{13}C is made available as a neutron source as a consequence of a process (Iben and Renzini 1983) to be described in a later section. The neutron densities which are created by the ^{13}C$(\alpha, n)^{16}$O source are too large to produce neutron-rich isotopes in a distribution resembling the solar system distribution of s-process isotopes. They are also too small to produce, in one pulse, the classical r-process distribution. It remains to be seen if the combination of an intermediate strength neutron flux and an exponential distribution of exposures may perhaps produce the classical r-process distribution.

Dredge-Up Following Thermal Pulses. The rapid expansion and cooling that is initiated in the helium-burning zone extends beyond this zone into the hydrogen-rich layers, with the result that hydrogen-burning is extinguished. There is a sharp but temporary drop in the surface luminosity as a consequence of this extinction. However, even as the total rate of energy generation by helium-burning drops dramatically from its peak values, more and more of the energy produced by helium-burning makes its way outward by radiative diffusion toward the base of the convective envelope and then through the convective envelope, until the major source of surface luminosity becomes helium-burning. Expansion and cooling in the helium-burning layers is eventually reversed and not only does the rate of energy generation by helium-burning increase for a time, but the flux of energy passing through the base of the convective envelope also increases. The net result is that the base of the convective envelope moves inward in mass and, in AGB models of sufficiently large core mass, this base extends into the region where ^{12}C and neutron-rich isotopes have been produced (Iben 1975a, 1976). The freshly made nucleosynthetic products are then convected to the surface. The minimum core mass at which this form of dredge-up occurs depends sensitively on the core mass, the total mass, and the metallicity of the model star as well as on the choice of mixing length/scale height used in the standard algorithm for modeling convective flow (Wood 1981).

For model stars of metallicity appropriate to the SMC, the minimum core mass is on the order of 0.6 $M_\odot$ (Iben and Renzini 1982a,b, Iben 1983) and a surface ratio C/O > 1 is achieved after the first or second dredge-up episode. At the end of this dredge-up episode, the stellar luminosity corresponds to $M_{BOL} \sim -5$, but the model then settles down into an extended phase of quiescent helium-burning during which the stellar luminosity drops slowly (over about 10 percent of the interpulse life-time) to about $M_{BOL} \sim -4$ and then rises slowly over the remainder of the interpulse lifetime to reach $M_{BOL} \sim -4.7$ just before the onset of another pulse. Thus, the models suggest that carbon star characteristics will first appear at $M_{BOL} \sim -4$ and that there should be an overlap in M_{BOL} between carbon stars (with $M_{CO} \gtrsim 0.6\ M_\odot$) that are decreasing in luminosity following a helium shell flash and M-stars (with $M_{CO} \lesssim 0.6\ M_\odot$) that are increasing in luminosity after reaching the luminosity mini-mum following a helium shell flash. These results are consistent with the distribu-tions of field M-stars and field C-stars in both the SMC and the LMC.

For stars of core mass larger than, say, 0.7 $M_\odot$, dredge-up occurs readily, regardless of metallicity, ℓ/H, and total stellar mass (Becker 1983). However, the larger M_{CO} is at the start of the thermally pulsing AGB phase, the larger is the main sequence mass of the progenitor model star. Thus, the larger M_{CO}^{start}, the more ^{12}C-rich material must be dredged-up to produce a surface C/O > 1. In models with $M_{CO}^{start} \gtrsim 0.8\ M_\odot$ (corresponding to a maximum luminosity prior to a thermal pulse of $M_{BOL} \sim -6$), roughly 10^6 yr of AGB evolution is required to achieve C/O > 1 (Iben and Truran 1978, Iben 1981, Renzini and Voli 1981).

Since M_{CO}^{start} decreases with decreasing main sequence mass (and increasing main sequence lifetime) of a progenitor star, the theory suggests that the minimum lumi-nosity at which C-star characteristics can occur in a population of a given age will decrease as the age of the population increases. This, of course, is what is known to be the case among globular clusters in the Magellanic Clouds (Frogel and Blanco 1983, Lloyd-Evans 1983). Not only is the agreement between theory and observation qualitatively satisfactory, it is also consistent quantitatively.

The absence of luminous single carbon stars in the oldest clusters in the Clouds can be understood in terms of the theoretical results that: (1) $M_{CO}^{start} \sim 0.53$ $M_\odot$ for stars of initial mass less than $\sim 2\ M_\odot$; (2) the oldest stars (age $\sim 10^{10}$ yr) have initial main sequences masses $\lesssim 0.8\ M_\odot$; (3) dredge-up does not occur until $M_{CO} \gtrsim 0.6\ M_\odot$; and (4) wind mass loss causes the hydrogen-rich envelope of an AGB star of such a low mass to evaporate before its luminosity exceeds the luminosity at the top of the first red giant branch in Galactic globular clusters ($M_{BOL} \gtrsim -3.5$,

$M_{CO} \lesssim 0.53$). Such stars have therefore at best just reached the thermally pulsing AGB phase just before ceasing to exist as AGB stars.

The observed fact that the frequency of carbon stars among AGB stars with intermediate age progenitors decreases with increasing Z can be understood as the consequence of two effects. The first effect is the obvious one that, the larger Z (and hence presumably the larger O is to begin with), the more C-rich material must be dredged-up before surface C > O. The second effect is that, the larger Z is, the more extended is the envelope and the smaller is the gas pressure relative to the radiation pressure. Since the radiative gradient at any point in the envelope is proportional to κ (L/M) (1 + P_{gas}/P_{rad}), the smaller P_{gas}/P_{rad}, the less likely is the adiabatic gradient to be smaller than the radiative gradient (Iben 1983a).

<u>Implication of the Paucity of Bright AGB Stars for s-Process Nucleosynthesis.</u> The absence in the Magellanic Clouds of carbon stars brighter than $M_{BOL} \sim -6$, coupled with the paucity of M-stars brighter than this, may have dramatic implications not only for our understanding of the origin of the s-process isotopes in the solar system but also for our understanding of Galactic nucleosynthesis in general. If one adopts the standard choice for the $^{22}Ne(\alpha,n)^{25}Mg$ reaction (Fowler, Caughlan, and Zimmerman 1975), then early studies suggest that ^{22}Ne is not converted substantially into ^{25}Mg and a neutron unless $M_{CO} > M_{CO}^{crit} \gtrsim 0.95$ M_0 (Iben 1976, Truran and Iben 1977, Iben and Truran 1978). Do enough stars achieve core masses this large and maintain themselves as thermally pulsing AGB stars long enough to produce s-process isotopes in sufficient quantity to account for the solar-system s-process distribution on the assumption that this distribution is equivalent to the average Galaxy-wide distribution of s-process isotopes?

From Becker and Iben (1979) one has that, when Z $\sim$ 0.01 and Y $\sim$ 0.28, M_{CO}^{start} and initial main sequence mass M_{MS} are related by $M_{CO}^{start} \sim 0.85 + 0.053$ (M_{MS} - 4) when $4 \lesssim M_{MS} \lesssim 8$. Masses and luminosities are here and hereinafter in solar units unless otherwise specified. For all $M_{MS} \lesssim 2$, $M_{CO}^{start} \sim 0.53$ and $M_{CO}^{start} \sim 0.53 + 0.16$ (M_{MS} - 2) for $2 \lesssim M_{MS} \lesssim 4$.

In the absence of convective dredge-up following thermal pulses, core mass grows according to $\dot{M}_{CO}$ (M_0 yr^{-1}) $\sim 10^{-6}$ (M_{CO} - 0.5) so that, after t_{AGB} years of evolution, $\Delta M_{CO} \sim [\exp (10^{-6} t_{AGB}) - 1]$ (M_{CO}^{start} -0.5). Since the maximum core mass is 1.4 M_0 and since, from the early studies, M_{CO} must exceed ~ 0.95 M_0 if s-process isotopes are produced in the solar system distribution, one has that the maximum time which a star can spend producing s-process isotopes in this distribution is $\sim 7 \times 10^5$ yr. The abundance of s-process isotopes relative to solar is about 200 (Truran and Iben 1977). Assuming that dredge-up brings up a fraction λ of the mass that has been added to the He-N zone between pulses, the total amount of dredged-up material is approximately $\Delta M_{dredge} \sim \lambda \Delta M_{CO}$, and λ is on the order of $\sim 1/3$ when

$M_{CO} \gtrsim 0.95.$

Let us next assume that the rate of star formation varies with mass according to $\frac{d}{dM}\left(\frac{dN}{dt}\right) = 1.3\,\frac{1}{M^{2.3}}$, where we have normalized to a total birthrate of 1 star per year over the interval $M = 1 \rightarrow \infty$. A measure of the contribution of all stars (still assuming no mass loss) to the Galactic nucleosynthesis of solar system s-process isotopes may now be written as

$$\Delta_s \sim 200\ \lambda\ 1.3\ \left\{ \int_1^{1.4} \frac{dM}{M^{2.3}}\,(M-0.95) + \int_{1.4}^8 \frac{dM}{M^{2.3}}\,0.45 \right\} \cong 68\ \lambda \sim 23.$$

If mass were not lost, all stars initially more massive than 1.4 M_0 would become supernovae and thus the SN rate would be on the order of $\nu_{SNI} = 1.3\int_{1.4}^8 \frac{dM}{M^{2.3}} = 0.58$ yr^{-1}, which is over 50 times the observed SNI rate. It is clear that most real stars of initial mass less than 8 M_0 must lose their hydrogen-rich envelopes before their cores reach the Chandrasekhar mass.

Before pursuing quantitatively the consequences of mass loss let us explore just a bit further the demands of Galactic nucleosynthesis. Type I supernovae are thought to be the consequence of binary star evolution, they occur at the rate $\nu_{SNI} \sim 10^{-2}\ yr^{-1}$ (see, e.g., Iben and Tutukov 1984), and they are thought to produce at least 0.8 M_0 of ^{56}Fe in the explosion (see, e.g., Woosley, Axelrod, and Weaver, 1983, Nomoto 1983). Since the overabundance (relative to solar) of ^{56}Fe in one gram of pure ^{56}Fe is $\sim 1000x$, a measure of the ^{56}Fe production rate in the Galaxy is

$$\Delta_{Fe} \sim 1000\ x\ 0.01\ x\ 0.8 \sim 8.$$

In the absence of mass loss, all intermediate mass stars of initial mass larger than 1.4 M_0 would also become SNeI and contribute 0.8 M_0 of ^{56}Fe each, leading to $\Delta_{Fe} \sim 58\ x\ 8 = 464.$ Thus, we have an essential paradox: the actual ^{56}Fe production rate is a factor of about 60 smaller than it would be if all intermediate mass stars of initial mass larger than 1.4 M_0 were to lose no mass until they become supernovae, and yet, the estimated rate of production of solar system s-process isotopes in the absence of mass loss is only four times larger than the actual rate of ^{56}Fe production by SNeI with binary star progenitors. That is, abolishing single star progenitors of SNeI also abolishes the source of s-process isotopes in the solar system distribution.

We are now in a position to make the paradox even more dramatic. Assuming that mass is lost by real AGB stars at some fraction of the rate given by the semi-empirical Reimers expression (1975), it may be shown that the typical lifetime of an AGB star is on the order of $\sim 10^6$ yr (Fusi-Pecci and Renzini 1976, Iben and Truran 1978, Renzini and Voli 1981). The paucity of real AGB stars brighter than $M_{BOL} \sim -6$ means that the actual lifetime of an AGB star with $M_{BOL} < -6$ must be much less than

10^6 yr and, from a comparison of the number of LPV's and Cepheids in the Magellanic Clouds one may estimate $t_{AGB} \lesssim 10^5$ yr for $M_{BOL} < -6$.

Note that, when $M_{CO} \sim 0.95$, the quiescent hydrogen-burning luminosity is $L \sim 6 \times 10^4$ (0.95-0.5), corresponding to $M_{BOL} \sim -6.3$. Thus, when M_{CO} exceeds ~ 0.95, the core can grow no more than $\Delta M_{CO} < 10^5 \times 10^{-6}$ (0.95-0.5) < 0.05 before the hydrogen-rich envelope evaporates. Being generous and assuming that all stars that develop an initial core mass as large as $M_{CO}^{start} \sim 0.9$ (corresponding to an initial main sequence mass of ~ 5 M_0) produce s-process isotopes in the solar system distribution we have that

$$\Delta_s' \lesssim 200 \lambda \times 0.05 \times 1.3 \int_5^8 \frac{dM}{M^{2.3}} \sim 0.2.$$

Since $\Delta_s' \ll \Delta_{Fe}$, we might conclude that, if the cross section for the $^{22}Ne(\alpha,n)^{25}Mg$ reaction is correct and if the early estimates of the maximum temperatures achieved in the convective helium-burning shell are correct, then either there is another process than the one we have envisioned for the Galaxy-wide production of s-process isotopes in the solar system distribution, or the solar system distribution of these isotopes has been produced by an isolated massive AGB star which injected its nucleosynthesis products into the matter out of which the Sun was born. This latter interpretation is consistent with the recent arguments of Olive and Schramm (1982) and, in the reluctant view of this author, this interpretation becomes more and more plausible with the passage of time.

An alternative interpretation is that the cross section for the $^{22}Ne(\alpha,n)^{25}Mg$ reaction has been significantly underestimated. Suppose, for example, that the $^{22}Ne(\alpha,n)^{25}Mg$ reaction goes to completion in all AGB stars with progenitor main sequence masses in the range $(1.5-3)M_0$ and that they produce s-process isotopes in the solar system distribution as core mass increases, on average, by ~ 0.3 M_0. Then

$$\Delta_s'' \sim 200 \times 1/3 \times 0.4 \times 1.3 \int_2^4 \frac{dM}{M^{2.3}} \sim 7.$$

The proximity of this number to Δ_{Fe} allows one to argue that the solar system distribution of s-process isotopes is equivalent to the Galactic one. To the best of the author's knowledge there has been no effort on the part of the experimental physicist to address the possibility that the $^{22}Ne(\alpha,n)^{25}Mg$ cross section at 30 keV has been underestimated by one or two orders of magnitude or even by a factor of 2!

Still another possibility is that the maximum temperatures reached at the base of the convective shell have been underestimated by a considerable amount. Certainly, these maximum temperatures have been rising as the total number of thermal pulses has slowly increased over the past few years (see the discussions by Becker 1981 and by Iben and Renzini 1983). For example, from a selection of the data available before 1976, Iben and Truran (1978) constructed the approximation $T_{CSB}^{max} \sim$

$[3.1 + 2.85 \ (M_{CO}-0.96)] \times 10^8$ K. It has now become clear, however, that, even for core masses as small as ~ 0.65 M_0, T_{CSB}^{max} exceeds 3×10^8 K after only a dozen or so pulses (Iben 1983b, Becker 1983) and that the maximum temperatures achieved after 20 or so pulses at larger core masses have still not reached asymptotic values (Becker 1983, Chieffi and Iben 1983). Considering the fact that, at $T \sim 3 \times 10^8$ K, the $^{22}Ne(\alpha,n)^{25}Mg$ reaction rate is proportional to $\sim T^{23.4}$, it is conceivable that, even with the currently accepted cross section, M_{CO}^{crit} may drop from the value $M_{CO}^{crit} \sim$ 0.95 M_0 suggested by data available five years ago. That is, only a ten percent increase in T_{CSB}^{max} will lead to a factor of 10 increase in the neutron producing reaction rate, and this might translate into $M_{CO}^{crit} \sim 0.75$ M_0. It is abundantly clear that progress in answering the crucial questions about nucleosynthesis in AGB model stars has been hampered by the lack of sufficient computer power in the hands of interested scientists.

Implication of the Paucity of Bright AGB Stars for the Nitrogen Abundance in Planetary Nebulae. The observed short lifetime of high luminosity AGB stars also has ramifications for the abundance of nitrogen in planetary nebulae. The first attempt to account for the observed relationship between N and He in planetary nebulae assumed that the effects of third dredge-up episodes could be neglected (Kaler, Iben, and Becker 1979) and found reasonable agreement between observation and theory. In this picture, most of the correlation between N and He is due to the effect of the second dredge-up episode which is experienced by massive intermediate mass stars. However, during the thermally pulsing AGB phase, third dredge-up episodes cause the abundance of nitrogen in the convective envelope to decrease, since there is no ^{14}N in the dredged-up material (Iben and Truran 1978, Becker and Iben 1980), and this decrease will persist unless enough of the simultaneously dredged-up ^{12}C is converted into ^{14}N by burning at the base of the convective envelope during the interpulse phase (Renzini and Voli 1981, Becker and Iben 1980).

The fact that there are no carbon stars brighter than $M_{BOL} \sim -6$ has on occasion been attributed to the effective burning of dredged-up carbon. However, in the absence of such burning, it still requires $\sim 5 \times 10^5$ yr for surface C to exceed surface O (Iben and Truran 1978, Becker and Iben 1980, Renzini and Voli 1981, Iben 1981) and, since the observations reduce the available lifetime to only $\sim 10^5$ yr, the absence of bright carbon stars may not be invoked as a demonstration of burning. It may very well be that the observed correlation between N and He in planetary nebulae is, after all, due almost entirely to the second dredge-up!

Activation of the $^{13}C(\alpha,n)^{16}O$ Neutron Source in Low Mass, Low Z AGB Stars. Although the maximum temperatures in the convective shells of low mass thermally pulsing AGB stars can reach 300×10^6 K when $M_{CO} \gtrsim 0.65$ M_0, the short duration of the high temperature phase prevents more than about a percent of the ^{22}Ne from being burned (Becker 1981, Iben 1982, 1983b), provided, of course, that the currently used

α-capture cross section is correct. However, even if this cross section were ten times larger than currently used, burning would be incomplete and the solar-system distribution of s-process isotopes would not result (Truran and Iben 1977).

Another potential source of neutrons is the $^{13}C(\alpha,n)^{16}O$ reaction (Cameron 1955) and, ever since the early work of Schwarzschild and Härm (1967) and of Sanders (1967), it has been assumed that somehow hydrogen may be injected into the ^{12}C-rich convective helium-burning shell and that the reaction $^{12}C(p,\gamma)^{13}N(\beta^+\nu)^{13}C$ will produce the ^{13}C that can then immediately act as a neutron source. Subsequent calculations have not provided support for this envisioned scenario (see, in particular, the entropy argument of Iben 1976, 1982), but another sequence of events has been found to occur in low-mass, low-metallicity AGB stars (Iben and Renzini 1983). In this sequence, semiconvective mixing during the dredge-up phase brings small amounts of hydrogen into the outer edge of the ^{12}C rich region after shell convection has died down. When hydrogen-burning is rekindled, the major result of burning in the region of initially small hydrogen abundance and large ^{12}C abundance is ^{13}C. Then, after the extended quiescent hydrogen-burning phase, when a new helium shell flash is triggered, the already prepared ^{13}C is engulfed by the convective shell. The effective rate of release of neutrons by the $^{13}C(\alpha,n)^{16}O$ reaction is governed more by the rate at which ^{13}C enters the convective shell than by the cross section for the α-capture reaction (Iben 1983a). The neutrons are released when the convective shell has attained only half of its maximum size and the temperature at the base of the shell is only about 150×10^6 K.

This process appears to work only in AGB models of low total mass, low core mass, and low Z (Iben 1983a,b), but considerably more theoretical exploration is required before its dependences on all parameters have been properly elucidated. As of this writing, it would appear that the ^{13}C neutron source will be highly variable from one AGB model to another and this should be reflected in a diversity in the distribution of neutron-rich isotopes from real AGB star to the next.

REFERENCES

Becker, S.A. 1981, in Physical Processes in Red Giants, eds. I. Iben, Jr. and A. Renzini (Dordrecht: Reidel), p. 141.

_____ 1983, in progress.

Becker, S.A. and Iben, I. Jr. 1979, Ap. J., 232, 831.

_____ 1980, Ap. J., 237, 111.

Blanco, V.M., Frogel, J.A., and McCarthy, M.F. 1981, P.A.S.P., 93, 532.

Blanco, B.M., McCarthy, M.F., and Blanco, V.M. 1978, Nature, 271, 638.

Blanco, V.M. and McCarthy, M.F. 1983, preprint.

Blanco, V.M., McCarthy, M.F., and Blanco, B.M. 1980, Ap. J., 242, 938.

Cahn, J.H. 1980, Space Sci. Rev., 27, 457.

Cameron, A.G.W. 1955, Ap. J., 121, 144.

Carbon, D.F., Langer, G.E., Butler, D., Kraft, R.P., Suntzeff, N.B., Kemper, E., Trezger, C.F., and Romanishin, W. Ap. J., 49, 207.

Chieffi, A. and Iben, I. Jr. 1983, in progress.

Cohen, J.G. 1982, Ap. J., 258, 143.

Cohen, J.G., Frogel, J.A., Persson, S.E., and Elias, J.H. 1981, Ap. J., 249, 481.

Frogel, J. and Blanco, V.M. 1983, in Observational Tests of Stellar Evolution Theory, eds. A. Maeder and A. Renzini (Dordrecht: Reidel), in press.

Frogel, J. and Cohen, J.G. 1982, Ap. J., 253, 580.

Frogel, J. A., Cohen, J. G., Persson, S. E., and Elias, J. H. 1981, in Physical Processes in Red Giants, eds. I. Iben, Jr. and A. Renzini (Dordrecht: Reidel), p. 159.

Frogel, J. and Whitford, A.E. 1982, Ap. J. Lett., 259, L7.

Fusi-Pecci, F. and Renzini, A. 1976, A. and Ap., 46, 447.

Iben, I. Jr. 1975a, Ap. J., 196, 525.

_____ 1975b, Ap. J., 196, 549.

_____ 1976, Ap. J., 208, 165.

_____ 1977, Ap. J., 217, 788.

_____ 1981, Ap. J., 246, 278.

_____ 1982, Ap. J., 260, 821.

_____ 1983a, Ap. J. Lett., 275, _____ .

_____ 1983b, in progress.

Iben, I. Jr. and Renzini, A. 1982, Ap. J. Lett., 263, L188.

_____ 1983, Ann. Rev. Ast. and Ap., 21, 271.

_____ 1984, Physics Reports, in press.

Iben, I. Jr. and Truran, J.W. 1978, Ap. J., 230, 980.

Kaler, J. B., Iben, I. Jr., and Becker, S. A. 1978, Ap. J. Lett., 224, L63.

Kraft, R. P. 1979, Ann. Rev. A. and Ap., 17, 309.

Kraft, R. P., Suntzeff, N. B., Langer, G. E., Carbon, D. F., Trefzger, Ch. F.,

Fried, E., and Stone, R. P. S. 1982, P.A.S.P., $\underline{94}$, 55.

Iben, I. Jr. and Tutukov, A. V. 1984, Ap. J. Suppl., Feb. 1 issue.

Little-Marenin, I.R. and Little, S.J. 1979, A. J., $\underline{84}$, 1374.

Lloyd-Evans, T. 1983, M.N.R.A.S., $\underline{205}$, _____.

McClure, R.D. 1983a, Ap. J., $\underline{268}$, 264.

_____ 1983b, preprint.

McClure, R.D., Fletcher, J.M., and Nemec, J.M. 1980, Ap. J. Lett., $\underline{238}$, L35.

Nomoto, K. 1983, in Stellar Nucleosynthesis, eds. C. Chiossi and A. Renzini (Dordrecht: Reidel), in press.

Olive, K. and Schramm, D.N. 1982, Ap. J., $\underline{257}$, 276.

Reimers, D. 1975, Mem. Soc. R. Sci. Liege, 6^e Ser., $\underline{8}$, 369.

Renzini, A. and Voli, M. 1981, A. and Ap., $\underline{94}$, 175.

Richer, H.B. 1981, Ap. J., $\underline{243}$, 744.

Sanders, R. H. 1967, Ap. J., $\underline{150}$, 971.

Schwarzschild, M. and Härm, R. 1967, Ap. J., $\underline{150}$, 961.

Searle, L., Wilkinson, A., and Bagnulo, W.G. 1980, Ap. J., $\underline{239}$, 803.

Sweigart, A. V. and Mengel, J. G. 1979, Ap. J., $\underline{229}$, 624.

Truran, J.W. and Iben, I. Jr. 1977, Ap. J., $\underline{216}$, 797.

Whitford, A.E. and Rich, M. 1983, Ap. J., in press.

Wood, P.R. 1981, in Physical Processes in Red Giants, eds. I. Iben, Jr. and A. Renzini (Dordrecht: Reidel), p. 135.

Wood, P.R., Bessel, M.S., and Fox, M.W. 1983, Ap. J., $\underline{272}$, 99.

Woosley, S. E., Axelrod, T. S., and Weaver, T. A. 1983, in Stellar Nucleosynthesis, eds. C. Chiosi and A. Renzini (Dordrecht: Reidel), in press.

CHROMOSPHERES IN METAL DEFICIENT FIELD STARS

A. K. Dupree, L. Hartmann and G. Smith
Harvard-Smithsonian Center for Astrophysics
60 Garden Street, Cambridge, MA 02138, U.S.A.

The chromospheric structure of red giant stars with metal defi-cient atmospheres presents a fundamental observational and theoretical problem for atmospheric theory. Moreover, the occurrence and rate of mass loss from these stars have consequences for the theory of stellar evolution of low mass stars in globular clusters. This paper reports some results from a systematic program to assess the presence and character of chromospheric line profiles and their variations, and to obtain quantitative measurements of radiative losses from chromo-spheres. Such data are needed to constrain semi-empirical models of these atmospheres.

Three metal deficient field stars have been observed both in the ultraviolet and optical spectral regions: HD 165195, HD 110281, and HD 232078. High dispersion spectra near the Hα transition were obtained with the echelle spectrograph and reticon detector at the F. L. Whipple Observatory of SAO. These stars were observed twice in 1983, separated by a two month interval. IUE spectra were also obtained using the long wavelength cameras in low dispersion mode. Observations of the three stars are discussed below.

<u>HD 165195</u> The Hα cores show strong asymmetry indicating outward motion both in the May 1983 and July 1983 spectra. The velocity of the core center is blue-shifted by -3.4 and -1.8 km s^{-1}, but these values differ from zero just at the 3 km s^{-1} uncertainty of the velo-city measurement. No emission is present in the May observations whereas emission on the red wing of the line (modified by telluric absorption) is present in July. This contrasts with Mallia and Pagel's (1978) discovery of a blue emission wing in August 1977. The IUE long wavelength, low resolution spectra: LWR 11703 (6 Oct. 1981); LWP 1909 (27 June 1983); and LWR 16543 (7 Aug. 1983) show broad absorption in the Mg II region ($\lambda 2800$), no sign of an emission core (although at low resolution, this would be hard to detect in such a hot star) and no indication of changes in the line profile beyond the noise in the spectrum. An upper limit to the Mg II flux is given in Table 1.

<u>HD 110281</u> This is a metal deficient giant as identified by Bond (1980), but no other observations have been made of this object. The Hα region is shown in Figure 1 where a broadening of the line wings and reduction of the emission asymmetry was present in July, as com-pared to the May observation. The line core is asymmetric although the measured wavelength of the core center is blue-shifted by less

than 1 km s^{-1} relative to the metal lines in the photosphere. The centroid of the emission peaks occurs at -35.6 and +40.1 km s^{-1} in May and -43.6 and +50.0 km s^{-1} in July. The Mg II region is shown in Figure 2, where chromospheric emission is clearly present. The flux of this emission is given in Table 1.

HD 232078 Figure 3 shows the Hα region, again with variable emission and an asymmetric core. The shift of the core is measured at -2.85 km s^{-1} relative to the photospheric velocity of -391 $\pm$ 2.9 km s^{-1}. Ramsey (1979) noted that the emission in the Hα profile differed from Cohen's (1976) observations in a previous year. Our measurements show that the time scale of variation can be on the order of a few months or less. The photospheric velocity (as indicated by the metal lines) and the Hα core center were unchanged in the two observations to within the 2.9 km s^{-1} precision of our radial velocity determination. The Mg II transition is in emission (Figure 2) and its flux is given in Table 2.

Discussion The Hα profiles can exhibit strong asymmetric emission that is quite different from luminous Population I stars. In the stars HD 232078 and HD 110281, these emission wings are in the sense red wing > blue wing indicative of outward flow. The absorption cores are frequently asymmetric, also a signature of mass outflow.

The Hα emission and core asymmetry can vary on a time scale of two months or less. Our observations substantially reduce the previous upper limit (Ramsey 1979) of one year or less for variability in field giants, and are not inconsistent with the time scale of a few days noted by Cacciari and Freeman (1983) for red giants in globular clusters. Substantial variability or even disappearance of the emission wings on a short time scale would make the circumstellar origin suggested by Cohen (1976) seem less likely.

The surface flux of the Mg II lines is similar to that of luminous Pop I stars (see Hartmann, Dupree, and Raymond 1982) in spite of the lower metal abundance. While the correction for reddening in the ultraviolet can be substantial, A$_\lambda$ (mag) = 6E(B-V) at λ2800 (Seaton 1979), and the distance and radii may be uncertain, the derived surface fluxes of these metal deficient giants are not notably discrepant from Pop I stars indicating fundamental differences in atmospheric structure between the two populations.

TABLE I

Fluxes of Mg II ($\lambda 2800$)

Quantity	HD 110281[a]	HD 165195[b]	HD 232078[c]
V	9.34	7.7	8.7v
M_v	−2.0:	−1.8	−2.0
$(B-V)_o$	1.5	1.1	1.8:
E(B-V)	0.03:	0.25	0.25:
$R_*(R_\odot)$	100.	75.	100.
d(pc)	1300.	460.	1400.:
Mg II (10^{-13} erg cm^{-2} s^{-1} at Earth)	1.19	<3.83	0.87
Mg II (erg cm^{-2} s^{-1} at stellar surface)	4.4(+4)	<1.1(+5)	1.2(+5)
$F_*/F_\odot$	0.04	<0.1	0.1

[a] Little studied star; colors from Bond (1980); reddening from galactic position ($\ell = 297°.4$; b = +62°.1) following technique of Bond (1980). Assumed absolute magnitude and radius.

[b] Color and M_v from Kraft et al. (1982); reddening and distance from Bond (1980); radius from Mallia and Pagel (1978).

[c] Observed color from Bond (1980); Reddening consistent with galactic position ($\ell = 53°.2$; b = -2°.3); Values of E(B-V) in literature vary from 0.0 (Jones and Dixon, 1972), and 0.06 (Sandage 1969) to 0.9 (Christiansen 1978). M_v and R taken to agree with values for globular cluster giants (Mallia and Pagel 1978).

REFERENCES

Bond, H. E. 1980, Ap. J. Suppl., 44, 517.
Cacciari, C. and Freeman, K. C. 1983, Ap. J., 268, 185.
Christiansen, C. G. 1978, A. J., 83, 244.
Cohen, J. G. 1976, Ap J. Letters, 203, L127.
Hartmann, L., Dupree, A. K., and Raymond, J. C. 1982, Ap. J., 252, 214.
Jones, D. H. P., and Dixon, M. E. 1972, Ap. J., 177, 665.
Kraft, R. P., Suntzeff, N. B., Langer, G. E., Carbon, D. F., Trefzger, C. F., Friel, E. and Stone, R. P. S. 1982, Pub. A.S.P., 94, 55.
Mallia, E. A., and Pagel, B. E. J. 1978, M.N.R.A.S., 184, 55p.
Ramsey, L. W. 1979, Pub. A.S.P., 91, 252.
Sandage, A. 1969, Ap. J., 158, 1115.
Seaton, M. J. 1979, M.N.R.A.S., 187, 73.

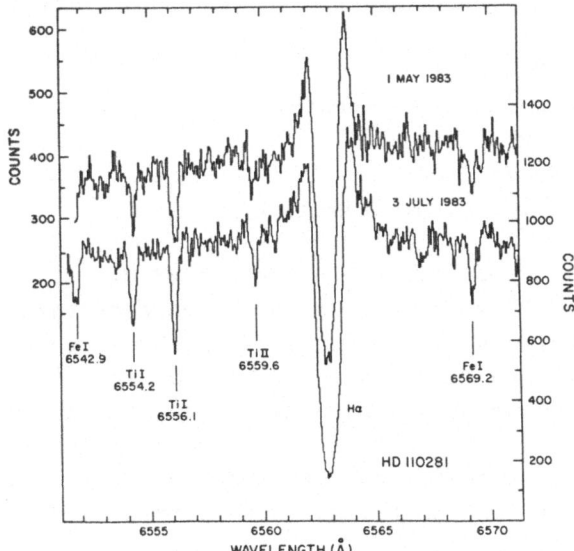

Figure 1. Hα region of HD 110281 showing broadening of the line wings in July as compared to May and indication of asymmetry of line core.

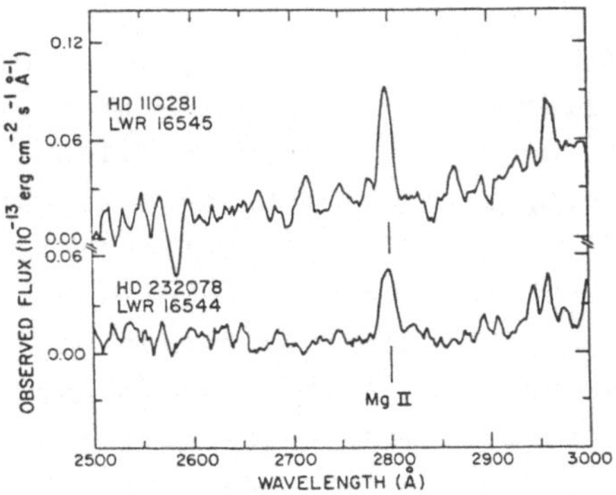

Figure 2. Mg II emission is prominent in both stars. HD 110281, being hotter than HD 232078 shows a continuum at long wavelengths.

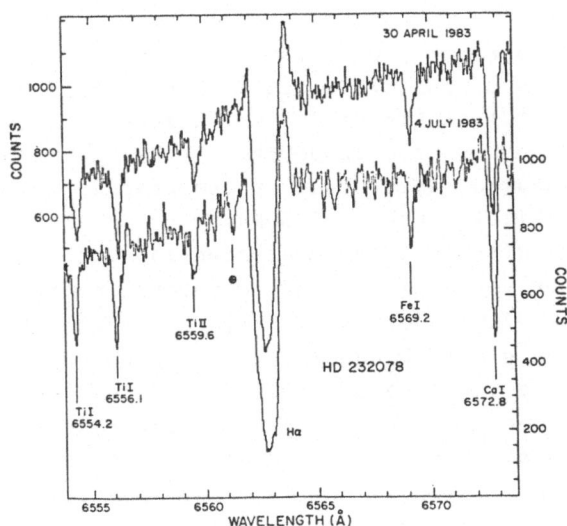

Figure 3. Hα region of HD 232078 showing both variable emission and an asymmetric absorption core.

INTERPRETATION OF THE OBSERVED MG II EMISSION FROM N-TYPE CARBON STARS

Eugene H. Avrett
Harvard-Smithsonian Center for Astrophysics
and
Hollis R. Johnson[1]
Astronomy Department, Indiana University

Johnson and O'Brien (1983) have reported the detection of Mg II h and k line emission, and C II $\lambda 2325$ emission, in the spectra of three N-type carbon stars: BL Ori, TX Psc, and T Ind, showing that these stars have chromospheres. Usually such chromospheric emission is accompanied by absorption or emission in the hydrogen Balmer lines, but it is puzzling that the Balmer lines are not seen in the spectra of these stars (Yamashita 1972, 1975). Here we investigate whether the presence of Mg II and C II lines and absence of Balmer lines can be explained by means of semi-empirical chromospheric models.

To construct a semi-empirical model we choose a temperature distribution, solve the equations of hydrostatic equilibrium, statistical equilibrium, and radiative transfer for H, H^-, H_2, Mg I, Mg II, C I, and C II, and compute flux profiles of the H, Mg II, and C II lines. We use the computational methods described by Vernazza, Avrett, and Loeser (1981), including partial frequency redistribution for the Mg II k line. The continuum opacity is calculated with an average line-opacity distribution function from Kurucz (private communication). We then study the computed results in order to revise the assumed temperature distribution and obtain line fluxes in a subsequent calculation that are in better agreement with observations.

We adopt a photospheric temperature distribution from Johnson (1982), with T_{eff} = 3000K, log g = 0.0, and C/O = 1.05. The photospheric model is plotted in the right panel of Figure 1. We assume a temperature minimum of 2000K and a chromospheric temperature rise beginning at the height 5×10^6 km (measured above the photospheric level corresponding to unit continuum optical depth at 5000A).

Two chromospheric temperature distributions are shown in the left panel of Figure 1. In the higher temperature case, the total hydrogen density n_H decreases less rapidly with increasing height, since the scale height is larger, and the electron density is enhanced, due to greater ionization. The temperature increase of ~ 500K from Model 1 to Model 2 causes the level-2 hydrogen number density n_2 to increase by a factor of 100. The optical thickness of the chromosphere in the center of the $H\alpha$ line is 0.008 for Model 1 and 0.8 for Model 2. There is no photospheric contribution to the $H\alpha$ line, because at the photospheric formation depth, where $T \sim 3000$K, the H^- continuum opacity is 10 to 100

---

[1] Guest Observer with the <u>International Ultraviolet Explorer Satellite</u>.

times larger than the Hα line center opacity. The chromospheric n_2 values for Model 1 are too small to produce an Hα feature, while the Model-2 values could produce either Hα absorption or emission, depending on whether the Hα chromospheric source function is smaller or larger than the corresponding photospheric continuum source function.

Since Model 1 clearly satisfies the requirement of no Hα flux, we consider whether the chromospheric temperature rise in this case is sufficient to produce the observed Mg II and C II lines. We show in Figure 1 the computed Model-1 number densities for Mg II and C II, obtained from a detailed solution of the ionization equilibrium equations for the three lowest Mg and C stages of ionization.

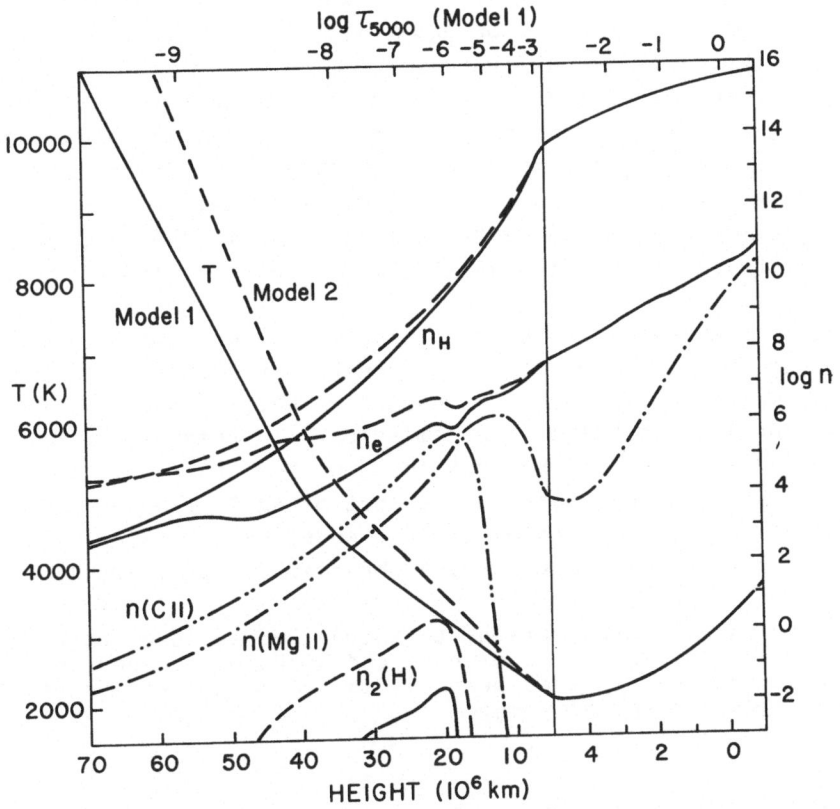

Figure 1 ---- Temperature as a function of height (measured above $\tau_{5000}=1$) for two chromospheric models, and the corresponding values of the total hydrogen density, the electron density, and the level-2 hydrogen density. The τ_{5000} scale and the Mg II and C II number densities for Model 1 are also shown.

The computed optical thickness of the chromosphere in the center of the Mg II k line (measured to the temperature minimum) is 6×10^5. The k-line source function reaches a maximum value just where n(Mg II) has a maximum (at a line-center optical depth 3×10^5). The computed maximum value of the line flux profile corresponds to a brightness temperature of about 2500K, while the continuum on either side of the line has a brightness temperature of 2200K. The integrated k-line emission flux at the stellar surface is computed to be ~ 100 ergs cm^{-2} s^{-1}. This value corresponds to a predicted flux at the Earth (for the combined h and k lines) about 2 or 3 times smaller than the value reported by Johnson and O'Brien, without including a possible correction for extinction.

The line-center chromospheric optical thickness for C II $\lambda 2325$ is only 4×10^{-3} for Model 1 (7×10^{-9} times the k line optical thickness), because the Einstein A coefficient for this intercombination line is 2×10^{-7} times the A coefficient of the k line, and because n(C II) drops rapidly for $T < 3000K$. Consequently, the calculated $\lambda 2325$ emission flux from Model 1 is negligible.

We have also calculated the emission flux from the C II resonance lines near 1335A. The line-center optical thickness in the strongest of these lines is 4×10^3, but the continuum flux at 1335A is also formed in the chromosphere and is comparable to the maximum line flux. Both have brightness temperatures $\sim 2700K$. We find that the 1335A flux is 10^{-3} times the continuum flux at 2325A, so that the resonance lines may be difficult to detect.

We plan to carry out further trial calculations with higher temperatures in the low-chromospheric height region $10-15 \times 10^6$ km ($T \sim 2500K$), which should lead to enhanced k-line and $\lambda 2325$ fluxes without an increase in Balmer line emission. In this way we hope to obtain better agreement with observations. On the basis of the results obtained so far, we feel that the presence of Mg II line emission can be understood in terms of our chromospheric model calculations, but it is not clear that we can account for the observed C II $\lambda 2325$ line.

Acknowledgements: We are grateful to Andrea Dupree, Lee Hartmann, and George O'Brien for useful comments, and to Rudolf Loeser for his help with the calculations reported here. This research was supported by NASA Grants NAGW-100 and NAG5-182.

References

Johnson, H.R. 1982, Ap.J., 260, 254.
Johnson, H.R. and O'Brien, G.T. 1983, Ap.J., 265, 952.
Vernazza, J.E., Avrett, E.H., and Loeser, R. 1981, Ap.J.Suppl., 45, 635.
Yamashita, Y. 1972, Ann. Tokyo Astr. Obs., 13, 169,
_____. 1975, Ann. Tokyo Astr. Obs., 15, 47.

INTRODUCTORY REMARKS TO THE SESSION ON
ALPHA ORIONIS AND RELATED TOPICS

L. Goldberg
Kitt Peak National Observatory
Tucson, AZ 85726

In opening this special session on Betelgeuse, I want to draw attention to a number of questions about the atmospheric structure of this and other red giant stars, which will probably have to be answered before we can reach an understanding of mass loss mechanisms and derive realistic rates of mass loss for these stars.

1). At great distances from the star, the emission from the circumstellar shell seems to be reasonably symmetric, but within the first five or ten stellar radii, the shells are strongly asymmetric. Nearly simultaneous interferometric observations in November, 1980 by Hege (see Goldberg et al., 1981) and by F. and C. Roddier (1983) show pronounced asymmetries at about the same position angle, despite the fact that the former observed through a narrow H-Alpha filter and the latter through a relatively broad-band filter transmitting visible continuum light. Does the continuum emission originate from scattering by dust, or from Rayleigh scattering by neutral hydrogen? What do the asymmetries tell us about the mass loss mechanism and how do they affect determinations of the mass loss rates?

2). The widths at half-intensity of weak Fraunhofer lines in the spectra of cool supergiants, including Betelgeuse, are about 27 km./sec. (Imhoff, 1976). What causes these great widths and what are their implications for models of these stars?

3). Similarly, the Fe+ emission lines near 3100A observed

by Boesgaard and Magnan, (1975) in the spectrum of Betelgeuse (and other red giants as well) show widths of 50-60 km./sec. and appear to arise from infalling matter close enough to the photosphere to mirror its radial velocity variations. How can these observations be incorporated into models of atmospheric structure?

4). The atmosphere of Betelgeuse shows evidence, from the radial velocities of H-Alpha (Weymann, 1962) and of the infrared triplet lines of Ca+ (Goldberg, 1979), of a layer, which is stationary with respect to the center of mass and is sufficiently far from the photosphere to be uncoupled from its radial velocity variations. A similar stationary region has been inferred from observations of CO in Chi Cygni and other Mira stars (Hinkle et al., 1982). The escape of mass from the stars appears to begin in this layer, which is clearly above the extensive regions in which the Fraunhofer and Fe+ emission lines are formed. Do their great widths imply the existence of a vast convective system which transports matter up to the stationary layer, where dust may form to provide the means for mass ejection?

5). Empirical determinations of mass loss rates from optical spectra of Betelgeuse and other red giants are based on one or both of two assumptions: **a)** the relative abundance of minor constituents and hydrogen are the same as in the sun, and **b)** the ionization theory used to calculate the relative amounts of neutral and singly-ionized atomic species is correct. The use of these assumptions has led to some rather absurd discordances of up to four orders of magnitude, as has been shown by Hagen (1978) and Bernat (1982). The reasons for these discrepancies need critical examination.

6). Finally, I wish to draw attention to the enormous disagreements between different authors seeking to infer the optical depth of the 10 mu silicate feature from the observed emission bump. For example, the values derived by Rowan-Robinson and Harris (1982, 1983) are smaller by one to two orders of magnitude than those deduced for the same stars by Hagen (1978). Part of the discrepancy can be attributed to inadequate handling of the radiative transfer problem, but by far the greatest source of uncertainty is introduced by assumptions made in calculating grain temperatures. We simply do not know enough about the composition and short-wave absorption efficiencies of the grains to derive reliable temperatures and rates of mass loss.

References

A. P. Bernat: 1982, Ap. J., 252, 644.

A. M. Boesgaard and C. Magnan: 1975, Ap. J., 198, 369.

L. Goldberg: 1979, Quart. J. Roy. Astron. Soc., 88, 660.

L. Goldberg, E. K. Hege, E. N. Hubbard, P. A. Strittmater and W. J. Cocke: 1981, in Second Cambridges Workshop on Cool Stars, Stellar Systems and the Sun. ed. M. S. Giampapa and L. Golub. SAO Special Report No. 392, vol. I, p. 131.

W. Hagen: 1978, Ap. J., Suppl., 38, 1.

K. H. Hinkle, D. N. B. Hall and S. T. Ridgway: 1982, Ap J., 252, 697.

C. L. Imhoff: 1976, Ap. J., 205, 455.

F. & C. Roddier: 1983, Ap. J., 270, L23.

M. Rowan-Robinson and S. Harris: 1982, Mon. Not. Roy. Astr. Soc., 200, 197.

M. Rowan-Robinson and S. Harris: 1983, Mon. Not. Roy. Astr. Soc., 202, 767.

R. J. Weymann: 1962, Ap. J., 136, 844.

MULTIBAND PHOTOELECTRIC PHOTOMETRY OF BETELGEUSE

Edward F. Guinan
Department of Astronomy
Villanova University
Villanova, PA 19085

I. Introduction

The semiregular (SRc) red supergiant Betelgeuse (Alpha Ori;
M2Iab) is at maximum brightness one of the brightest stars north of
the celestial equator. The physical properties of the star have been
discussed and summarized by Weymann (1977) and Goldberg (1980, 1984).
An oversimplified but useful physical model is one that consists of
an unstable star (T$\simeq$3500K) with a radius of R$\sim$600R$_\odot$, surrounded
by an extended "warm" chromosphere (T$\sim$8000K) and an extensive circumstellar shell of gas and dust that has been traced out to 9500 AU from
the star by Honeycutt et al. (1980).

Systematic photoelectric photometry of Betelgeuse was carried out
by Stebbins (1931), in the pioneering days of the technique, during
most years from 1916 to 1931. At nearly the same time Jones (1928)
and Sanford (1933) made systematic radial velocity measurements of
the star. The brightness and photospheric radial velocity were found
to vary with a period of about 5.78 years, and with mean amplitudes
of about 0.4 mag in brightness and $\sim$6 km/s in radial velocity. These
long-term variations are accompanied by apparently random variations
of brightness and in radial velocity with time-scales of several weeks
to several months. The long-period changes in light and velocity indicate that light maximum occurs near minimum radius and vice versa.
However, the short-term light and radial velocity variations are usually uncoupled, except for a few noteworthy occasions when larger than
usual outward motions were accompanied by fairly rapid increases in
brightness (Sanford 1933). More recently Goldberg (1983) has shown
that the radial velocity measurements made subsequent to Sanford's
study, generally appear to follow the form and phasing indicated by
Sanford. Because of its brightness and relatively large angular diameter (α =0.035-0.060), Betelgeuse has been a favorite target for
interferometric angular diameter measurements. Although changes in
its diameter apparently occur, it is difficult to assess the precise
magnitude and time-scale of these changes due to problems in defining
the disk of star and to differences in technique and instrumentation.
(cf. Balega et al. 1979 and White 1980).

No systematic photoelectric measurements of Betelgeuse were made
until the B and V observations made by Krisciunas (1982,a,b) from 1979
to 1982, and those presented here, which started during 1981. However,
beginning in 1979 (and still going strong) Hayes (1980, 1981, 1984)
has made extensive systematic polarization measures of Betelgeuse
which reveal significant ordered changes, having characteristic time
scales of several months. The observed changes in the intrinsic polarization (of up to$\sim$0.8%) imply significant deviations from spherical
symmetry. Hayes has suggested that the changes in polarization arise
from the growth, the decay, as well as the motion of large-scale elements of the star's inner atmosphere. In apparent accord with this
model, is the recent discovery of variable asymmetries in the bright-

ness distribution of the circumstellar shell, which is indicated by speckle observations of Goldberg et al. (1981) and Roddier and Roddier (1983).

The present study of multi-band, high precision photometry was undertaken to complement the ongoing polarimetric program of Hayes and also the high resolution spectroscopic program of Baliunas, Guinan and Hartmann (1983). In addition, a number of other investigations are currently being carried out or planned for Betelgeuse in the UV, IR and radio regions.

II. Observations

The photoelectric photometry of Betelgeuse commenced during 1981 November and is continuing at the present time. The observations were made with the 38-cm reflector of Villanova University. A description of the instrumentation is given elsewhere (Guinan et al. 1982). A pair of narrow and intermediate-band interference filters centered near the Balmer Hα line, and an intermediate-band blue (λ4530) filter were used. Because of the extreme brightness of Betelgeuse, a neutral density filter (with a reduction factor of $\sim$5.0 mag) was used when observing the variable star. Differential photometry was made with respect to comparison and check stars that are within 1°.5 of the variable. In previous photometric studies the comparison stars have been up to 20° away from the variable. In this study BD+6°1051 (m_v=+7°.0;G5) was the comparison star and BD+6°1056 (m_v=+ 7°.3; G0) served as the check star. No significant light variations have been detected between the two stars. Although the Hα filter pair was originally designed to measure the Hα line strength in B to G stars, in the case of Betelgeuse and other very cool stars, the Hα line is relatively weak and the TiO bands are the dominant spectral features. It turns out that relatively strong TiO bands at $\lambda\lambda$· 6650-6800 fall within the spectral response of the Hα intermediate band filter while the narrow-band filter contains only relatively weak spectral features. Thus, the α-index which is conventionally defined by

$$\alpha- \text{index} = -2.5 \log \text{FN/FI}$$

where FN and FI are the fluxes through the narrow and intermediate band filters, yields chiefly a measure of the TiO band strength for M-type stars such as Betelgeuse. The α-index correlates well with spectral type. Thus, the α-index for M stars is chiefly a measure of the TiO band strength and thus can be used to determine spectral type and stellar temperature. Furthermore, over 1981-1983, spectroscopic observations made by Baliunas, Guinan, and Hartmann (1983) indicate no significant variation in the strength of the Hα feature in Betelgeuse. In addition to the α-index, two other differential photometric indices were formed from the differential magnitudes. These are the ($\Delta m \lambda$4530-$\Delta m \lambda$6600)=and ($\Delta m\lambda$4530-$\Delta m\lambda$6568) -indices which are measures of the TiO line strength and color temperature, respectively. The differential λ4530 magnitudes,$\Delta\alpha'$(TiO) -indices and the two color indices are plotted in Fig. 1 against time. Also plotted in Fig. 1 are polarization measures in the B-bandpass of Betelgeuse, obtained by Hayes (1984) during 1981-1983. In addition, the intermediate-band blue observations were transformed to Johnson B-magnitudes and combined with the B magnitudes of Kriscuinas (1982a, b). The transformation of the intermediate-band blue data was made possible by the fact that Betelgeuse was observed contemporaneously in B during 1981 December through 1982 February. By combining the wide-and intermediate-band blue data, comparison can be made between the long-term changes in polarization and brightness over 1979-1983.

The net polarization (P%) in the B-bandpass determined by Hayes, and the B-magnitudes are plotted in Fig. 2. Following the same procedure of Stebbins, mean seasonal magnitudes were computed from the data and these are shown in the figure as large open circles.

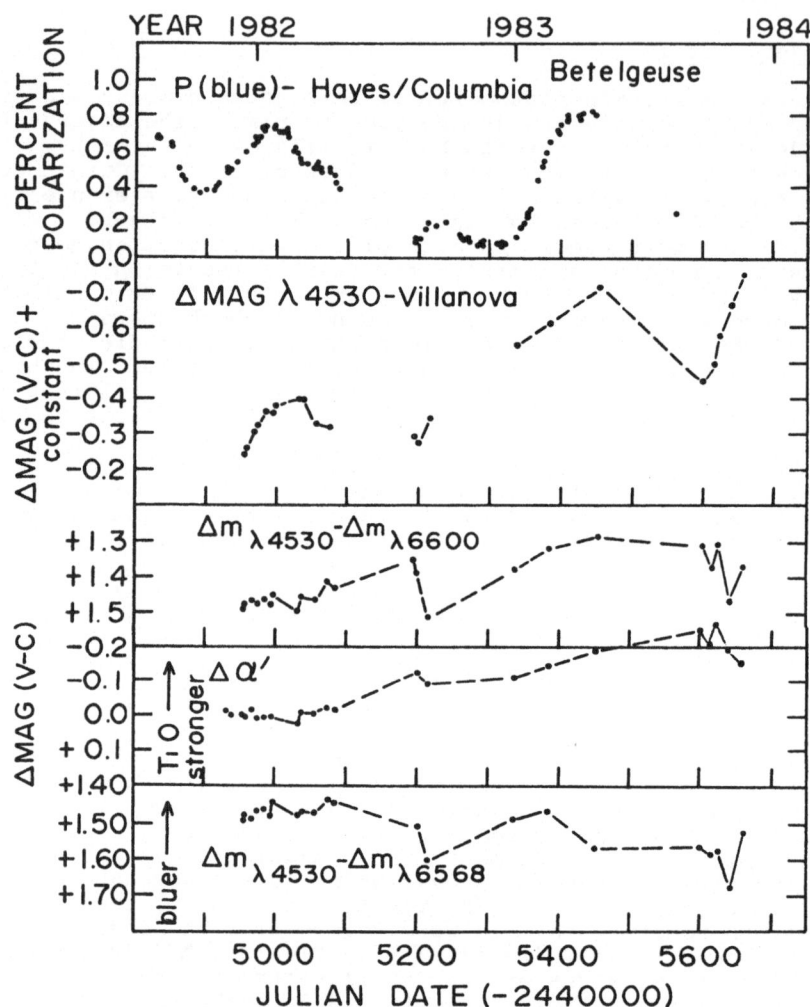

Figure 1. The top panel is a plot of Hayes' polarization measures. The second panel from the top shows a plot of the differential intermediate-band blue magnitudes obtained at Villanova University. The lower three panels show the various differential photometric indices of Betelgeuse.

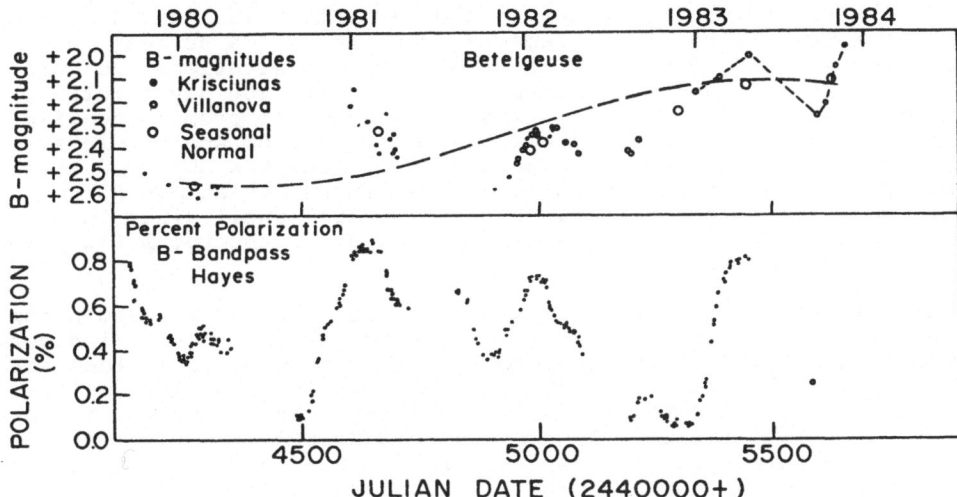

Figure 2. Plot of the B-magnitudes and polarization measures of Betelgeuse. The large open circles are seasonal means and the solid curve shown in the top panel is the mean blue light curve of Stebbins.

III. Results and Conclusions

In the following the initial results of the photometric study of Betelgeuse are presented. A fuller analytical treatment of the data and a more thorough discussion are planned after the 1983-84 observing season.

An examination of the blue measurements plotted in Figs. 1 and 2 indicates that light variation of Betelgeuse is essentially the same as observed by Stebbins over 50 years ago. As shown in Fig. 2, the seasonally averaged B-magnitudes appear to show the long term ∿5.78 year light variation with a minimum occurring during mid-1980 and a maximum occurring during 1983. The broken curve shown in Fig. 2 is the mean light curve satisfying Stebbins' data which has been brought forward in time, and adjusted in its brightness level, to satisfy the observed minimum and maximum of the seasonal means. Although the fit is by no means perfect, the seasonal means generally still appear to follow the ∿5.78 year period, both in its phasing and light ampli- tude. The short-term, irregular variations with characteristic rise or decline times of 100-200 days are present and superposed on the long-term periodic light variation. As shown in the figure, there is no evidence of the long-term variation apparent in the polari- zation data. However, there appears to be a rough correlation be- tween the percent polarization and the short-term light variation in the sense that the polarization maxima usually (but not always) occur during the maxima of the short-term light variations and vice versa. Furthermore, the characteristic time scales of the short-term light and polarization changes are also similar. If the long-term light variation arises from the radial pulsation of the star, there is no reason to expect concurrent polarization changes since there should be little or no deviation of the star from spherical symmetry.

If the short-term brightness and polarization variations arise from large-scale convective elements over the star's surface, as first suggested by Schwarzschild (1975), the observations indicate that they are large, few in number, and have 100-200 day lifetimes. However, the growth and dissipation of large-scale prominence-like features above the star's surface (as reported from speckle interferometry by Goldberg et al.) could also produce the observed short-term light and polarization changes. The light variations do not appear to arise in outer portions of the star's shell since the outer shell contributes only a small fraction to the total flux (Roddier and Roddier 1983).

Additional and important information about the origin and nature of the light variations can be deduced from the photometric indices formed from the multi-band photometry. The photometric indices are plotted in Fig. 1, where the $\Delta\alpha'$(TiO)-index yields a measure of the TiO band strength and the $(\Delta m \lambda 4530 - \Delta m \lambda 6568)$-index yields a measure of the color temperature. The $(\Delta m \lambda 4530 - \Delta m \lambda 6600)$-index is more difficult to interpret, but for M-stars is chiefly a measure of the TiO bands and color temperature. Although the indices show short-term variations, the seasonal means show a clear indication of long-term variations that appear to coincide with the net increase in mean brightness observed from 1981-82 to late 1983. For example, the seasonal mean $\Delta\alpha$(TiO)-index shows a net change from $\sim 0^{m}.00$ in 1981-82 to $\sim -0^{m}.20$ during late 1983. Calibration of these indices with spectral type indicates that the spectral type of Betelgeuse has varied from $\sim$M1.5 during 1981-82 to $\sim$M2.5 during 1983. (Similar changes in spectral type have been reported earlier by White and Wing (1980) over 1969-1978). According to the spectral type-Teff tabulation for supergiants given by Novotny (1973), these inferred spectral types indicate an overall decrease in temperature from T=3550K to T=3400K from 1981-82 to late 1983. An overall increase in mean brightness of about 0.3 mag occurs over the same interval, which when combined with the inferred temperature change, indicates an increase in the star's radius of about 25%. This indicates that the star attains its maximum radius near the time when it is coolest and brightest. This tentative result is contrary to what was inferred earlier from Sanford's combined radial velocity-luminosity study. An increase in the star's radius of $\sim$25% should be detectable by interferometric techniques, but no recent measurements are available at present.

Finally there appears to be little or no correlation between the short-term variations in the TiO-band strength and the short-term light and polarization changes. This suggests that the mechanism responsible for the short-term light enhancements is not linked to temperature increases as would be expected from ascending giant convective cells. The apparent lack of a correspondence between temperature and luminosity for the short-term light variations seems to favor the hypothesis that the brightness enhancements are produced above the star's surface - perhaps related to the atmospheric structures suggested by the interferometric measures by Goldberg et al. More observations are necessary to resolve this problem. Although it may be wishful thinking, with contemporaneous photometric, polarimetric, spectroscopic and interferometric measurements, it should now be possible to make considerable progress in understanding the complex processes taking place in the interior, on the surface, and in the extended atmosphere of this red supergiant star.

Acknowledgements

I wish to thank Dr. Daniel Hayes of Columbia University for encouraging me to undertake the photometry. I also wish to thank Mrs. Hildred Nason and Ms. Carol McMenamin for preparing the manuscript for publication. The observations obtained during Fall 1983 were made by the following Astronomy undergraduates at Villanova University: Edwin Dombrowski, Robert Donahue, Sue Draus, and Scott Wacker.

References:

Baliunas, S.L., Guinan, E.F. and Hartmann, L. 1983, in preparation.

Balega, Y., Blazit, A., Bonneau, D., Koechlin, L., Foy, R., and Labeyvie, A. 1982, Astr. Ap., 115, 253.

Goldberg, L. 1980, Quart. J.R.A.S., 20, 361.

Goldberg, L. 1983, private communication.

Goldberg, L. 1984, (in this volume).

Goldberg, L., Hege, E., Hubbard, E., Strittmatter, P., and Cocke, W., 1982, in Second Cambridge Workshop on Cool Stars, Stellar Systems, and the Sun, ed. M. Giampapa and L. Golub, SAO Special Report No. 392, Vol. 1, p. 131.

Guinan, E.F., McCook, G.P., Fragola, J.L., O'Donnell, W.C., Tomczyk, S., and Weisenberger, A.G. 1982, Astron. J., 87, 893.

Hayes, D.P. 1980, Ap.J. (Letters), 241, L165.

Hayes, D.P. 1981, Publ. A.S.P., 93, 752.

Hayes, D.P. 1984, (in this volume).

Honeycutt, R.K., Bernat, A.P., Kephart, J.E., Gow, C.E., and Sanford, M.T. 1980, Ap.J., 239, 565.

Jones, H.S. 1928, Mon. Nat. Roy. Astr. Soc., 88, 660.

Krisciunas, K. 1982a, I.A.U., Info. Bull. Var. Stars, No. 2104.

Krisciunas, K. 1982b, I.A.U. Info. Bull. Var. Stars, (in press).

Novotny, E. 1973, Introduction to Stellar Atmospheres and Interiors (Oxford Univ. Press, New York) p. 12.

Roddier, C. and Roddier, F. 1983, Ap.J. (Letters), 270, L23.

Sanford, R.F. 1933, Ap.J. 77, 110.

Schwarzschild, M. 1975, Ap.J., 195, 137.

Stebbins, J. 1931, Publ. Washburn Obs., Univ. of Wisconsin, 15, 177.

Weymann, R. 1977, in IAU Colloq. 42, The Interaction of Variable Stars With Their Environment, ed. R. Kippenhahn, J. Rahe, and W. Strohmuer (Bamberg:Remeis-Sternwarte).

White, N.M. 1980, Ap.J., 242, 646.

White, N.M. and Wing, R.F. 1978, Ap.J. 222, 209.

LOWER ATMOSPHERIC CHANGES IN
BETELGEUSE DEDUCED FROM
OPTICAL CONTINUA POLARIZATION
MEASUREMENTS

Daniel P. Hayes
Astronomy Department
Columbia University
New York, New York 10027

Abstract

Observations of Betelgeuse's optical continua linear polarization are reviewed. Arguments are presented that the observed ordered large-scale polarization variations emanate from this star's lower atmosphere, i.e., at or near the photosphere. Polarization-producing mechanisms are indicated, the most promising being large-scale surface convective features.

I. Introduction

The optical continua polarization of the semiregular (SRc) light varying, late-type supergiant Betelgeuse (α Ori, M2 Iab) was initially observed by (at least) the following: Behr (1959), Dyck (1968), and Dyck and Jennings (1971). Intercomparison of their data reveals variability (and thus that this star is intrinsically polarized). But the initial data had several drawbacks: the relatively few observations were scattered over a long time base, and the available instrumentation did not have the requisite stability for carrying out precise long-term observations. The first systematic study of this star using contemporary-technology instrumentation was carried out by the late Khzysztop Serkowski, the father of modern astronomical polarimetry. Serkowski (1971) presented convincing evidence for long-term polarization variations and demonstrated that this star's polarization showed both a Rayleigh scattering and a dust component. The next major long-term studies were carried out by Hayes (1980), and by Tinbergen et al. (1981) who consolidated their own data with Serkowski's published and unpublished results. The remainder of this paper will be devoted to reporting on observations of Betelgeuse carried out as part of an ongoing Columbia University Astronomy Department program to determine the precise morphology of polarization variations in a wide variety of intrinsically polarized stars.

II. Columbia Polarization Survey of Betelgeuse

The Columbia University wide-band (B-filter) linear polarimetry survey of Betelgeuse was carried out at the Cassegrain focus of the Harriman Observatory's 61-cm telescope.

The instrumentation and observing procedures are discussed in detail by Hayes (1975, 1978). A total of 232 observations were made over four consecutive observing seasons (Years 1979-1983). In this report the amounts (P) and the directions (θ) of linear polarization are expressed in percentages and equatorial coordinates, respectively. All observations have Poisson photon-count standard deviations of 0.015% for both P as well as the Q and U Stokes parameters (with $Q = P \cos 2\theta$ and $U = P \sin 2\theta$). The standard deviation of θ is given by $28°7 (\sigma_p/P)$. The photon-count errors always exceeded the contribution of sky and any moonlight background. The years 1979-1980 and 1980-1981 observations have previously been reported by Hayes (1980) and Hayes (1981), respectively. Throughout these observations several standard polarized stars were frequently monitored to test for polarimeter precision. Chi-square (χ^2) tests showed both a high degree of system precision and that Poisson photon-count statistics adequately account for instrumentation variations over the course of the observations. Quantitative test results are reviewed in detail by Hayes (1983).

Each of the four observing seasons is separately plotted in the equatorial Q - U Stokes parameter reference frame in Figures 1 a-d (where the data trains are broken into monthly segments). While the data are herein presented in a rectangular cartesian coordinate system, a polar coordinate plot offers the following representation: the distance from the origin is the percentage polarization P, while the angular coordinate is the double position angle 2θ as measured counterclockwise from the positive Q axis (which corresponds to the equatorial North-South direction). These plots reveal that ordered (as opposed to stochastic) changes occurred over the course of each season. Scrutiny of Fig. 1 reveals that with the possible exception of one relatively quiescent interval (1980 February-April), the polarization appears to be displaying ordered changes throughout the course of these observations. Whether or not the polarization was quiescent during the spring and summer interregnums when observing was precluded is a matter that can not be satisfactorily resolved.

Modulation by a rotating atmosphere can be discounted as an explanation for the observations. Assigning Betelgeuse a radius of 633 solar radii (Weymann 1977) and an equatorial rotational velocity of 10 km s^{-1} (probably a very generous upper limit) yields a stellar rotational period of ≈8.8 yr. Since such a period is appreciably longer than the observed polarization duty-cycles, rotational modulation can be dismissed as an explanation. The time-scales of the polarization changes being reported here also enable us to discount changes in the circumstellar envelope due to mass-loss as being responsible for such variations. For it would take ≈1.4 yr for a fluid element traveling at a representative mass-loss velocity of 10 km s^{-1} to traverse even one stellar radius. And such a distance is still only a small portion of Betelgeuse's circumstellar envelope. Therefore it would appear unlikely that the polarization changes can be attributed to the effect of material transport over a significant portion of the envelope. On the contrary, it would appear likely that the origin of the observed polarization variations is ultimately seated in the lower atmosphere - either being directly produced at or near the photosphere, or being indirectly produced through photospheric processes which control temporal variations in the anisotropic illumination of circumstellar polarization-producing material. In addition to these relatively short-term components, the polarization may also have a temporally invariant interstellar component and a circumstellar component which would be expected to undergo relatively long-term variations.

It is proposed that the polarization variations being reported here arise from the presence of one or at most a few large-scale convective cells on the stellar surface. The cell(s) would break the azimuthal symmetry of the stellar surface--resulting in a net polarization. The changes being reported here could arise from the waxing and waning of such photospheric features. Harrington (1969) has explicitly shown how polarization could be produced by Rayleigh scattering in the photospheres of late-type stars. A somewhat allied but less-favored explanation entails large-scale magnetically-controlled surface activity complexes akin to those reported in the Sun by Bumba and Howard (1965). An alternative explanation would be that the polarization emanates from a region located above but still relatively near the photosphere. Recent preliminary interferometric measurements give some tentative indications for the existance of features located at or near the surface as well as for features located at greater distances (e.g., Goldberg et al. 1982, Roddier and Roddier 1983).

The results of this study may be summarized as follows. The late-type supergiant Betelgeuse has evinced ordered (as opposed to stochastic) changes in its linear polarization B-band continuum over four consecutive observing seasons (Years 1979-1983). The origin of such changes has been shown to be ultimately seated in or near the photosphere. Such polarization changes are likely to be a manifestation of the waxing and waning of large-scale convective cells. The data presented here in concert with that obtained using other observational techniques should form the basis for constructing definitive atmosphere models of this and related stars.

References

Behr, A. 1959, Nachr. Akad. Wiss. Gottingen, Math.-Phys. Kl., 7, 185; Veroff, Gottingen, No. 126.

Bumba, V., and Howard, R. 1965, Ap. J., 141, 1502.

Dyck, H. 1968, A. J., 73, 688.

Dyck, H., and Jennings, M. 1971, A. J., 76, 689.

Goldberg, L., Hege, E., Hubbard, E., Strittmatter, P., and Cocke, W. 1982, in Second Cambridge Workshop on Cool Stars, Stellar Systems, and the Sun, ed. M. Giampapa and L. Golub, Smithsonian Astrophysical Observatory Special Report No. 392, Vol. I, p. 131.

Harrington, J. 1969, Ap. Letters, 3, 165.

Hayes, D. 1975, Ap. J. (Letters), 197, L55.

Hayes, D. 1978, Ap. J., <u>219</u>, 952.

Hayes, D. 1980, Ap. J. (Letters), <u>241</u>, L165.

Hayes, D. 1981, Pub. A.S.P., <u>93</u>, 752.

Hayes, D. 1983, submitted for publication.

Roddier, C., and Roddier, F. 1983, Ap. J. (Letters), <u>270</u>, L23.

Serkowski, K. 1971, <u>Kitt Peak Obs. Contr. 554</u>, p. 107.

Tinbergen, J., Greenberg, J., and de Jager, C. 1981, Astr. Ap., <u>95</u>, 215.

Weymann, R. 1977, in IAU Colloquium 42, <u>The Interaction of Variable Stars With Their Environment</u>, ed. R. Kippenhahn, J. Rahe and W. Strohmeier (Bamberg: Remeis-Sternwarte).

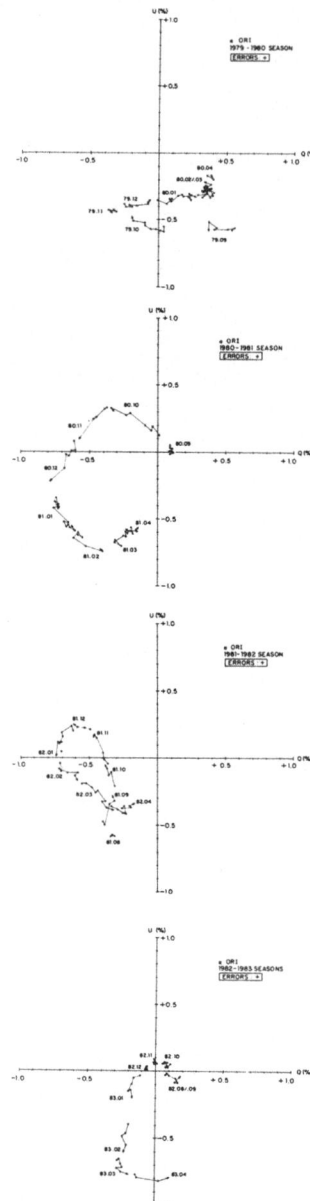

Figures 1 a-d. Linear polarization observations of Betelgeuse over four consecutive observing seasons: 1979-1980, 1980-1981, 1981-1982, and 1982-1983. The polarization is plotted in the equatorial Stokes parameter (Q-U) reference frame, with the data trains being broken into monthly segments. In a polar coordinate system the distance from the origin is the percentage polarization P while the angular coordinate is the double position angle 2θ as measured counterclockwise from the positive Q axis (which corresponds to the equatorial North-South direction). The cited errors are the 2σ standard deviations derived from Poisson photon-counts.

PHOTOIONIZATION OF THE CIRCUMSTELLAR SHELL OF BETELGEUSE BY CHROMOSPHERIC RADIATION

A.E. Glassgold and P.J. Huggins
New York University Physics Department
New York, NY, 10003

INTRODUCTION

The occurrence of a chromosphere can have a profound effect on the cool circumstellar envelope surrounding a star such as α Ori. Passage of a wind through the warm, partially ionized chromosphere reduces the gas to atoms or atomic ions. The dominant ion in the chromosphere is H^+, which recombines when photoionization from excited states freezes out. The gas does not become neutral, however, because the chromospheric UV radiation is sufficiently strong to maintain a significant fraction of the heavy ions almost fully ionized throughout the bulk of the CS shell. We have analyzed the physical processes which enter into the dynamic balance between photoionization and recombination in the CS shell of α Ori by solving appropriate time-dependent equations which generalize the original treatment of this problem by Weymann (1962). In addition to describing the physical processes affecting the ionization of the CS shell, we present preliminary results on the interpretation of the measured spatial profile of the KI density and on the determination of the mass loss rate, previously discussed by Jura and Morris (1981).

PHOTOIONIZATION AND RECOMBINATION

Models of the chromosphere (e.g. Avrett and Hartmann 1983) suggest that the photoionization rate of H becomes less than the wind expansion rate just outside the position of maximum chromospheric temperature. Beyond this point, our solutions of the time dependent recombination equation show that the electron fraction decays rapidly to x_i, the abundance of heavy ions. The

ionization of the heavy atoms is maintained by the strong
chromospheric UV radiation, which has been measured by various
satellites, as discussed previously by Clegg et al. (1983). We
have calculated the photoionization rates for the most abundant
atomic species and the atoms and ions of diagnostic interest; at
the surface of the star they are typically 1 million times larger
than for the interstellar radiaton field, in agreement with Clegg
et al. For elements with high or low ionization potentials, the
rates may be a factor of 10 smaller or larger, as in the cases of
C and K. The calculations imply that the electron fraction is
constant throughout a large portion of the outer envelope of α
Ori, and that it is independent of the initial value in the
chromosphere.

KI DENSITY PROFILE

The analysis of time-dependent photoionization and
recombination theory for KI shows that the KI density can be
represented by

$$n_{KI}(r) = B \frac{\alpha(T(r)) n_e(r)}{G_a(r)} \xi_K n(r)$$

where n(r)=c/r^2 is the hydrogen density for a uniformly expanding
wind (c = 3(37)cm^{-1} M$_{-5}$/v$_6$), T is the gas temperature, α is the K
recombination coefficient, n$_e$ is the electron density, G$_a$(r) is
the ionization rate due to chromospheric radiation, and ξ_K is the
gaseous abundance of potassium. Throughout most of the envelope
where chromospheric ionization from the ground state of KI
dominates, the factor B(r) = 1. When the radial dependence of all
the quantities are inserted in this equation, we find that

$$n_{KI}(r) \sim \alpha(T(r)) r^{-2}$$

i.e., the KI density will depart from an inverse square law
because of the temperature dependence of the recombination
coefficient. Little is known about the temperature profile of
the outer envelope of α Ori, but we can estimate the effect by
considering adiabatic cooling. In this case T(r) $\propto$ r$^{-4/3}$, and
$n_{KI} \propto r^{-1.1}$. Honeycutt et al. (1980) have measured the KI profile

from 15" to 50", and deduced a slope of 1.65±0.2, which is significantly steeper than the prediction for an adiabatic envelope. This suggests that the temperature profile is determined by a balance between heating and cooling, along the lines discussed by Goldreich and Scoville (1976).

At smaller (< 10 R_*) and larger (> 1000 R_*) distances, the KI density decreases more rapidly than inverse square due to increased ionization and photoionization, respectively. Close to the chromosphere it is essential to include collisional and photoionization from the ground state and potoionization from the first excited state. Near the dedge of the shell, photoionization by the interstellar radiation field dominates, as discussed by Jura and Morris (1981).

An estimate of the mass loss rate can be made using the above theory and assuming that the K abundance is solar. If the distance to the star is 200 pc, the extreme range permitted by various uncertainties is 1(-6) to 1(-5) $M_\odot$/yr, and our best current estimate is 3(-6) $M_\odot$/yr to within a factor of 2. This value is 5 times smaller than the result of Jura and Morris.

This research has been supported by NASA grant NGR-016-33-196 (AEG), NSG grant AST 82-16484 (PJH), and an Alfred P. Sloan Foundation Research Fellowship to PJH.

REFERENCES

Avrett, E. and Hartmann. L. 1983, submitted to Ap.J.

Clegg, R. E. et al. 1983, A.A. 203, 125.

Honeycutt, R.K., et al. 1980, Ap.J. 203, 565.

Jura, M. and Morris, M. 1981, Ap.J. 251, 181.

Weymann, R.J. 1962, Ap.J. 136, 844.

BETTER DETERMINATIONS OF MASS LOSS RATES FOR RED GIANTS AND SUPERGIANTS

Stephen A. Drake and Jeffrey L. Linsky[*]
Joint Institute for Laboratory Astrophysics, University of Colorado and National
Bureau of Standards, Boulder, Colorado 80309

I. INTRODUCTION

Mass loss in the form of stellar winds is of great relevance to the study of stellar atmospheres, stellar structure and evolution, and the interstellar medium, and progress in these areas would greatly benefit from reliable estimates of mass loss rates $\dot{M}$ for many stars. Deutsch (1956) first showed that significant mass loss does occur for cool, luminous stars, but order of magnitude or more disagreements in the value of $\dot{M}$ between different studies of the same star are fairly typical. Goldberg (1979) and Zuckerman (1980) have reviewed the optical and infrared techniques by which the vast majority of $\dot{M}$ estimates have been obtained. In this paper we discuss the "new" methods which have become available in the last decade or so utilizing other wavelength regions. We will limit this study to giants and supergiants of spectral type G to mid M, and thus we will not mention $\dot{M}$ measurements in late M stars using molecular emission lines such as CO(J=2-1). We will also exclude techniques that are only appropriate for binary systems of known orbital parameters (e.g. Reimers 1977).

II. ULTRAVIOLET TECHNIQUES

The circumstellar features in lines like Ca II K, Mg II k, and Fe II UV1 that are often observed in cool, evolved stars can be modeled in a manner similar to the optical lines. The major advantages of these UV lines over the optical are that they are resonance lines of the dominant stages of ionization of relatively abundant elements, and that, in the case of Mg II, blue-shifted absorption features are observed in giants as hot as K2, whereas the optical circumstellar lines are normally seen only in M giants. One disadvantage, principally in analyzing the Ca and Mg lines, is the presence of complicating chromospheric emission and central self-reversal.

Wilson (1960) argued from the absence of any blue-shifted Ca II absorption in most K giants, that $\dot{M} \lesssim 10^{-10}$ $(R_*/20\ R_\odot)$ $M_\odot$ yr^{-1}. Reimers (1973) derived $\dot{M}$ estimates for several G-K supergiants and M giants from their observed Ca II symmetries. Drake and Linsky (1982) have presented preliminary results of a full model atmosphere calculation to model the Mg II k profile in α Boo (K2 IIIp) by solving the radiative transfer in the co-moving frame incorporating expansion, spherical symmetry, and partial redistribution effects. We found $\dot{M} \simeq 10^{-10}$ $M_\odot$ yr^{-1} for this star, in good agreement with Wilson's estimate quoted above.

---

[*]Staff Member, Quantum Physics Division, National Bureau of Standards.

III. MICROWAVE CONTINUUM AND LINE TECHNIQUES

Continuum free-free radiation is emitted from the partially ionized warm plasma in a stellar wind. Assuming a spherical, constant velocity (V_W), optically thick ionized wind, the mass loss rate in $M_\odot$ yr^{-1} is

$$\dot{M}_{FF} = 4 \times 10^{-9} \left(\frac{S_\nu}{mJy}\right)^{0.75} D^{1.5} \left(\frac{V_W}{km\ s^{-1}}\right)^{1.0} \left(\frac{\nu}{4.9\ GHz}\right)^{-0.45} \left(\frac{T_e}{10^4\ K}\right)^{-0.075} , \quad (1)$$

(Wright and Barlow 1975), where S_ν is the observed flux and D is the distance in kiloparsecs. For a cool star, $\dot{M}_{FF}$ is the mass loss rate of the <u>ionized</u> material only. Extensive surveys at cm wavelengths with ~10 mJy sensitivity detected only one cool, luminous, non-binary star, α Ori (M2 Iab) (Altenhoff <u>et al</u>. 1979). VLA observations at these wavelengths in the last six years with a sensitivity of 0.1 mJy at 6 cm have added a handful more of detected red giants (Hjellming and Newell 1983; Drake and Linsky 1983) including α^1Sco (M1.5 Ib) and α^1Her (M5 II). Mass loss rates derived from these observations are presented in Table 1. Because of the high sensitivity of the VLA, many of the non-detections yield significant upper limits to $\dot{M}$, particularly for the nearest giants. The major sources of error in these estimates are probably ' the assumptions of constant velocity and ionization fraction: the observed radio spectra of α Ori and α^1Sco have spectra significantly steeper than the $\nu^{0.6}$ predicted by eq. (2), indicating the need for a more realistic radial dependence in the models.

From a spherical constant velocity <u>neutral</u> wind, atomic hydrogen in the ground state will emit at 21 cm due to the well-known spin flip transition. Assuming that the radiation is optically thin and that the inner radius of the H I emitting region is small compared to the outer radius, it can be shown that:

$$\dot{M}_{H\ I}/M_\odot\ yr^{-1} = 2 \times 10^{-4} \left(\frac{S_\nu}{mJy}\right)^{1.0} \left(\frac{D}{kpc}\right)^{1.0} \left(\frac{V_\omega}{km\ s^{-1}}\right)^{2.0} \left(\frac{\phi}{maS}\right)^{-1.0} \quad (2)$$

where ϕ is the observed angular diameter for a detected source and the beam size (HPBW) for a non-detection. No cool, luminous star has been detected as a HI source (e.g. Knapp and Bowers 1983; Zuckerman <u>et al</u>. 1980) and most of the upper limits to $\dot{M}$ are uninteresting, except in the case of the M supergiants like α Ori (see Table 1), where they can be significantly less than the largest optically obtained mass loss rates.

IV. X-RAY CONSTRAINTS ON CORONAL MASS LOSS

From an observed Einstein X-ray luminosity or upper limit, one can calculate the volume emission measure of coronal ($T \gtrsim 10^6$ K) material, assuming that the emission is due to thermal processes only. Making the <u>maximal</u> assumption that all the X-rays are coming from a stellar wind region (which, of course, is very unlikely given the fact that the bulk of the X-radiation from the Sun comes from essentially static active regions), a strong <u>upper</u> limit can be obtained for the mass loss rate of "coronal" material, if one adopts the usual assumptions of spherical symmetry and constant wind speed. Applying this technique to the Sun, we obtain $\dot{M}_{X-ray} \lesssim 10^{-11} M_{\odot}$ yr^{-1}, two orders of magnitude above the usually quoted solar rate. In Table 1, we tabulate other X-ray upper limits to $\dot{M}$.

V. CONCLUSIONS

We have summarized the large variety of methods now available to measure mass loss from red giants. In Table 1 we have assembled mass loss rates for a sample of stars, obtained using both the techniques discussed here and the more standard optical and infrared methods. We also tabulate the values of $\dot{M}$ predicted by the Reimers' scaling law for mass loss from cool giants (Reimers 1975,1977). Until more data are obtained (e.g. by making observations and/or model calculations to replace the question marks in Table 1 with numbers), it is difficult to know how accurate, for example, the widely used Reimers' scaling formula is. (The data for β Dra, for example, suggest that the Reimers' law may not be appropriate for G stars.) We believe that ultraviolet and radio methods hold great promise for producing reliable mass loss rates for these stars.

REFERENCES

Altenhoff, W. J., Oster, L., and Wendker, H. J. 1979, Astr. Ap., 73, L21.
Deutsch, A. J. 1956, Ap. J., 123, 210.
Drake, S. A. and Linsky, J. L. 1982, Bull. A.A.S., 14, 894.
_____. 1983, Ap. J. (Lett.), in press.
Goldberg, L. 1979, Quant. J. R. A. S., 20, 361.
Hjellmung, R. M. and Newell, R. T. 1983, Ap. J. (Lett.), in press.
Knapp, G. R. and Bowers, P. F. 1983, Ap. J., 266, 701.
Reimers, D. 1973, Astr. Ap., 24, 79.
_____. 1975, Mem. Roy. Soc. Sci Liege, 8, 369.
_____. 1977, Astr. Ap., 61, 217 [1978, Astr. Ap., 67, 161 (Erratum)].
Wilson, O. C. 1960, Ap. J., 132, 136.
Wright, A. E. and Barlow, M. J. 1975, M.N.R.A.S., 170, 41.
Zuckerman, B. 1980, Ann. Rev. Astr. Ap., 18, 263.
Zuckerman, B., Terzian, Y., and Silverglate, P. 1980, Ap. J., 241, 1014.

Table 1. Comparison of Mass Loss Rates Obtained by Various Techniques

Star	Spectral Type	$\dot{M}_{opt+IR}$	$\dot{M}_{Reimers}$	$\dot{M}_{X-ray}$	$\dot{M}_{UV}$	$\dot{M}_{6\ cm}$	$\dot{M}_{HI}$
β Dra	G2 Ib-II	---	1.9(-8)	≲6.9(-10)	---	≲5.8(-9)	---
β Gem	K0 III	---	1.2(-10)	≲3.4(-12)	---	1.2(-10)	---
β Cet	K0 III	---	1.0(-10)	≲6.3(-11)	---	≲1.8(-10)	---
α Boo	K2 IIIp	---	1.4(-9)	≲1.3(-12)	1.0(-10)	7.5(-11)	?
ε Peg	K2 Ib	2→9(-6)	5.7(-8)	≲7.3(-11)	?	≲3.2(-9)	?
ι Aur	K3 II	?	6.1(-9)	?	1→2(-8)	≲2.0(-9)	?
α Tau	K5 III	?	5.3(-9)	≲5.0(-12)	?	≲1.5(-10)	≲1.8(-7)
α¹ Sco	M1.5 Iab	1(-7)→2(-6)	9.8(-7)	≲1.3(-10)	6.7(-6)	8.7(-9)	?
α Ori	M2 Iab	1(-7)→3(-5)	1.2(-6)	≲2.1(-10)	?	8.3(-9)	≲5.0(-6)
μ Cep	M2 Ia	1(-6)→4(-4)	6.8(-6)	?	?	≲3.0(-8)	≲1.2(-4)
α Her	M5 II	2(-8)→1(-6)	1.2(-7)	?	?	1.7(-9)	?

SULPHUR I EMISSION LINES IN THE EUV SPECTRUM OF β GRU (M3 II)

P.G. Judge

Department of Theoretical Physics, 1 Keble Road, Oxford.

Summary. The excitation mechanisms producing the strong S I lines observed in the EUV spectra of late-type giants are discussed. For the case of β Gru it is shown that the broad H Ly-α line plays an important role in producing the observed multiplets.

Introduction

S I was first noted as a strong contributor to the EUV flux in late type giants by Brown and Jordan (1980 - henceforth BJ). Their short wavelength IUE spectra of β Gru and α Tau (K 5 III) (BJ figs. 1,2) indicate the importance of the S I lines in these stars, which show no evidence for temperatures greater than 10^4 K. Brown, Ferraz and Jordan (1980) noted that UV(1) was optically thin, whilst UV(2) was optically thick, which places limits on the S I column density. (See fig. 1 for labled multiplets). Stencel *et al.* (1981) measured the electron density using the C II multiplet at ~ 2325 Å. Thus:

$$10^{14} \text{ cm}^{-2} < \int n_{SI} \, dh < 10^{17} \text{ cm}^{-2} \tag{1}$$

$$n_e \sim 10^8 \text{ cm}^{-3} \tag{2}$$

Here we consider processes which can account for the observed absolute and relative strengths of the S I multiplets including fluxes obtained from a new high resolution spectrum (Engvold & Stencel, private communication).

Calculations

The table lists atomic data from Wiese, Smith and Miles (1969) and observed fluxes $F_\oplus$ from the new observations:

Multiplet	λ average [Å]	gf	Flux at earth $F_\oplus$ $(10^{-13} \text{erg cm}^{-2} \text{ sec}^{-1})$
UV(1)	1907	2.3×10^{-4}	7.5
UV(2)	1814	1.1	23
UV(3)	1478	0.85	11.6
UV(9)	1299	1.3	~12
All others			< 1.0

The line flux is given by equation (3), where the symbols have their usual meanings (Jordan and Brown, 1981). The bracketed expression is deduced and compared to the theoretical rates calculated below.

353

$$F_\phi = \frac{1}{2} \frac{hc}{\lambda} \frac{1}{4\pi} \Omega_* \int_{\Delta h} n_{SI} \frac{[n_2 \, A_{21}]}{n_{SI}} \, dh \qquad (3)$$

1. *Excitation by Electron Collisions*. The semiempiricle formula of Van Regemorter (1962) for neutrals has been used for the allowed transitions. All others are estimated using a collision strength of 1. Results show that the total rates are rather low, and that UV(3) is underpopulated relative to the other multiplets by a factor of ~ 100. This discrepancy cannot be explained by uncertainties in collision strengths.

2. *Photoexcitation to high-n states by H Ly-α*. As suggested in BJ, the effects of H Ly-α on n~12 states observed by Tondello (1972) have been studied, a typical pumped transition being shown in fig. 1. Several hundred oscillator strengths have been compiled in order to obtain net excitation rates for UV(2) and UV(5). The physics of the Ly-α line has not yet been investigated, but a minimum for the photoexcitation rates has been calculated using the *observed* Ly-α flux, and this yields values comparable with the collisional rates above. Fig. 2 shows the combined effects of the two processes and how the S I lines might be used as diagnostics once the atomic data and Ly-α physics are better known. A rough upper limit obtained by taking Ly-α to be thermalised at ~ 6000 K, yields rates of which are too high be 2 orders of magnitude.

3. *Excitation by Recombinations*. Fig. 2 shows estimates of the radiative rates (Aldrovandi & Péquignot 1973) and dielectronic rates for UV(3) (in the low temperature regime of Storey, 1981) where the ion balance is determined by Ly-α pumping levels above the $(^4S^o)$ limit (see fig. 1). Details will be published elsewhere.

Conclusions

The important excitation processes for all the S I EUV multiplets have been identified. Modelling work using these must await determination of the necessary collision and oscillator strengths.

Aldrovandi S.M.V., Péquignot D., *Astron. Astrophys.* 25, 321 (1973)
Brown A., Ferraz M., Jordan C., in *The First Two Years of IUE*, NASA Conference
    Publication, 2171, 297 (1980)
Brown A., Jordan C., MNRAS 191, 37P (1981)
Jordan C., Brown A., in *Solar Phenomena in Stars and Stellar Systems*, Proceedings of
    NATO Advanced Study Institute, Bonas, p.199, pub. D. Reidel, 1980
Stencel R.E., Linsky J.L., Brown A., Jordan C., Carpenter K.G., Wing R.F., Czyzak S.,
    MNRAS 195 27P (1981)
Storey P.J., MNRAS 195 27P (1981)
Tondello G., Ap.J. 172, 771 (1972)
Wiese W.L., Smith M.W., Miles B.M., NSRDS-NBS 22 (1969)
Van Regemorter H., Ap.J. 132, 906 (1962)

Figure 1

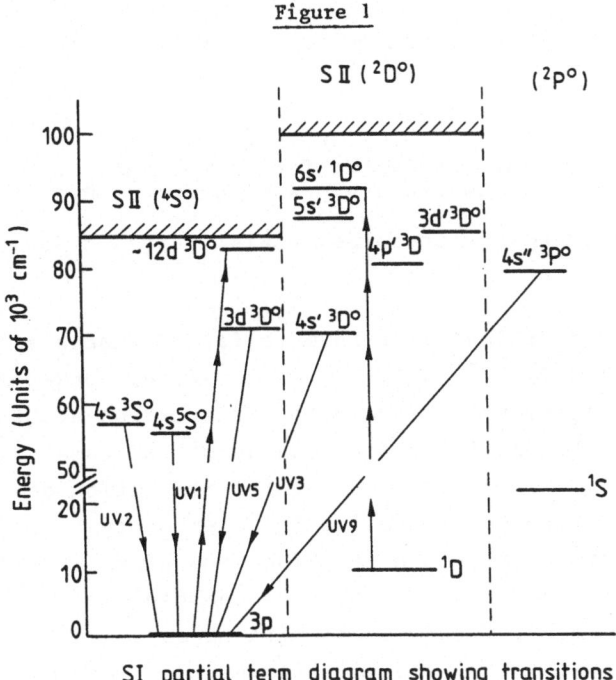

SI partial term diagram showing transitions

Figure 2

Key:

– – **2** – –	Combined collisional + photo rate for UV(2) etc.
—————	Radiative recombination rate to all upper levels
–·–**3**–·–	Estimate of dielectronic rate for UV(3)

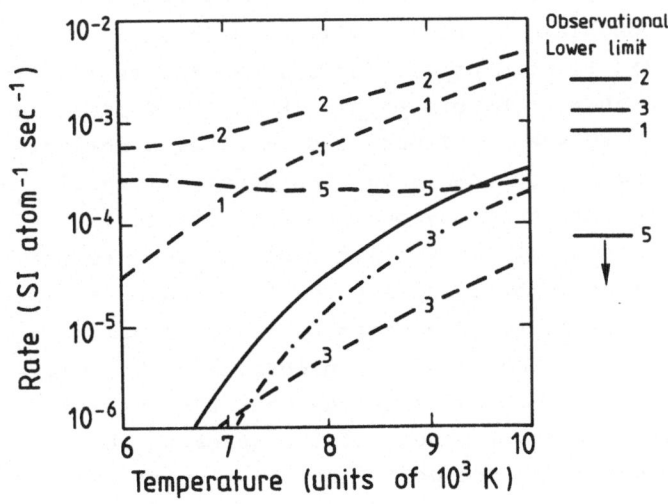

OH MASER EMISSION AS A PROBE OF CIRCUMSTELLAR ENVELOPE STRUCTURE

P. F. Bowers
E. O. Hulburt Center for Space Research
Naval Research Laboratory, Washington, D. C.   20375
and
Sachs/Freeman Associates, Bowie, Maryland 20715

Interferometric observations of circumstellar OH masers associated with cool, high mass loss ($\gtrsim 10^{-6}$ M_o y^{-1}), oxygen-rich, long-period variables can, in principle, provide valuable information about the geometry and kinematics of the envelopes and thus about the mass loss process in these evolved stars. Previous high angular resolution data could be obtained only with Very Long Baseline Interferometry. These data were useful to determine the relative locations and sizes of individual, bright maser spots, but could not provide unambiguously the size, geometry, or kinematics of the maser region because of the limited spatial sensitivity (most of the emission usually was over-resolved) and the limited spectral sensitivity (only the brightest, most complex objects could be observed easily). Newly developed intermediate resolution ($\sim$ 1"), spectral line interferometers such as the VLA and MERLIN can detect weak, extended emission and (for the VLA) can determine accurate absolute positions of all emission features. In this paper I briefly summarize recent results in the context of what maser data can or cannot tell us about circumstellar structure.

VLA or MERLIN maps of OH/IR stars with the classic double-peaked velocity structure have been obtained by Baud (1981), Norris et al. (1982), Bowers et al. (1983), Diamond et al. (1983), and Bowers (1984). The most extensive study is that of Bowers et al. (BJS) who mapped the 1612 MHz OH distribution for 20 stars. Figure 1 summarizes a comparison of their data to the commonly assumed expanding spherical shell (ESS) model. This model predicts that the peak emission features should be positionally coincident at the stellar position because the line-of-sight velocity gradient is smallest and the maser gain is largest in the direction of the star. The emission region should appear as ring-like structures at intermediate velocities with a maximum size at the stellar velocity V_o. BJS conclude that the ESS model appears to be valid, to first order, for most of the stars in their sample. This conclusion is strongest for well resolved sources in which a circular outline of the complete shell is seen near the

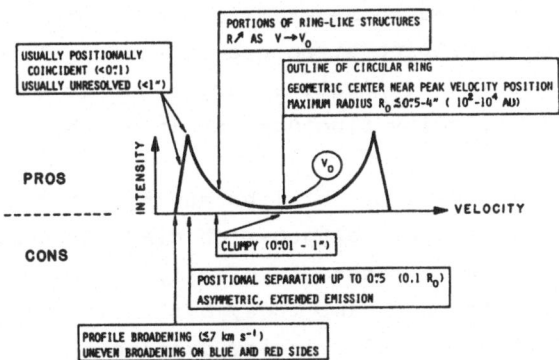

Figure 1. Points of agreement ("Pros") and disagreement ("Cons") in comparison of BJS data to the ESS model for sources whose underlying geometry is clearly spherical based on the presence of a circular ring structure detected near the stellar radial velocity V_o.

stellar velocity. About half the stars in their sample are unresolved or marginally resolved. In these cases the underlying geometry cannot be examined, but in virtually every case the positional coincidence to within 0".1 of peak or extreme velocity features is consistent with the ESS model.

Nevertheless, there can be significant deviations from the simple ESS model, as indicated in Figure 1. Each of these deviations could be caused by a geometric or kinematic effect, and the interpretation of the data can lead to significantly different conclusions. For example, positional offsets of low and high velocity emission and the presence of emission at extreme velocities outside the peaks can be produced in a tilted expanding disk configuration (Morris and Bowers 1980) or in an ESS model where there is velocity streaming (BJS); the implications for mass loss mechanisms thus can be quite different. In addition, maser emission from some sources is extremely clumpy or asymmetric, providing valuable clues but not unambiguous answers about the underlying geometry (Bowers 1984).

Because the deviations listed in Figure 1 are present in sources whose underlying geometry is clearly spherical, BJS suggest that density clumping ($\sim 10^{15}$-10^{16} cm) and velocity streaming (~ 1 to 2 km s^{-1}) play an important role in determining the observed OH distribution. Density/velocity perturbations may be produced by asymmetries in the mass ejection from the star, by interaction of the circumstellar material with the ambient medium, or by anistropies in the interstellar

UV radiation field. It can be difficult to distinguish between density and velocity effects based on maser data alone, thus making it difficult to deduce the underlying structure of the envelope. An asymmetric emission distribution may indicate a larger density in one portion of the shell or larger velocity fluctuations in another portion.

Density/velocity perturbations notwithstanding, this discussion does not imply that all late-type, oxygen-rich stars lose mass in a spherical outflow expanding at a constant velocity in the outer envelope. Expansion is clearly the dominant mode, but small rotational components (Norris et al. 1982) or radial velocity gradients (Bowers 1984) might be present in some stars. OH maser data provide evidence for non-spherical (possibly ellipsoidal) expanding geometries for at least two stars: the peculiar supergiant VY CMa (BJS) and the bipolar nebula OH231.8+4.2 (Morris et al. 1982; Bowers and Morris 1984). Diamond et al. (1983) also propose a disk-like geometry for the peculiar supergiant IRC+10420, but more extensive OH observations by Bowers (1984) do not support their model.

The OH(1612) masers usually are located at radii $\gtrsim$ 1000 AU. The apparent validity of the ESS model at this distance suggests that the mass loss mechanism for most oxygen-rich red giants does not produce strong deviations from spherical outflow when averaged over time. This does not eliminate possibilities such as sporadic mass loss which distributes material in different directions at different times (Bowers 1984) or mass loss confined to a modest (but not small) opening angle. Investigations to examine the inner structure of the envelope by means of maser transitions associated with H_2O (Johnston et al. 1984) and SiO (Lane 1984) are in progress.

REFERENCES

Baud, B. 1981, Ap. J. (Letters), 250, L79.
Bowers, P. F. 1984, Ap. J., 279, in press.
Bowers, P. F., Johnston, K. J., and Spencer, J. H. 1983, Ap. J., 274, in press (BJS).
Bowers, P. F. and Morris, M. 1984, Ap. J., 276, in press.
Diamond, P. J., Norris, R. P., and Booth, R. S. 1983, Astr. Ap. Letters, 124, L4.
Johnston, K. J., Spencer, J. H., and Bowers, P. F. 1984, in preparation.
Lane, A. P. 1984, I.A.U. Symp. #110, in press.
Morris, M. and Bowers, P. F. 1980, A. J., 85, 724.
Morris, M., Bowers, P. F., and Turner, B. E. 1982, Ap. J., 259, 625.
Norris, R. P., Diamond, P. J., and Booth, R. S. 1982, Nature, 299, 131.

PRELIMINARY RESULTS FROM STUDIES OF HIGH RESOLUTION LINE SPECTRA OF LATE-TYPE GIANT AND SUPERGIANT STARS

O.Engvold[1], O.Kjeldseth Moe[1], E.Jensen[1], C.Jordan[2],
R.Stencel[3], and J.Linsky[4].
[1]Institute of Theoretical Astrophysics, University of Oslo
[2]Department of Theoretical Physics, Oxford University
[3]NASA Headquarters, Washington D.C. [4]JILA, University of Colorado

High resolution EUV emission line spectra ($\lambda\lambda$1200–2100Å) of four giant and supergiant stars of spectral types K1–M3 have been obtained through joint NASA-ESA observations with the International Ultraviolet Explorer (IUE) satellite (Boggess et al., 1978). The exposure times were 13–16 hours in order to bring out a number of weak emission lines in this spectral region. The high resolution mode of the instrument is indispensable in order to make reliable line identifications. Confidence in the identification of weak lines is enhanced by an intercomparison of equal quality spectra of reasonably similar stars. For this reason we include here the spectra of three more stars (αBoo, αTrA, αTau) which have been discussed earlier (cf. Brown and Jordan, 1980). Their exposure times were shorter and range from 6 to 9 hours.

The aim of this initial discussion of our data is to test line excitation mechanisms in the atmospheres of giant and supergiant stars. The property of supergiants that sets them apart from dwarfs and giants is their low gravities which leads to lower chromospheric densities. This effect is partly offset by the much larger column mass densities in which the supergiant chromospheres are embedded.

The stars of spectral type K radiate strongly in the resonance triplet O I λ1302–06Å uv 1. The lines appear especially bright in Arcturus (αBoo). One assumes that the lines are formed by resonance fluorescence pumped by H Lyβ (Haisch et al., 1977). It is expected that in the chromospheres of a wide range of stars the O I resonance lines will be formed by the same mechanism, and that the line strengths provide an indirect measure of the Lyβ radiation field. The O I lines weaken noticeably in the M-stars and they are virtually absent in the spectrum of αOri.

In the stars of spectral type K2 III and later one notes a bright line around λ1640–41Å coinciding with the intersystem O I λ1641.3Å (Brown and Jordan, 1980). The O I line at λ1641Å has its upper level in common with the strong uv 1 triplet. The intersystem line weakens noticeably from K to M-stars, yet somewhat less than the uv 1 lines.

An abrubt decline of the O I triplet line flux from K5 III to M3 II is evidently caused primarily by a decrease in the Lyβ radiative

Line flux in SWP/HI IUE spectra of late type stars (10^{-14} erg cm^{-2} s^{-1})

Line (Å)	βCet K1 III	αBoo K2 III	αTrA K4 III	αTau K5 III	λVel K5 Ib	αOri M2 Iab	βGru M3 II
1206.522 Si III	<3	–	–	–	–	–	–
1215.670 H I	190	980	–	>280	–	128	–
1294.534 Si III	5	–	–	–	–	–	–
1295.653 S I	18	95	73	41	85	10	20
1296.174 S I	6	80	52	28	56	11	36
1296.726 Si III	30	–	–	–	–	–	–
1302.169 O I	40	(340	(114	(131	185	–	43
1302.337 S I	17				–	5	–
1302.863 S I	4	70	36	26	59	–	14
1303.111 S I	(2)	(2)	–	–	–	–	–
1304.858 O I	101	660	205	224	180	–	54
1305.883 S I	–	–	–	–	–	–	–
1306.029 O I	111	833	214	228	169	–	59
1334.532 C II	14	11	–	–	–	–	–
1335.708 C II	–	10	–	–	–	–	–
1355.598 O I	21	85	18	24	–	–	–
1358.512 O I	6	31	–	–	–	–	–
1393.755 Si IV	30	–	–	–	–	–	–
1402.770 Si IV	14	–	–	–	–	–	–
1474.380 S I	7	12	6	–	–	28	53
1533.432 Si II	5	–	–	–	–	–	–
1548.185 C IV	9	–	–	–	–	–	–
1550.774 C IV	19	–	–	–	–	–	–
1640.474 He II	29	–	–	–	–	–	–
1641.31 O I	–	76	35	37	47	51	29
1785.262 Fe II	~10	~17	~12	~12	10	25	11
1786.738 Fe II	–	–	–	–	–	23	12
1788.072 Fe II	~10	~15	~12	~13	11	22	13
1807.311 S I	–	12	12	13	21	280	60
1808.012 Si II	50	78	23	38	20	40	20
1816.928 Si II	156	20	(83	74	61	(130)	66
1817.451 Si II	25	47		24	–	–	6
1820.323 S I	–	15	17	37	31	430	60
1826.245 S I	–	11	(30)	(9)	27	540	87
1892.030 Si III	83	16	33	–	–	–	–
1900.286 S I	2	45	16	19	19	178	44
1908.734 C III	17	23	45	–	–	–	–
1914.698 S I	–	30	17	(10)	11	164	26
1993.620 C I	6	91	17	25	17	124	26

flux, and secondly by self absorption in the extended atmospheres. The O I λ1303Å of the triplet lines may be attenuated by interstellar oxygen (Basri et al., 1981).

The "solar type" βCet seems to contain possibly a very weak trace of the O I λ1641.3Å line. βCet shows instead the line of He II λ1640.474Å. This line is an indicator of the strength of the stellar coronal XUV flux (Hartman et al., 1980). The absence of the He II λ1640Å line in the other stars of our sample implies that these stars have not developed a hot corona (Linsky, 1980).

The Fe II λ1785-88Å uv 191 triplet appears commonly in the spectra of binaries of red supergiant and hot B dwarf (Stencel et al., 1979; Hagen et al., 1980). Highly excited Fe II lines are then formed through scattering and fluorescence of the B star radiation in the atmosphere of the red star. Possible mechanisms for the formation of the Fe II uv 191 triplet in the atmospheres of single, cool supergiants, and a puzzling absence of the central component in the K-star spectra, are discussed by Engvold et al., (1983).

The S I λ1807-26Å uv 2 multiplet is strong and becomes the dominant spectral feature for M-stars. These triplet lines become increasingly saturated with increasing column mass of the stellar atmospheres. The flux of the optically thick S I uv 2 multiplet increases relative to the optically thin intersystem doublet S I uv I λ1900Å and λ1914Å by a factor of 3 from early K to early M-stars. The observed variation may reflect differences in relative statification in the chromospheres going from K to M-stars.

The S I lines at λ1295.635Å, λ1296.174Å, and λ1302.863Å of multiplet uv 9 appear in the spectra of all the stars of our sample. Two other lines of this multiplet are located within two of the O I uv 1 triplet lines, and the S I uv 9 multiplet is pumped by O I, which itself is excited by Lyβ.

The flux in the Si II uv 1 triplet λ1808-17Å is noticeably reduced in the spectra of M-stars. The Si II λ1808Å line is barely detectable in αOri. The strong emission feature around λ1817Å in the spectrum of αOri cannot be attributed to Si II alone.

Centrally reversed, strongly asymmetric O I uv 1 triplet lines are evidence for an outward directed mass flow in the outher atmospheric layers. The shift of the central absorption components corresponds to relative velocities in the range 10 to 30 km/s for the K5 III-M3 II stars.

References

Allen,C.W.:1973, Astrophysical Quantities, 3rd edn. Athlone Press, London.
Ayres,T.R., Linsky,J.L., Basri,G.S., Landsman,W., Henry,R.C., Moos,H.W. and Stencel,R.E.:1982, Astrophys.J. 256, 550.
Basri,G.S., Linsky,J.L. and Eriksson,K.:1980, Astrophys.J. 251, 162.
Bogess et al.:1978, Nature 275, 377.
Brown,A. and Jordan,C.:1980, M.N.R.A.S. 191, 37P.
Engvold,O., Jensen,E. and Kjeldseth Moe.O.:1983, Paper presented at Nordic Astronomy Meeting, 15-17 August, University of Oslo.
Hagen,W., Black,J.H., Dupree,A.K. and Holm,A.V.:1980, Astrophys.J. 238, 203.
Haisch,B.M., Linsky,J.L., Weinstein,A. and Shine,R.A.:1977, Astrophys.J. 214, 785.
Hartmann,L., Dupree,A.K. and Raymond,J.C.:1980, Astrophys.J. 236, L143.
Linsky,J.L.:1980, Ann.Rev.Astr.Ap. 18, 439.
Stencel,R.E., Kondo,Y., Bernat,A.P. and McCluskey,G.E.:1979, Astrophys.J. 233, 621.

DISCOVERY OF A HOT, INTERACTIVE COMPANION TO THE S STAR HD 35155

H.R. Johnson[a]

Indiana University, Bloomington, Indiana

NAS-NRC Senior Fellow, NASA Ames Research Center, Moffett Field, California

and

T.B. Ake[a]

Computer Sciences Corporation, NASA-Goddard Space Flight Center
Greenbelt, Maryland

[a]Guest Observer with the International Ultraviolet Explorer Satellite

Abstract

During an IUE investigation of the chromospheres of S stars, we detected strong emission lines of such ions as C IV and Si IV and excess continuous emission in an SWP exposure of HD 35155 (S3,2). These emission lines indicate a hot gas with a state of ionization exceeding that found in the recently studied Ba star systems of Zeta Cap and 56 Peg, both of which have white dwarf companions. IUE observations 7 months later showed the continuum had decreased by about 0.5 mag. and the emission lines by 15-50%, the high excitation lines decreasing most rapidly.

Introduction and Observations

HD 35155 is a cool giant star of type S3,2 (Keenan and Boeshaar 1980) or S2.5 Zr 1 Ti 3 (Ake 1979). Broad-band photometry has been obtained by Eggen (1972), who also estimated E(B-V) = 0.03. The star is present in the IRC catalog (Wing and Yorka 1977). Eggen found HD 35155 to be one of only two photometrically constant S stars among the sixteen he studied although Wing (1967) claimed variations. A fairly secure value of T_{eff} = 3660 K has been recently obtained (Bregman et. al. 1983) from model fitting to infrared spectrophotometry and from the Blackwell-Shallis method. A complete low-resolution IUE spectrum of HD 35155, made of nearly simultaneous LWR and SWP spectrograms, is shown in Figure 1. A portion of a high-resolution LWR spectrum near 2800 A is shown in Figure 2.

Results and Discussion

The energy flux of the strong emission lines in Figure 1 (in units of 10^{-13} erg/cm^2 s) are: N V (1237 A): 1.2; OI (1301 A): 4.1; CII (1335 A): 1.4; Si IV (1400 A): 6.5; C IV (1550 A): 29.4; Si III (1892 A): 2.0; C III (1909 A): 3.5; C II (2325 A): 2.5; Mg II (2800 A): 30.0. The surprising weakness of He II 1640 suggests that the flux of soft X-rays, which is the source of He excitation, is low. The strength of C III relative to Si III indicates that the lines are emitted in a region of relatively low density. Ionization in HD 35155 significantly exceeds that found in the Ba star systems Zeta Cap (Bohm-Vitense 1980) and 56 Peg (Schindler et al. 1982), each of which

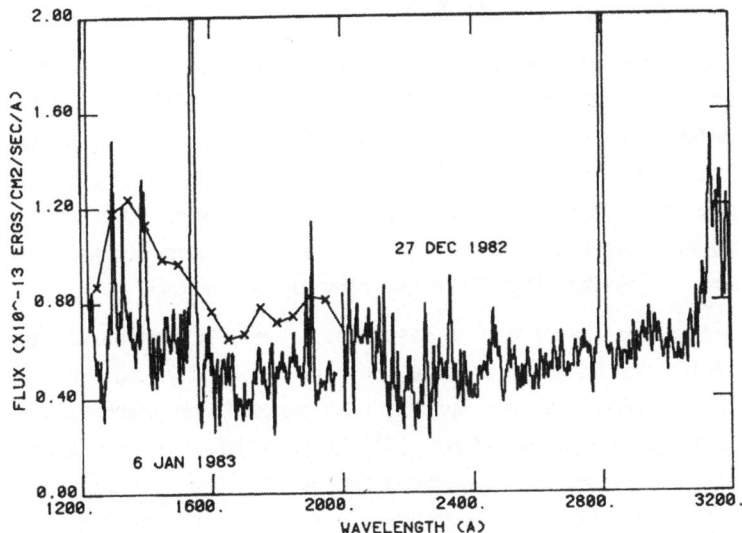

Figure 1. LWR and SWP spectra of HD 35155 taken on the dates shown. The SWP spectrum taken on 27 December 1982 is weakly exposed, with only C IV at 1550 A clearly visible, and the flux (neglecting C IV) has therefore been recorded in bins 50 A wide, and these are shown by crosses joined with straight-line segments.

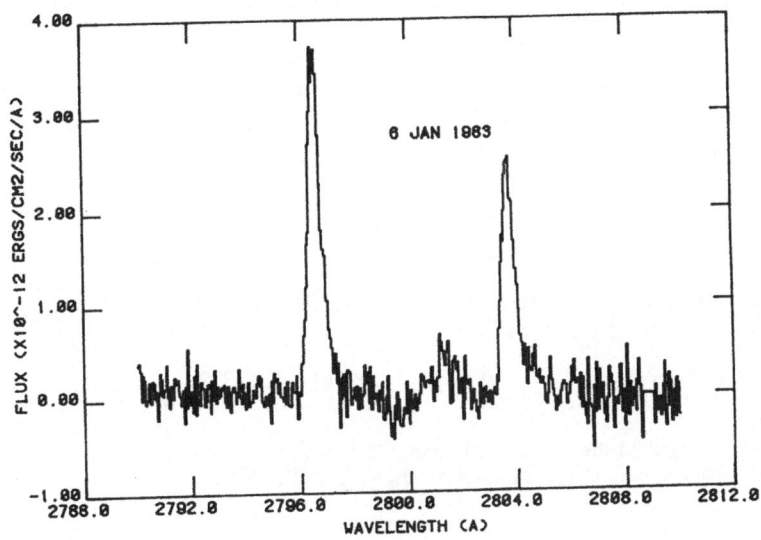

Figure 2. LWR high-resolution spectrum of HD 35155 on 6 January 1983. Note that the blue wings of the Mg II h and k lines are entirely obliterated by a dense shell of overlying, outflowing material.

consists of a cool giant and a white dwarf.

Close examination of the asymmetries and of the apparent wavelengths of the Mg II h and k lines (whose rest wavelengths are 2795.5 and 2802.7 A) in Figure 2 indicates that only the redward remnant of each line is visible, except for a trace of the blue edge of the k line. The entire blueward portion of the line is obliterated by overlying, outflowing gas. This outflowing gas has a velocity of 30-60 km/s and a large optical depth at line center. A high rate of mass loss seems probable, but no quantitative estimate is possible until the geometry is understood.

Observations to date are inadequate to distinguish between various stellar configurations; possibilities for HD 35155 include, but are not limited to, the following. (1) It is a single S star with an unusual, time varying chromosphere–corona which accounts for all the observed emission. (2) It is a binary system consisting of an S star and a white dwarf, whose photosphere cannot be seen. The observed SWP emission arises either from an accretion disk around the WD or a dense stellar wind from the S star which is excited by the WD. Alternately, the WD excites the outer layers of the S star. (3) It is a binary system consisting of an S star and a compact companion (WD?) which undergoes outbursts of some sort (as in some symbiotics or in the recurrent nova T CrB). (4) It is an eclipsing binary system consisting of an S star and a hot companion (main sequence or subdwarf) undergoing mass exchange. The system may be similar to certain W Ser systems. (5) It is a binary system in which the cool giant, ordinarily a slow rotator with a weak chromosphere, has been spun up by its companion until it is now very active.

The authors gratefully acknowledge a NASA grant for IUE observations and stimulating discussions with several colleagues. One author (H.R.J.) acknowledges an NAS–NRC senior fellowship and the hospitality of NASA Ames.

References

Ake, T.B. 1979, Ap.J. 234, 538.
Bregman, J.D., Johnson, H.R., Augason, G.C., and Witteborn, F.C. 1983, Ap.J. (Submitted).
Bohm–Vitense, E. 1980, Ap.J. 239, L79.
Eggen, O.J. 1972, Ap.J. 177, 489
Keenan, P.C. and Boeshaar, P.C. 1980, Ap.J. Suppl. 43, 379.
Schindler, M., Stencel, R.E., Linsky, J.E., Basri, G.S. and Helfand, D.J. 1982, Ap.J. 263, 269.
Wing, R.F. 1967, Unpublished Ph.D. Thesis, University of California
Wing, R.F. and Yorka, S. 1977, Mon. Note R.A.S. 178, 383.

A.M.Fridman, V.L.Polyachenko

Physics of Gravitating Systems

Volume 1

Equilibrium and Stability of Gravitating Systems

Translated from the Russian by A.B.Aries,
I.N.Poliakoff

1984. 86 figures. Approx. 480 pages
ISBN 3-540-11045-3

Contents: Introduction. – Theory: Equilibrium and
Stability of a Nonrotating Flat Gravitating Layer.
Equilibrium and Stability of a Collisionless Cylinder.
Equilibrium and Stability of Collisionless Spherically
Symmetrical Systems. Equilibrium and Stability of
Collisionless Ellipsoidal Systems. Equilibrium and
Stability of Flat Gravitating Systems.

Volume 2

**The Nonlinear Theory of Collective Processes in
a Gravitating Medium: Astrophysical Applications**

Translated from the Russian by A.B.Aries,
I.N.Poliakoff

1984. 64 figures. Approx. 384 pages
ISBN 3-540-13103-5

Contents: Non-Jeans Instabilities of Gravitating
Systems. – Problems of Nonlinear Theory. — Astro-
physical Applications: General Remarks. Spherical
Systems. Ellipsoidal Systems. Disk-Like Systems;
Spiral Structure. Other Applications. – Appendix. –
References. ↘Additional References. – Index.

C.Hoffmeister, G.Richter, W.Wenzel

Variable Stars

Translated from the German by S.Dunlop

1984. Approx. 310 pages. ISBN 3-540-13403-4

C.Hoffmeister, G.Richter, W.Wenzel

Veränderliche Sterne

2.Auflage. 1984. ISBN 3-540-13396-8

Springer-Verlag
Berlin
Heidelberg
New York
Tokyo

Lecture Notes in Physics

Selected Issues from
Lecture Notes in Mathematics